Differential Forms on Singular Varieties

De Rham and Hodge Theory Simplified

MONOGRAPHS AND TEXTBOOKS IN PURE AND APPLIED MATHEMATICS

Recent Titles

V. Lakshmikantham and D. Trigiante, Theory of Difference Equations: Numerical Methods and Applications, Second Edition (2002)

T. Albu, Cogalois Theory (2003)

A. Bezdek, Discrete Geometry (2003)

M. J. Corless and A. E. Frazho, Linear Systems and Control: An Operator Perspective (2003)

I. Graham and G. Kohr, Geometric Function Theory in One and Higher Dimensions (2003)

G. V. Demidenko and S. V. Uspenskii, Partial Differential Equations and Systems Not Solvable with Respect to the Highest-Order Derivative (2003)

A. Kelarev, Graph Algebras and Automata (2003)

A. H. Siddiqi, Applied Functional Analysis: Numerical Methods, Wavelet Methods, and Image Processing (2004)

F. W. Steutel and K. van Harn, Infinite Divisibility of Probability Distributions on the Real Line (2004)

G. S. Ladde and M. Sambandham, Stochastic versus Deterministic Systems of Differential Equations (2004)

B. J. Gardner and R. Wiegandt, Radical Theory of Rings (2004)

J. Haluska, The Mathematical Theory of Tone Systems (2004)

C. Menini and F. Van Oystaeyen, Abstract Algebra: A Comprehensive Treatment (2004)

E. Hansen and G. W. Walster, Global Optimization Using Interval Analysis, Second Edition, Revised and Expanded (2004)

M. M. Rao, Measure Theory and Integration, Second Edition, Revised and Expanded (2004)

W. J. Wickless, A First Graduate Course in Abstract Algebra (2004)

R. P. Agarwal, M. Bohner, and W-T Li, Nonoscillation and Oscillation Theory for Functional Differential Equations (2004)

J. Galambos and I. Simonelli, Products of Random Variables: Applications to Problems of Physics and to Arithmetical Functions (2004)

Walter Ferrer and Alvaro Rittatore, Actions and Invariants of Algebraic Groups (2005)

Christof Eck, Jiri Jarusek, and Miroslav Krbec, Unilateral Contact Problems: Variational Methods and Existence Theorems (2005)

M. M. Rao, Conditional Measures and Applications, Second Edition (2005)

K. M. Furati, Zuhair Nashed, and Abul Hasan Siddiqi, Mathematical Models and Methods for Real World Systems (2005)

Vincenzo Ancona and Bernard Gaveau, Differential Forms on Singular Varieties: De Rham and Hodge Theory Simplified (2005)

Differential Forms on Singular Varieties

De Rham and Hodge Theory Simplified

Vincenzo Ancona

University of Firenze
Florence, Italy

Bernard Gaveau

University Pierre et Marie Curie
Paris, France

CRC Press
Taylor & Francis Group
Boca Raton London • New York

CRC Press is an imprint of and
Taylor & Francis Group, an Informa business

FIRST INDIAN REPRINT, 2015

Published in 2006 by
CRC Press
Taylor & Francis Group
6000 Broken Sound Parkway NW, Suite 300
Boca Raton, FL 33487-2742

International Standard Book Number-10: 0-8493-3739-9 (Hardcover)
International Standard Book Number-13: 978-0-8493-3739-0 (Hardcover)
Library of Congress Card Number 2005048467

Library of Congress Cataloging-in-Publication Data

Ancona, Vincenzo.
Differential forms on singular varieties : De Rham and Hodge theory simplified / Vincenzo Ancona, Bernard Gaveau.
p. cm. – (Monographs and textbooks in pure and applied mathematics ; 273)
Includes bibliographical references and index.
ISBN 0-8493-3739-9
1. Differential forms. 2. Hodge theory. 3. Singularities (Mathematics) 4. Geometry, Algebraic. I. Gaveau, Bernard. II. Title. III. Series.

QA381.A53 2005
515¢.37–dc22 2005048467

T&F informa

Taylor & Francis Group
is the Academic Division of T&F Informa plc.

Visit the Taylor & Francis Web site at
http://www.taylorandfrancis.com

and the CRC Press Web site at
http://www.crcpress.com

Printed and bound in India by Bhavish Graphics.

FOR SALE IN SOUTH ASIA ONLY

Introduction

The theory and applications of differential forms have been central themes of algebraic (and analytic) geometry for the last two hundred years. At the beginning of the 19th century, Abel defined and started the classification of abelian integrals, namely integrals of differential forms of the type $R(x, y)dx + S(x, y)dy$, where R and S are rational functions and x and y are variables related by a polynomial relation $f(x, y) = 0$. In modern terminology, Abel was defining and studying meromorphic 1-forms on a projective curve. The properties and the explicit construction of these forms "attached" to a curve was the subject of intensive study during the 19th century: they were classified as forms of the 1st, 2nd and 3rd species, depending on the nature of their singularities. The main results were the Abel-Jacobi inversion theorem and the construction of the jacobian variety of a curve, the theorem of Riemann-Roch relating the number of functions and forms having given singularities and the theory of linear series by the italian school of algebraic geometry (in modern language, the theory of line bundles).

A different but related direction came from the theory of residues by Cauchy which was also used to study global properties of algebraic curves and evidently was the central result of the theory of holomorphic functions.

Picard and Poincaré attempted to generalize these theories to algebraic surfaces; they started the study of meromorphic forms of degree 1 and 2 on algebraic surfaces, they tried to extend Abel-Jacobi theory as well as the notion of linear series, they started the study of divisors and Picard proved the first version of Lefschetz theorem on surfaces. But at the end of the 19th century, algebraic topology was not sufficiently developped and further progresses in algebraic geometry had to wait for corresponding progresses in algebraic topology by Lefschetz and others in the 1920's, as well as the development of exterior differential calculus by Goursat, E. Cartan. This culminated in the statement and proof of the Lefschetz theorems about hyperplane sections of projective manifolds and the rigorous proof of De Rham theorems on general compact manifolds, stating that the De Rham cohomology, that is the closed forms modulo the exact forms, is dual to the singular homology by integration of forms on cycles: a closed form which has null periods on cycles is exact, and one can find a closed form which has given periods on the homology classes.

The problem which was posed by the definition of the De Rham cohomology was to find a representative in a given cohomology class. This problem was solved by De Rham: in a given cohomology class one specifies a representative using other equations than the d-equation satified by this representative. The method is to introduce a riemannian metric and to choose in a given cohomology class a closed form which is harmonic, namely, that it is in the kernel of the adjoint of the exterior differential d, the adjoint being taken with respect to the chosen riemannian metric. It turns out that such harmonic forms are exactly the forms in the kernel of a second order elliptic system of equations, which reduces to the standard Laplace equation for the case of functions. Then, the cohomology class space is isomorphic to the space of harmonic forms. In particular the space of harmonic forms does not depend on the chosen metric. As a consequence, it became possible to relate the topology of a compact manifold to its metric properties, roughly speaking, positive curvature implies a vanishing of cohomology. This was generalized to forms with coefficients in a vector bundle.

But the main application concerned the case of compact complex Kählerian manifolds. In the 1930's and 1940's, applying the methods of harmonic forms to the case of compact complex kählerian manifolds, Hodge proved that the cohomology of such a manifold can be decomposed in a direct sum of subspaces of harmonic forms of well-defined types, which later, around 1950, were identified by Dolbeault as cohomology spaces of the sheaves of holomorphic forms. This implies that the Hodge decomposition of the cohomology did not depend of the chosen kählerian metric, but only of the complex structure of the manifold. Moreover, Kodaira used systematically the method of harmonic forms to prove vanishing theorems for the cohomology with coefficients in certain line bundles, "sufficiently positive," in the sense that their curvature forms (or their Chern class) is sufficiently positive as a hermitian form. As a consequence, it was possible to give an analytical proof of the Lefschetz theorem on hyperplane sections of a projective manifold and to give a characterization of the class of projective manifolds in term of the existence of a positive line bundle whose sections define a projective embedding.

At about the same time, Leray had started new foundations of algebraic topology, based on sheaf theory that he had invented while he was a prisoner in Austria during the second world war. The three main notions that Leray introduced were the definition of sheaf cohomology using fine resolutions; fine resolutions constitute a very broad generalization of the notion of differential forms and it was proved that the cohomology of the complex of global sections of a fine resolution of a sheaf does not depend of that resolution. The second notion was the long exact sequence of cohomology and the third one was the concept of spectral sequence, as an approximation of the cohomology of a space. Sheaf theory allowed the rapid development of analytic geometry, in particular for analytic spaces with singularities, for which differential geometric notions like differential forms has no well-defined meaning a priori. Moreover, at the end of the 1950's, starting from the work of Poincaré on residues in two complex variables, Leray defined a general theory of residues for differential forms with what are now called "logarithmic singularities".

In the 1960's, Grothendieck used systematically sheaf cohomology in algebraic geometry; his final achievement was the introduction of the concept of "motif", as a way of considering an algebraic variety as made of basic building blocks which were treated as being "added" together. This led him to the notion of the so-called "weight filtration." The weight filtration was finally used by Deligne to prove the existence of a mixed Hodge structure on the cohomology (with constant coefficients) of algebraic varieties. The main tools to define this structure, were the desingularization theorems of Hironaka, the Leray spectral sequence and the notion of "descente cohomologique", as well as the theory of residues for forms with logarithmic singularities on a divisor with normal crossing. It is possible to explain easily what is the mixed Hodge structure on the cohomology of an algebraic variety. Recall that on a compact complex Kählerian manifold, the cohomology can be naturally decomposed as a direct sum of subspaces, each of which are cohomologies of other sheaves. This decomposition is in fact the graduation associated to the so-called "Hodge filtration" on the differential forms of the manifold. This filtration is defined by the number of dz which appear in the differential form in a local system of complex coordinates. In this situation, one says that the cohomology carries a pure Hodge structure, and this implies, among many other things, that the odd dimensional Betti numbers are even. Now, this last statement is clearly wrong for a singular space: for example, take a projective space of dimension 1 with two points identified. Its homology is generated by a single cycle, which can be represented by the curve joining the two identified points, so the cohomology of degree 1 has dimension 1. Thus there is no hope in general that the cohomology of a singular space carries a pure Hodge structure. Instead, it carries a mixed Hodge structure, namely the weight filtration of Deligne and Grothendieck induces a graduation on the cohomology, and each quotient space of this graduation carries a pure Hodge structure: this is the definition of the mixed Hodge structure. It is quite difficult to see how this mixed Hodge structure is made, the main difficulty coming from the weight filtration, which is itself constructed by Deligne using the "descente cohomologique".

Before outlining our method to deal with these questions, we start with general comments about the problem of the passage from a local situation to global statements. Usually, one starts with some kind of local calculations, for example in differential geometry, and these calculations are used as a basis for an integration by part, leading to a variational problem or to the definition of self-adjoint extensions. The theory of harmonic forms on a compact manifold is a typical example of this situation: using a metric as a supplementary structure, one can calculate the De Rham-Laplace operator, verify that it is an elliptic system and then use the global theory of self-adjoint extensions to deduce global results. Moreover, in the Kählerian situation, local calculations, although not so easy ones, prove that the De Rham-Laplace operator for the d, ∂ and $\bar{\partial}$ complexes are identical, so that the associated global theories of Green operators, spectral decompositions and harmonic forms are also identical: this leads immediately to the pure Hodge theory on compact, complex, Kählerian manifolds. Another different exemple is the standard theory of residues in complex analysis, where the extension from local to global is a consequence of Stokes formula, and the analysis of the local situation is the decomposition of a function near its poles and the calculation of the integral of dz/z along a circle. A second method for extending local results to global results is the theory of sheaves: in fact the notion of cohomology is exactly the obstruction of extending locally defined sections in globally defined ones. Here, again, one starts with local calculations or local solutions of partial differential equations like the d- or $\bar{\partial}$- equations (Poincaré or Dolbeault theorems) to prove that a complex of fine sheaves is a resolution of the constant sheaf or the holomorphic functions sheaf. This identifies the cohomology of these sheaves in term of cohomologies of complexes of global sections of the sheaves of the resolution. The theory of harmonic forms can be used, in combination with curvature properties, to prove vanishing theorems of certain cohomology spaces (again by integration by part), then the exact sequence of cohomology allows one to prove global extensions results. Another method is the Leray spectral sequence of hypercohomology: in this case, the idea is to approximate the cohomology (more precisely the graded space of the cohomology for a certain filtration of an underlying complex), by simpler spaces, with a systematic rule for going from one space of the approximation to the next one. The second term of the spectral sequence is thus a cohomology of sheaves of local cohomologies. These local cohomologies being local objects, can be calculated by local methods, at least in principle. For example, this is how one can identify the global cohomology of an open algebraic manifold using complexes of differential forms with logarithmic poles at infinity: the Leray spectral sequence reduces everything to local (and easy) calculations of residues around a point situated on a transverse intersection of complex hyperplanes.

Given the importance of differential forms in algebraic geometry, it was natural to try to extend the definition of forms on singular spaces. One possibility, investigated by Grauert and Grothendieck, was to embed the singular space in a space $\mathbb{C}^n$, and to restrict the sheaf of differential forms of the ambient space, modulo forms which are identically zero when restricted to the smooth part of the singular space; unfortunately, this does not give a resolution of the constant sheaf. After various attempts, which provided resolutions of the constant sheaf, but which were not functorial with respect to morphisms of complex spaces, we realized that we could not deal with a single sheaf but that we needed a whole family of sheaves of differential forms. There are many reasons for this new situation: first, we use a resolution of singularities which is not uniquely defined, second, the image of singular points under a morphism need not be situated in the singular points of the image space.

Let X be a (possibly) singular space; we fix a resolution of singularities of

X, that is a diagram

$$\begin{array}{ccc} \tilde{E} & \xrightarrow{i} & \tilde{X} \\ {\scriptstyle q}\downarrow & & \downarrow{\scriptstyle \pi} \\ E & \xrightarrow{j} & X \end{array}$$

where $E \subset X$ is a nowhere dense closed subspace, containing the singularities of X, $j \colon E \to X$ is the natural inclusion, $\tilde{X}$ is a smooth manifold and π is a proper modification inducing an isomorphism $\tilde{X} \setminus \tilde{E} \simeq X \setminus E$.

Thus, we replace the singular space by three spaces: the desingularized space $\tilde{X}$, the subspace E containing the singularities and its inverse image $\tilde{E}$ in the desingularization; the last two spaces may be singular in general, but they are of lower dimension than X; the desingularized space $\tilde{X}$ has the same dimension as X, but it is smooth. Associated to this diagram, one can construct by recursion complexes of differential forms, which coincide with the usual notion of differential forms on the regular points of the space. Forms, in this sense, are formal triples of forms on the three spaces of the diagram. This is possible because we can use a recursion argument on the dimension of the spaces, namely we assume that the complexes of forms have already been constructed on lower dimensional spaces. The differential of the complex is constructed in such a way that the vanishing of the differential of forms indicates both a kind of compatibility condition for the forms of the triple and the closedness of forms in the usual sense. To construct it, we need to define "pullback" of forms by a morphism of analytic spaces between complexes of forms carried by the spaces related by the given morphism; these pullbacks generalize the familiar pullback of forms by morphisms between manifolds.

We use complexes of differential forms to give a complete treatment of Deligne theory of mixed Hodge structures on the cohomology of singular spaces. The advantages of this approach are the following:

1. We can use recursive arguments on dimension, and we do not introduce spaces of higher dimension than the initial space.

2. The weight filtrations can be easily identified: in the compact situation, the weight is zero for the forms on the desingularization $\tilde{X}$, and is inductively defined for the forms carried by E and $\tilde{E}$ (for $\tilde{E}$ a shift by 1 is needed).

3. The Hodge filtration is the usual Hodge filtration for the forms on the desingularization $\tilde{X}$, and is inductively defined for the forms carried by E and $\tilde{E}$.

4. We avoid the use of cohomological descent theory [SD], which is a very difficult topic, especially for a reader not much acquainted with derived categories and derived functors.

Of course we are mainly indebted to the fundamental work of Deligne [D], [D1], [D2]; but also to other papers containing very interesting approaches to the theory, like [A], [E], [GNPP1], [GNPP2], [GS], [N].

Contents

Part I

Classical Hodge theory

Chapter 1

Spectral sequences and mixed Hodge structures

1.1 Introduction

In this chapter we collect some algebraic preliminaries, especially on spectral sequences and pure and mixed Hodge structures.

1.2 Filtrations

Let E be a vector space. An increasing filtration is an increasing sequence of subspaces

$$\cdots \subset W_m E \subset W_{m+1} E \subset \cdots$$

with

$$\bigcap_m W_m E = (0)$$

We will consider only *finite* filtrations, that is $W_m E \neq (0)$ *and* $\neq E$ *only for a finite number of* m.

A decreasing filtration F^p on E can be defined in an obvious way. We deal mainly with increasing filtrations. All the properties can be translated to the decreasing case F^p, just observing that the formula $W_m = F^{-m}$ transforms a decreasing filtration to an increasing one, and conversely.
One defines for every $e \in E$

$$w(e) = \inf \left\{ n \in \mathbb{Z} \;\middle|\; e \in W_n E \right\} \tag{1.1}$$

in particular $w(0) = -\infty$. If $F \subset E$ is a subspace, W_m induces a filtration by the formula

$$W_m F = W_m E \cap F$$

On the quotient E/F there is also a natural induced filtration

$$W_m\,(E/F) = W_m E / W_m F$$

If A, B are vector spaces with increasing filtrations $W_m A$ and $W_m B$ and corresponding functions w_A and w_B as in (1.1), a morphism of filtered spaces is a linear mapping $\phi\colon A \to B$ such that $\phi(W_m A) \subset W_m B$, which is equivalent to $w_B(\phi(a)) \leq w_A(a)$ for every $a \in A$.

1.3 Strict morphisms

Let $\phi\colon A \to B$ be a morphism of filtered spaces. On the subspace $\operatorname{im}\phi$, one can define two filtrations, namely

(i) the filtration on $\operatorname{im}\phi$ as a subspace of B

$$W_m \operatorname{im}\phi = W_m B \cap \operatorname{im}\phi$$

(ii) the quotient filtration defined by the function w_ϕ

$$w_\phi(b) = \inf\{\, n \in \mathbb{Z} \mid \exists a \in W_n A \text{ with } b = \phi(a) \,\} \tag{1.2}$$

Because $\phi(W_n A) \subset W_n B$, it follows that

$$w_B(b) \leq w_\phi(b) \quad \text{for } b \in \operatorname{im}\phi$$

One says that ϕ is a *strict morphism* if

$$w_B(b) = w_\phi(b) \quad \text{for } b \in \operatorname{im}\phi$$

Lemma 1.1. *A morphism $\phi\colon A \to B$ of filtered spaces is strict, if and only if*

$$\phi(W_n A) = \operatorname{im}\phi \cap W_n B \tag{1.3}$$

Proof. In general $\phi(W_n A) \subset \operatorname{im}\phi \cap W_n B$ so that $w_B \leq w_\phi$.

(i) If ϕ is strict, namely $w_B = w_\phi$, and if $b \in \operatorname{im}\phi$ with $w_B(b) = n$, by definition (1.2) there exists an $a \in A$ with $w_A(a) = n$, $b = \phi(a)$ so $b \in \phi(W_n A)$.

(ii) Conversely, if (1.3) holds and if $b \in \operatorname{im}\phi$ with $w_B(b) = n$, one has $b \in \operatorname{im}\phi \cap W_n B$ and there exists an a with $b = \phi(a)$, $a \in W_n A$, so that $w_B(b) \leq n$ and $w_\phi \leq w_B$. Thus $w_B = w_\phi$ (because $w_B \leq w_\phi$ is always true). □

The following lemma is easy to prove.

Lemma 1.2. *Let E be a vector space with an increasing filtration W_n, F a subspace. The quotient mapping $E \to E/F$ is a strict morphism and the filtration W_n is associated to the function*

$$w([e]) = \inf\{\, n \in \mathbb{Z} \mid \exists e' \in [e], e' \in W_n E \,\}$$

1.4 Filtered complexes

Let $(L^{\cdot}, d)$ be a complex of vector spaces with $d\colon L^k \to L^{k+1}$. We say that it is an increasing (resp. decreasing) filtered complex if each L^k has an increasing (resp. decreasing) filtration $W_m L^k$ such that

$$d\colon W_m L^k \to W_m L^{k+1}$$

i.e. d is a morphism of filtered spaces. We will consider only *regular* filtrations, that is the filtrations $W_m L^k$ are finite for every k.

The cohomology of the complex is defined as usual

$$H^k\left(L^{\cdot}\right) = \frac{\ker\left\{ d\colon L^k \to L^{k+1} \right\}}{dL^{k-1}}$$

and the filtration W_m induces a natural filtration on $H^k\left(L^{\cdot}\right)$ by

$$W_m H^k\left(L^{\cdot}\right) = \frac{\ker\left\{ d\colon L^k \to L^{k+1} \right\} \cap W_m L^k}{(dL^{k-1}) \cap W_m L^k} \tag{1.4}$$

If $x \in L^k$, we say that its cohomological degree is k. If $w(x) = m$, we say that its formal degree is m.

1.5 Spectral sequences

For the spectral sequences (associated to a filtration) we use notations which are different from those which appear in many papers and books. The advantage is that in our notation $E_r^{m,k}$, the indices have a more clear algebro-geometric meaning: m is the degree of the filtration and k is the degree of the complex (the degree of differential forms in the case of the De Rham complex).

The reader who is willing to work with the more classical indices $E_r'^{p,q}$ can use the following dictionary:

$$E_r^{m,k} = E_r'^{-p,p+q}$$
$$E_r'^{p,q} = E_r^{-m,k+m}$$

We define for any $r \geq 0$, the approximate cocycles

$$Z_r^{m,k} = \left\{ x \in W_m L^k \,\middle|\, dx \in W_{m-r} L^{k+1} \right\}$$

There we have

$$Z_0^{m,k} = W_m L^k$$
$$Z_s^{m,k} \subset Z_r^{m,k} \quad (s \geq r)$$
$$dZ_r^{m,k} \subset W_{m-r} L^{k+1}$$
$$dZ_r^{m,k} \subset Z_s^{m-r,k+1} \quad \text{for } s \geq 0$$
$$dZ_{r-1}^{m+r-1,k} \subset Z_s^{m,k+1} \quad \text{for } s \geq 0$$
$$Z_{r-1}^{m-1,k} \subset Z_r^{m,k} \cap W_{m-1} L^k$$

Let us remark that there exists r_0 such that for $r \geq r_0$ the approximate cocycles are usual cocycles: $x \in Z_r^{m,k}$ implies $dx = 0$; in fact there exists r_0 such that $W_{m-r}L^{k+1} = 0$ for $r \geq r_0$. One defines

$$E_r^{m,k} = \frac{Z_r^{m,k}}{dZ_{r-1}^{m+r-1,k-1} + Z_{r-1}^{m-1,k}}$$

We denote $[x]_r$ the class of $x \in Z_r^{m,k}$ in the quotient $E_r^{m,k}$.

We give the main properties of spectral sequences. First, we remind the reader that the indices are not the usual ones, for obvious reasons of simplicity: m is the *formal degree* (or degree of filtration), k is the *cohomological degree.*

Second, the differential d induces a differential d_r

$$d_r \colon E_r^{m,k} \to E_r^{m-r,k+1}$$

with $d_r^2 = 0$.

Third, $E_r^{m,k}$ is the cohomology of the complex

$$\cdots \longrightarrow E_{r-1}^{m+r-1,k-1} \xrightarrow{d_{r-1}} E_{r-1}^{m,k} \xrightarrow{d_{r-1}} E_{r-1}^{m-r+1,k+1} \longrightarrow \cdots$$

namely

$$E_r^{m,k} = \frac{\ker\left\{ d_{r-1} \colon E_{r-1}^{m,k} \to E_{r-1}^{m-r+1,k+1} \right\}}{\operatorname{im}\left\{ d_{r-1} \colon E_{r-1}^{m+r-1,k-1} \to E_{r-1}^{m,k} \right\}}$$

We will often write (with a slight abuse) that $E_r^{m,k}$ is the cohomology of the complex $\left(E_{r-1}^{m,\cdot}, d_{r-1}\right)$.

1.6 The first term of the spectral sequence

We define the graded complex

$$\operatorname{grad}_m L^k = \frac{W_m L^k}{W_{m-1} L^k}$$

with the differential

$$d_0\colon \operatorname{grad}_m L^k \to \operatorname{grad}_m L^{k+1}.$$

Then one has

$$E_1^{m,k} = H^k\left(\operatorname{grad}_m L^{\cdot}\right).$$

1.7 The graded cohomology

For every $r \geq 0$ there are natural morphisms $W_m H^k\left(L^{\cdot}\right) \to E_r^{m,k}$, sending to zero the subspace $W_{m-1} H^k\left(L^{\cdot}\right)$. Hence we obtain morphisms

$$\frac{W_m H^k\left(L^{\cdot}\right)}{W_{m-1} H^k\left(L^{\cdot}\right)} \to E_r^{m,k} \tag{1.5}$$

There is an r_0 such that for $r \geq r_0$ the approximate cocycles are cocycles. Hence for $r \geq r_0$ the above morphisms (1.5) are isomorphisms and $d_r = 0$, so that $E_r^{m,k} = E_{r_0}^{m,k}$.

We say that the spectral sequence $E_r^{m,k}$ converges, for $r \to \infty$, to the graded cohomology defined by (1.4):

$$\bigoplus_m \frac{W_m H^k\left(L^{\cdot}\right)}{W_{m-1} H^k\left(L^{\cdot}\right)}$$

and that it degenerates at $E_{r_0}^{m,k}$, or at the level r_0. Then we define

$$E_\infty^{m,k} = E_{r_0}^{m,k} = E_{r_0+1}^{m,k} = \cdots = E_r^{m,k} = \cdots$$

and we write

$$E_r^{m,k} \Longrightarrow \frac{W_m H^k\left(L^{\cdot}\right)}{W_{m-1} H^k\left(L^{\cdot}\right)}$$

Summarizing:

Proposition 1.1. *There exists r such that $d_s \equiv 0$ for $s \geq r$, if and only if the natural morphisms* (1.5) *are isomorphisms:*

$$\frac{W_m H^k\left(L^{\cdot}\right)}{W_{m-1} H^k\left(L\right)} \simeq E_r^{m,k} \tag{1.6}$$

Then $E_\infty^{m,k} = E_r^{m,k}$.

1.8 Pure Hodge structures

Let A be a complex vector space endowed with a conjugation $a \to \bar{a}$ and a decreasing filtration F^pA of A

$$\cdots \subset F^{p+1}A \subset F^pA \subset \cdots$$

We note $\bar{F}^q = \overline{F^q}$ the conjugate filtration. We denote also

$$\phi(a) = \sup \{ p \in \mathbb{Z} \mid a \in F^pA \} \qquad (\phi(0) = \infty).$$

Definition 1.1. We say that A has a pure Hodge structure of weight N if

(i) A is the direct sum of its graded spaces for the filtration F^p

$$A = \bigoplus_{p=0}^{N} A^p, \quad A^p = \frac{F^pA}{F^{p+1}A} \tag{1.7}$$

(ii) The conjugation $a \mapsto \bar{a}$ is an isomorphism from A^p to A^{N-p}, i.e. $\overline{A^p} = A^{N-p}$.

From (1.7) we deduce

$$F^pA = \bigoplus_{s \geq p} A^s \tag{1.8}$$

Using the conjugation, we also deduce

$$\frac{\bar{F}^qA}{\bar{F}^{q+1}A} = \overline{\left(\frac{F^qA}{F^{q+1}A}\right)} = \bar{A}^q = A^{N-q} \tag{1.9}$$

$$\bar{F}^qA = \bigoplus\nolimits_{t \geq q} \bar{A}^t = \bigoplus\nolimits_{t \geq q} A^{N-t} \tag{1.10}$$

Proposition 1.2. *Let A be a complex vector space having a pure Hodge structure of weight N. Then*

$$F^p\bar{F}^qA = F^pA \cap \bar{F}^qA = 0 \quad \textit{for } p+q > N \tag{1.11}$$

Let

$$A^{p,q} = F^p\bar{F}^qA = F^pA \cap \bar{F}^qA \quad \textit{for } p+q = N \tag{1.12}$$

Then A decomposes as a direct sum

$$A = \bigoplus_{p+q=N} A^{p,q} \tag{1.13}$$

and the filtrations F^pA, $\bar{F}^qA$ are given by

$$\begin{cases} F^pA = \displaystyle\bigoplus_{s \geq p, s+q=N} A^{s,q} \\ \bar{F}^qA = \displaystyle\bigoplus_{t \geq q, p+t=N} A^{p,t} \end{cases} \tag{1.14}$$

Proof. We note that $A^s \cap A^{N-t} \neq \emptyset$ only if $s+t=N$. If $p+q>N$, s and t in the formulas (1.8) and (1.10) satisfy $s+t \geq p+q > N$, hence we find $F^p\bar{F}^qA = F^pA \cap \bar{F}^qA = 0$. By (1.9) the graded spaces for the filtrations F and $\bar{F}$ are the same. So, an $a \in A^p$ has degree p for the filtration F and degree $N-p$ for the filtration $\bar{F}$, and we can write

$$A^p = F^p\bar{F}^{N-p}A = F^pA \cap \bar{F}^{N-p}A = A^{p,N-p}$$

from which (1.13), (1.14) follow. □

1.9 Morphisms of pure Hodge structures

Let

$$A = \bigoplus_{p=0}^{N} A^p \tag{1.15}$$

be a graded vector space (where the A^p are subspaces of A). Then we define on A the filtration

$$F^pA = \bigoplus_{s \geq p} A^s \tag{1.16}$$

so that the graded spaces of the filtration are exactly the A^p:

$$\operatorname{grad}_F^p = A^p$$

For a graded space (1.15) we will always refer to the induced filtration (1.16).

Definition 1.2. Let A, B be filtered vector spaces, and $f\colon A \to B$ a linear mapping.

(i) f is said a morphism of filtered spaces if $f(F^pA) \subset F^pB$.

(ii) f is said a strict morphism if $f(F^pA) = \operatorname{im} f \cap F^pB$.

(iii) If A and B are graded vector spaces, f is said a graded morphism if $f(A^p) \subset B^p$.

Lemma 1.3. *Let $f\colon A \to B$ be a graded morphism of graded spaces. Then*

(i) f is a strict morphism of filtered spaces.

(ii) We have

$$\ker f = \bigoplus_p (\ker f \cap A^p)$$

$$\operatorname{im} f = \bigoplus_p (\operatorname{im} f \cap B^p)$$

$$A/\ker f = \bigoplus_p A^p/(\ker f \cap A^p)$$

$$B/\operatorname{im} f = \bigoplus_p B^p/(\operatorname{im} f \cap B^p)$$

Proof.

(i) Clearly f is a morphism of filtered spaces. Let $b \in \operatorname{im} f \cap B^p$, so $b = f(a)$. Now $b = \sum_{s \geq p} b^s$, $b^s \in B^s$ and $a = \sum_r a^r$, $a^r \in A^r$. Then

$$\sum_{s \geq p} b^s = \sum_r f(a^r)$$

so $f(a^r) = 0$ for $r < p$, $f(a^r) = b^r$ for $r \geq p$, hence $b = f\left(\sum_{r \geq p} a^r\right)$.

(ii) If $a = \sum_p a^p$ is in $\ker f$, $f(a) = \sum_p f(a^p) = 0$, $f(a^p) \in B^p$, so each $a^p \in \ker f$. In the same way, if $b = f(a)$, $b^p = f(a^p)$ for all p. Finally, if $a = \sum_p a^p$, to take the class modulo $\ker f$ is equivalent ot take the class of each a^p modulo $\ker f \cap A^p$. □

Definition 1.3. Let A, B be two spaces having pure Hodge structures, and r an integer. A morphism of pure Hodge structures of degree r is a linear mapping $f \colon A \to B$ such that

(i) $f(\bar{a}) = \overline{f(a)}$ for $a \in A$;

(ii) for every p, $f(F^p A) \subset F^{p+r} B$

A morphism of pure Hodge structures is by definition a morphism of pure Hodge structures of degree 0.

Clearly (i) and (ii) of the above definition imply that $f\left(\bar{F}^p A\right) \subset \bar{F}^{p+r} B$.

Proposition 1.3. *Let $f \colon A \to B$ be a morphism of pure Hodge structures. Then f is a morphism of filtered spaces for $\bar{F}^q$, is a morphism of graded spaces and is strict.*

Proof. We already know that f is a morphism of filtered spaces for $\bar{F}^q$. We use (1.11)

$$f\left(A^{p,q}\right) = f\left(F^p A \cap \bar{F}^q A\right) \subset F^p B \cap \bar{F}^q B = B^{p,q}$$

so that f is a morphism of graded spaces, and f is strict by lemma 1.3. □

Proposition 1.4. *Let $f\colon A \to B$ be a morphism of pure Hodge structures of respective weights N and M.*

(i) If $N = M$, $\ker f$, $\operatorname{im} f$, Cokerf and Coimf carry pure Hodge structures of weight $N = M$.

(ii) If $N > M$, f is zero.

Let us prove (ii). We have $A = \bigoplus_{p+q=N} A^{p,q}$. If $x \in A^{p,q}$ then $f(x) \in F^pB \cap \bar{F}^qB$, which is zero by the proposition 1.2 because $p + q > M$.

1.10 Mixed Hodge structures

Definition 1.4. Let A be a complex vector space endowed with a conjugation $a \to \bar{a}$. We say that A has a mixed Hodge structure if

(i) there is an increasing filtration W_mA in A;

(ii) there is a decreasing filtration F^pA on A and the corresponding conjugate filtration $\bar{F}^qA$;

(iii) on each graded space $\frac{W_mA}{W_{m-1}A}$ the quotient filtration of F^p induces a pure Hodge structure of weight m, such that

$$\frac{W_mA}{W_{m-1}A} = \bigoplus_{p+q=m} \left(\frac{W_mA}{W_{m-1}A}\right)^{p,q}$$

where

$$\begin{aligned}\left(\frac{W_mA}{W_{m-1}A}\right)^{p,q} &= F^p\bar{F}^q\left(\frac{W_mA}{W_{m-1}A}\right) = \\ &= F^p\left(\frac{W_mA}{W_{m-1}A}\right) \cap \bar{F}^q\left(\frac{W_mA}{W_{m-1}A}\right)\end{aligned} \tag{1.17}$$

Remark. To be precise, the above definition refers to the notion of mixed Hodge structures over $\mathbb{C}$, which is suitable for the purposes of this book.
In [D], Deligne deals with mixed Hodge structures over $\mathbb{Z}$; in this case, the vector space A is an extension $H \otimes_{\mathbb{Z}} \mathbb{C}$ of a $\mathbb{Z}$-module H of finite type, and the filtration W is an extension of a filtration defined on the $\mathbb{Q}$-vector space $H \otimes_{\mathbb{Z}} \mathbb{Q}$.

As a matter of fact the vector spaces we are interested in are cohomology vector spaces, which fall within Deligne framework. For them, the results about mixed Hodge structures over $\mathbb{C}$ could be stated also as results on mixed Hodge structures over $\mathbb{Z}$.

Definition 1.5. A linear mapping $f\colon A \to B$ of vector spaces with mixed Hodge structures is called a morphism of mixed Hodge structures if

(i) $f(\bar{a}) = \overline{f(a)}$ for $a \in A$

(ii) f is a morphism of filtered spaces for the filtrations W_m and F^p.

Proposition 1.5. *Let $f\colon A \to B$ be a morphism of mixed Hodge structures. Then*

(i) f is strict for the filtration W_m, and f induces morphisms of graded spaces for the filtrations W_m

$$f_m\colon \frac{W_m A}{W_{m-1}A} \to \frac{W_m B}{W_{m-1}B} \tag{1.18}$$

which are morphisms of the pure Hodge structures of these spaces.

(ii) f is strict for the filtrations F^p and $\bar{F}^q$.

(iii) If f is an isomorphism of vector spaces A onto B, each f_m is an isomorphism of the corresponding pure Hodge structures, hence f is an isomorphism of mixed Hodge structures.

Proof. (i) Clearly f induces morphisms f_m of the W-graded spaces, and each of them commutes with the conjugation and respects the filtration F^p on the graded spaces. Hence f_m is a morphism of pure Hodge structures on the graded spaces $\frac{W_m A}{W_{m-1}A} \to \frac{W_m B}{W_{m-1}B}$.

Let us prove that f is strict for the filtrations W_m. Let $b \in \operatorname{im} f \cap W_m B$ so that $b = f(a)$ and assume that $a \in W_s A$, with $s > m$. Let $[a]_s$ be the class of a in $\frac{W_s A}{W_{s-1}A}$. Then

$$f_s([a]_s) = [b]_s = 0$$

because $b \in W_m B \subset W_{s-1}B$. Now

$$[a]_s = \sum_{p+q=s} [a^{p,q}]_s \tag{1.19}$$

with

$$a^{p,q} \in F^p W_s A \cap \bar{F}^q W_s A$$

$$[a^{p,q}]_s \in F^p \bar{F}^q \frac{W_s A}{W_{s-1}A}$$

so that each $f_s([a^{p,q}]_s) = 0$, hence $f(a^{p,q}) \in W_{s-1}B$. But $f(a^{p,q})$ is also in $F^p\bar{F}^q B$. So, because $\frac{W_{s-1}B}{W_{s-2}B}$ is a direct sum of $F^p\bar{F}^q$ with $p+q = s-1$, this implies that $f(a^{p,q}) \in W_{s-2}B$ etc..., so finally $f(a^{p,q}) = 0$. Let us define

$$a_1 = a - \sum_{p+q=s} a^{p,q}$$

Then

$$f(a_1) = f(a) = b$$
$$[a_1]_s = 0$$

(because of (1.19)) so that $a_1 \in W_{s-1}A$ etc ..., until finally we find an element $a_{s-m} \in W_mA$ with $f(a_{s-m}) = b$.
(ii) We prove that f is strict for F^p; the result for $\bar{F}^q$ will follow by conjugation. It is enough to prove, by induction on m, the following statement. Let $x \in W_mA$ such that $f(x) \in F^pB$; there exists $y \in F^pW_mA$ with $f(y) = f(x)$. For $m \ll 0$, $W_mA = 0$, and the statement is obvious. Let us assume that the statement holds for $m-1$. The morphism $\frac{W_mA}{W_{m-1}A} \to \frac{W_mB}{W_{m-1}B}$ induced by f is a morphism of pure Hodge structures, hence by proposition 1.3 is strict for the filtration induced by F^p; there exists $u \in F^pW_mA$ such that $z = u - x \in W_{m-1}A$ and $t = f(u) - f(x) \in W_{m-1}B$; then $f(z) = t \in F^pW_{m-1}B$, so that by induction there exists $\zeta \in F^pW_{m-1}A$ with $f(\zeta) = f(z)$. Then $y = u - \zeta \in F^pW_mA$ and $f(y) = f(x)$.
(iii) We assume that $f\colon A \to B$ is an isomorphism, and we want to prove that each f_m is an isomorphism. First, we prove that f_m is injective. Let $[a]_m \in \frac{W_mA}{W_{m-1}A}$ with $f_m([a]_m) = 0$. We write

$$[a]_m = \sum_{p+q=s} [a^{p,q}]_m$$

with

$$a^{p,q} \in F^pW_mA \cap \bar{F}^qW_mA$$

Because f_m is a morphism of graded spaces (see (i)), $f_m([a^{p,q}]_m) = 0$, so that $f(a^{p,q}) \in W_{m-1}B$; but $f(a^{p,q})$ is also in $F^pW_mA \cap \bar{F}^qW_mA$ so that $f(a^{p,q})$ induces an element of

$$F^p\bar{F}^q\left(\frac{W_{m-1}B}{W_{m-2}B}\right) = \left(\frac{W_{m-1}B}{W_{m-2}B}\right)^{p,q}$$

which is 0 because $p+q=m$, while the non zero $F^p\bar{F}^q\frac{W_{m-1}B}{W_{m-2}B}$ correspond to $p+q=m-1$. It follows that $f(a^{p,q}) \in W_{m-2}B$ etc ... until finally $f(a^{p,q}) = 0$ which implies $a^{p,q} = 0$ because f is injective. Then $[a]_m = 0$ and f_m is injective.

Now, because f_m is injective

$$\dim \frac{W_mA}{W_{m-1}A} = \dim f_m\left(\frac{W_mA}{W_{m-1}A}\right) \leq \dim\left(\frac{W_mB}{W_{m-1}B}\right) \tag{1.20}$$

We sum the above inequalities over m, to obtain

$$\dim A = \sum_m \dim \frac{W_mA}{W_{m-1}A} \leq \sum_m \dim \frac{W_mA}{W_{m-1}B} = \dim B$$

But $\dim A = \dim B$ because f is an isomorphism so that necessarily, we have equalities for all m in (1.20) so that f_m is onto. So f_m is an isomorphism. □

1.11 Exact sequences of mixed Hodge structures

Lemma 1.4. *Let*

$$\cdots \longrightarrow A \longrightarrow B \longrightarrow C \longrightarrow \cdots$$

be an exact sequence of vector spaces with filtrations W_m, so that each morphism is a strict morphism of filtered spaces. Then, one has exact sequences

$$\cdots \longrightarrow W_m A \longrightarrow W_m B \longrightarrow W_m C \longrightarrow \cdots \tag{1.21}$$

and

$$\cdots \longrightarrow \frac{W_m A}{W_{m-1}A} \longrightarrow \frac{W_m B}{W_{m-1}B} \longrightarrow \frac{W_m C}{W_{m-1}C} \longrightarrow \cdots \tag{1.22}$$

Proof. It is clear that the sequence of the W_m is a complex of vector spaces, because each morphism is a morphism of filtered spaces.
Let $b_m \in \ker\{W_m B \to W_m C\}$. Then $b_m \in \operatorname{im}\{A \to B\} \cap W_m B$, and since the morphisms are strict, $b_m \in \operatorname{im}\{W_m A \to B\}$, thus the sequence (1.21) is exact.
The same remark applies for the sequence (1.22). □

Lemma 1.5. *Let*

$$\cdots \longrightarrow A \longrightarrow B \longrightarrow C \longrightarrow \cdots$$

be an exact sequence of vector spaces with mixed Hodge structures so that each morphism of the sequence is a morphism of mixed Hodge structures. Then, the sequences (1.21) *and* (1.22) *are exact as well as the sequences*

$$\cdots \longrightarrow \left(\frac{W_m A}{W_{m-1}A}\right)^{p,q} \longrightarrow \left(\frac{W_m B}{W_{m-1}B}\right)^{p,q} \longrightarrow \left(\frac{W_m C}{W_{m-1}C}\right)^{p,q} \longrightarrow \cdots \tag{1.23}$$

for all (p,q) with $p+q=m$.

Proof. Because of proposition 1.5, each morphism is strict for the filtrations W_m, so that the sequences (1.21) and (1.22) are exact, by lemma 1.4. Moreover the morphisms of the sequence (1.20) are morphisms of the pure Hodge structures of these quotient spaces, and so they are strict, and also morphisms of graded spaces (proposition 1.3), so that the sequences are all exact (for $p+q=m$) and are all 0 for $p+q \neq m$. □

1.12 Shifted complexes and shifted filtrations

Let A be a vector space filtered with a (for example, decreasing) filtration F. For any integer r we define the *shifted filtration* ${}^{(r)}F$ on A by

$$ {}^{(r)}F^p A = F^{p+r} A $$

If the filtration F, together with its conjugate $\bar{F}$, induces on A a pure Hodge structure of weight N, the shifted filtrations ${}^{(r)}F$ and ${}^{(r)}\bar{F}$ induce a pure Hodge structure of weight $N - 2r$.

Let $\left(L^{\cdot}, d^{\cdot}\right)$ be a complex, r an integer. The *r-shifted complex* $\left(L^{\cdot}(r), d(r)^{\cdot}\right)$ is given by

$$ L^{\cdot}(r)^k = L^{k+r} $$

and

$$ d(r)^k = d^{k+r} $$

Let us suppose that $\left(L^{\cdot}, d\right)$ is filtered with a (for example, decreasing) filtration F. We define the *shifted filtration* ${}^{(r)}F$ *on the shifted complex* $L^{\cdot}(r)$ by

$$ {}^{(r)}F^p L^k(r) = F^{p+r} L^{k+r} $$

1.13 The strictness of d and the degeneration of the spectral sequence

The following theorem relates the strictness of the differential in a filtered complex with the degeneration of the spectral sequence at the level 1.

Theorem 1.1. *Let $\left(L^{\cdot}, d\right)$ be a filtered complex of vector spaces. The following properties are equivalent.*

(i) The spectral sequence associated to the filtration degenerates at E_1: $d_r = 0$ for $r \geq 1$.

(ii) For every k, $d\colon L^k \to L^{k+1}$ is strict.

Proof. Let us suppose for example that the filtration W_m on the complex is increasing.
(i) $\Rightarrow$ (ii) The degeneration of the spectral sequence at E_1 means that there are isomorphisms (1.6) with $r = 1$:

$$ E_1^{m,k} \simeq \frac{W_m H^k\left(L^{\cdot}\right)}{W_{m-1} H^k\left(L^{\cdot}\right)} \tag{1.24} $$

If $x \in W_m L^k$ and $dx \in W_{m-1}L^{k+1}$, then there exists $z \in W_{m-1}L^k$ with $dz = dx$. In fact by the isomorphism (1.24) there is $y \in W_m L^k$, $dy = 0$, such that

$$x - y = dt + z, \quad t \in W_{m-1}L^{k-1}, \quad z \in W_{m-1}L^k$$

hence $dx = dz$.
Step by step, if $x \in W_m L^k$ and $dx \in W_{m-s}L^{k+1}$, we find $z \in W_{m-s}L^k$ with $dz = dx$.
(ii) $\Rightarrow$ (i) If d is strict, it is easy to prove that there are isomorphisms (1.24), so that the spectral sequence degenerates at E_1. □

1.14 Flat modules

Let A be a commutative ring with unit. An A-module M is *flat* if for any short exact sequence of A-modules

$$0 \longrightarrow E_1 \longrightarrow E_2 \longrightarrow E_3 \longrightarrow 0 \tag{1.25}$$

the sequence obtained by tensoring with M

$$0 \longrightarrow E_1 \otimes M \longrightarrow E_2 \otimes M \longrightarrow E_3 \otimes M \longrightarrow 0 \tag{1.26}$$

remains exact.
If (1.25) is an exact sequence of A-modules, and E_3 is flat, the sequence (1.26) remains exact for any A-module M.
The following result is well known.

Lemma 1.6. *Let*

$$0 \longrightarrow M_1 \longrightarrow M_2 \longrightarrow M_3 \longrightarrow 0$$

be a short exact sequence of A-modules. If M_2 and M_3 are flat, M_1 is flat.

As a consequence we can state the following proposition.

Proposition 1.6. *Let*

$$0 \longrightarrow M_1 \longrightarrow M_2 \longrightarrow \cdots \longrightarrow M_k \longrightarrow 0 \tag{1.27}$$

be an exact sequence of flat A-modules M_i. For any A-module F the sequence obtained by tensoring with F

$$0 \longrightarrow M_1 \otimes F \longrightarrow M_2 \otimes F \longrightarrow \cdots \longrightarrow M_k \otimes F \longrightarrow 0 \tag{1.28}$$

remains exact.

In fact we cut the sequence (1.27) into short exact sequences

$$0 \longrightarrow E_{j-1} \longrightarrow M_j \longrightarrow E_j \longrightarrow 0$$

Starting with $E_k = M_k$ we prove by decreasing induction, by means of the lemma 1.6, that each E_j is flat. Hence the sequences

$$0 \longrightarrow E_{j-1} \otimes F \longrightarrow M_j \otimes F \longrightarrow E_j \otimes F \longrightarrow 0$$

are exact, which implies that (1.28) is exact.

1.15 Connecting homomorphisms

Let us consider a commutative diagram of morphisms of vector spaces (or modules over a ring...)

$$\begin{array}{ccccccccc}
 & & 0 & \longrightarrow & 0 & \longrightarrow & 0 & & \\
 & & \downarrow & & \downarrow & & \downarrow & & \\
0 & \longrightarrow & \mathcal{C}_1^0 & \xrightarrow{f_1^0} & \mathcal{C}_2^0 & \xrightarrow{f_2^0} & \mathcal{C}_3^0 & \longrightarrow & 0 \\
 & & {\scriptstyle d_1^0}\downarrow & & {\scriptstyle d_2^0}\downarrow & & {\scriptstyle d_3^0}\downarrow & & \\
0 & \longrightarrow & \mathcal{C}_1^1 & \xrightarrow{f_1^1} & \mathcal{C}_2^1 & \xrightarrow{f_2^1} & \mathcal{C}_3^1 & \longrightarrow & 0 \\
 & & {\scriptstyle d_1^1}\downarrow & & {\scriptstyle d_2^1}\downarrow & & {\scriptstyle d_3^1}\downarrow & & \\
 & & \vdots & & \vdots & & \vdots & &
\end{array} \tag{1.29}$$

where the columns are complexes, and the rows are exact. Then there is a long exact sequence for the cohomologies of the complexes:

$$\begin{gathered}
0 \longrightarrow H^0\left(\mathcal{C}_1^{\cdot}\right) \xrightarrow{F_1^0} H^0\left(\mathcal{C}_2^{\cdot}\right) \xrightarrow{F_2^0} H^0\left(\mathcal{C}_3^{\cdot}\right) \xrightarrow{\partial^0} H^1\left(\mathcal{C}_1^{\cdot}\right) \longrightarrow \cdots \\
\cdots \longrightarrow H^k\left(\mathcal{C}_1^{\cdot}\right) \xrightarrow{F_1^k} H^k\left(\mathcal{C}_2^{\cdot}\right) \xrightarrow{F_2^k} H^k\left(\mathcal{C}_3^{\cdot}\right) \xrightarrow{\partial^k} H^{k+1}\left(\mathcal{C}_1^{\cdot}\right) \longrightarrow \cdots
\end{gathered} \tag{1.30}$$

where the morphisms $\partial^k : H^k\left(\mathcal{C}_3^{\cdot}\right) \to H^{k+1}\left(\mathcal{C}_1^{\cdot}\right)$ are called connecting homomorphisms, and the F_j^k are natural homomorphisms induced by the morphism of complexes $f_j^{\cdot}$, $j = 1, 2$.

We give a partial sketch of the proof. First we define ∂^k. Let $x \in \mathcal{C}_3^k$ with $d_3^k(x) = 0$. Since f_2^k is surjective, there is $y \in \mathcal{C}_2^k$ with $f_2^k(y) = x$. It follows $f_2^{k+1}\left(d_2^k(y)\right) = d_3^k\left(f_2^k(y)\right) = d_3^k(x) = 0$, so that (by the exactness of the k-th row in the diagram (1.29)) there exists $z \in C_1^{k+1}$ with $f_1^{k+1}(z) = d_2^k(y)$. Then $f_1^{k+2}\left(d_1^{k+1}(z)\right) = d_2^{k+1}\left(f_1^{k+1}(z)\right) = d_2^{k+1}\left(d_2^k(y)\right) = 0$; since f_1^{k+2} is injective, it follows $d_1^{k+1}(z) = 0$, hence z induces a cohomology class in $H^{k+1}\left(C_1^{\cdot}\right)$. By definition, ∂^k takes the class of x in $H^k\left(C_3^{\cdot}\right)$ to the class of z in $H^{k+1}\left(C_1^{\cdot}\right)$ (one verifies that the class of z depends only on the class of x). Let us suppose that the class of z is zero. This means that there is $t \in C_1^k$ such that $z = d_1^k(t)$. We have $d_2^k\left(f_1^k(t)\right) = f_1^{k+1}\left(d_1^k(t)\right) = f_1^{k+1}(z) = d_2^k(y)$; thus $d_2^k\left(f_1^k(t) - y\right) = 0$, i.e. $f_1^k(t) - y$ induces a cohomology class in $H^k\left(C_2^{\cdot}\right)$; finally $f_2^k\left(-f_1^k(t) + y\right) = f_2^k(y) = x$. This proves the exactness of the sequence (1.30) at ∂^k. The rest of the proof is left to the reader.

Chapter 2

Complex manifolds, vector bundles, differential forms

2.1 Introduction

In this chapter, we recall the main definitions concerning complex manifolds and vector bundles. In particular, we introduce the decomposition of differential forms in types, the decomposition of the exterior differential in the operators ∂ and $\bar{\partial}$. The main results are the Poincaré lemma and Dolbeault lemma. This chapter should be considered mainly as a dictionary to fix notations.

2.2 Complex manifolds

2.2.1 Definitions of complex coordinates and manifolds

Let M be a topological space. A system of complex coordinates on M is an open covering $\mathcal{U} = (U_a)$ of M by open sets U_a, together with homeomorphisms

$$z_a \colon U_a \to z_a(U_a) \subset \mathbb{C}^n$$

on open sets $z_a(U_a)$ of $\mathbb{C}^n$, such that the mapping $z_b \circ z_a^{-1} \colon z_a(U_a \cap U_b) \to z_b(U_a \cap U_b)$ is biholomorphic. The set of numbers $\left(z_a^1(m), \ldots, z_a^n(m)\right) \in \mathbb{C}^n$ is called the complex coordinates of $m \in U_a$ in U_a.

Let $U' = (U'_{a'})$ be another system of complex coordinates, with coordinates $z'_{a'} : U'_{a'} \to z'_{a'}(U'_{a'})$. We say that the system (U_a, z_a) is equivalent to the system $(U'_{a'}, z'_{a'})$ if the changes of coordinates $z'_{a'} \circ z_a^{-1}$ are biholomorphic whenever defined.

A complex structure is an equivalence class of systems of complex coordinates, and M is then called a complex manifold.

A local chart on M is a couple (U, z) where U is an open set of M and $z \colon U \to z(U) \subset \mathbb{C}^n$ is a homeomorphism on an open set of $\mathbb{C}^n$, such that $z \circ z_a^{-1}$ is holomorphic on $z_a(U_a \cap U)$.

A complex manifold of dimension n is a C^∞ real manifold of real dimension $2n$; the real coordinates $\left(x_a^1, x_a^2, \ldots, x_a^{2n-1}, x_a^{2n}\right)$ are defined by

$$z_a^j = x_a^{2j-1} + i x_a^{2j}. \tag{2.1}$$

If M is a complex manifold, and (U_a, z_a) is a system of complex coordinates on M, then, the coordinates $\left(z_b^k\right)_{k=1,\ldots,n}$ are holomorphic functions of the $\left(z_a^j\right)_{j=1,\ldots,n}$ in $z_a(U_a \cap U_b)$, such that the jacobian matrix $\left(\frac{\partial z_b^k}{\partial z_a^j}\right)_{k,j}$ is invertible. Moreover, we have immediately

$$\frac{\partial z_b^k}{\partial \bar{z}_a^j} = 0 \qquad \text{(Cauchy-Riemann equations)} \tag{2.2}$$

because the z_b^k are holomorphic functions. Then, one deduces:

Lemma 2.1. *On a complex manifold the ordering* $\left(x_a^1, x_a^2, \ldots, x_a^{2n-1}, x_a^{2n}\right)$ *of the real coordinates induces a natural orientation.*

Proof. It is sufficient to prove that the jacobian of the change of coordinates is positive. But

$$\det\left(\frac{\partial x_b^k}{\partial x_a^j}\right)_{k,j=1,\ldots,2n} = \left|\det\left(\frac{\partial z_b^k}{\partial z_a^j}\right)_{k,j=1,\ldots,n}\right|^2$$

as it is checked easily using (2.2) and manipulations of determinants. □

2.2.2 Tangent vectors

A real (or complex) tangent vector at a point $m \in M$, with $m \in U_a$ is

$$v = \sum_{j=1}^{2n} \xi_a^j \frac{\partial}{\partial x_a^j} \tag{2.3}$$

where ξ_a^j are real (or complex). Because

$$\frac{\partial}{\partial x_a^j} = \sum_{k=1}^{2n} \frac{\partial x_b^k}{\partial x_a^j} \frac{\partial}{\partial x_b^k} \tag{2.4}$$

one sees that the tangent vector (2.3) in the coordinate system $\left(x_b^j\right)$ has components

$$\xi_b^k = \sum_{j=1}^{2n} \frac{\partial x_b^k}{\partial x_a^j} \xi_a^j. \tag{2.5}$$

We denote $V_m(M)$ the vector space of complex tangent vectors at m.

Recall that by definition

$$\frac{\partial}{\partial z_a^j} = \frac{1}{2}\left(\frac{\partial}{\partial x_a^{2j-1}} - i\frac{\partial}{\partial x_a^{2j}}\right)$$
$$\frac{\partial}{\partial \bar{z}_a^j} = \frac{1}{2}\left(\frac{\partial}{\partial x_a^{2j-1}} + i\frac{\partial}{\partial x_a^{2j}}\right).$$

Any complex vector field v as in (2.3) can be written as

$$\begin{aligned} v &= \sum_{j=1}^{n} \zeta_a'^j \frac{\partial}{\partial z_a^j} + \sum_{j=1}^{n} \zeta_a''^j \frac{\partial}{\partial \bar{z}_a^j} \\ \zeta_a'^j &= \xi_a^{2j-1} + i\xi_a^{2j} \\ \zeta_a''^j &= \xi_a^{2j-1} - i\xi_a^{2j}. \end{aligned} \tag{2.6}$$

Because

$$\begin{aligned} \frac{\partial}{\partial z_a^j} &= \sum_{k=1}^{n} \frac{\partial z_b^k}{\partial z_a^j}\frac{\partial}{\partial z_b^k} \\ \frac{\partial}{\partial \bar{z}_a^j} &= \sum_{k=1}^{n} \overline{\left(\frac{\partial z_b^k}{\partial z_a^j}\right)}\frac{\partial}{\partial \bar{z}_b^k} \end{aligned} \tag{2.7}$$

v can be also written in the coordinates $\left(z_b^k\right)$ as

$$v = \sum_{k=1}^{n} \zeta_b'^k \frac{\partial}{\partial z_b^k} + \sum_{k=1}^{n} \zeta_b''^k \frac{\partial}{\partial \bar{z}_b^k}$$

with

$$\begin{aligned} \zeta_b'^k &= \sum_{j=1}^{n} \frac{\partial z_b^k}{\partial z_a^j}\zeta_a'^j \\ \zeta_b''^k &= \sum_{j=1}^{n} \overline{\left(\frac{\partial z_b^k}{\partial z_a^j}\right)}\zeta_a''^j. \end{aligned} \tag{2.8}$$

We denote $T_m(M)$ the vector space of complex tangent vectors at m of type

$$v = \sum_{j=1}^{n} \zeta_a^j \frac{\partial}{\partial z_a^j} \tag{2.9}$$

and $\overline{T}_m(M)$ the vector space of complex tangent vectors at m of type

$$v = \sum_{j=1}^{n} \zeta_a^j \frac{\partial}{\partial \bar{z}_a^j}$$

so that

$$V_m(M) = T_m(M) \oplus \overline{T}_m(M). \tag{2.10}$$

2.2.3 Holomorphic functions

A holomorphic function $f\colon U \to \mathbb{C}$ is a continuous differentiable function such that in $U \cap U_a$, $f \circ z_a^{-1}$ is holomorphic in $z_a(U \cap U_a)$. In other words, in $U \cap U_a$, f becomes a holomorphic function of the coordinates z_a^j, or

$$\frac{\partial f}{\partial \bar{z}_a^j} = 0. \tag{2.11}$$

This definition does not depend on the choice of the complex coordinates, because

$$\frac{\partial}{\partial \bar{z}_b^k} = \sum_{j=1}^{n} \overline{\left(\frac{\partial z_a^j}{\partial z_b^k}\right)} \frac{\partial}{\partial \bar{z}_a^j}.$$

As a consequence, we deduce immediately:

Lemma 2.2. *On a compact connected complex manifold, a holomorphic function $f\colon M \to \mathbb{C}$ is constant.*

Proof. f has a maximum at a certain point $m \in M$. Suppose that $m \in U_a$, then on $z_a(U_a)$, $f(z_a)$ is a holomorphic function which has a maximum at $z_a(m)$, and thus it is constant in a neighborhood of m. Because any holomorphic function is real analytic, it is then globally constant. □

We denote by $\mathcal{O}(U)$ the algebra of holomorphic functions on an open set U of M.

2.2.4 Complex submanifolds

Let M be a complex manifold of dimension n. A complex submanifold W of dimension p is a real submanifold of real dimension $2p$ which is a complex manifold for the induced complex structure. This means that one can find a system of complex coordinates (U_a, z_a) for M such $U_a \cap W$, if it is non empty, is defined by the $n-p$ equations

$$z_a^{p+1}(m) = 0, \ldots, z_a^n(m) = 0.$$

Then the $\left(U_a \cap W, \left(z_a^1, \ldots, z_a^p\right)\right)$ define a system of complex coordinates on W. The tangent space at m to W is a subspace

$$V_m(W) \subset V_m(M)$$

defined by the vectors

$$v = \sum_{j=1}^{p} \zeta_a'^j \frac{\partial}{\partial z_a^j} + \sum_{j=1}^{p} \zeta_a''^j \frac{\partial}{\partial \bar{z}_a^j}$$

and one has also

$$T_m(W) \subset T_m(M),\ \overline{T}_m(W) \subset \overline{T}_m(M).$$

The holomorphic functions on W are obtained as functions $f\colon W \to \mathbb{C}$, so that in $U_a \cap W$, f is a holomorphic function of $(z_a^1, \ldots, z_a^p)$. Then f extends as a holomorphic function in U_a.

Let $f_1, \ldots, f_{n-p}$ be holomorphic functions in an open set U of M, and

$$W = \{\, m \in U \mid f_1(m) = 0, \ldots, f_{n-p}(m) = 0 \,\}$$

The implicit function theorem says that if the jacobian matrix

$$\left(\frac{\partial f_k}{\partial z_a^j}\right)_{\substack{k=1,\ldots,n-p\\ j=1,\ldots,n}}$$

has rank $n-p$ at every point $m \in W$, then W is a submanifold of U of dimension p. In particular, if $W = \{\, m \in U \mid f(m) = 0 \,\}$ $(f \in \mathcal{O}(U))$ is a hypersurface, it is a complex submanifold of dimension $n-1$, provided $df(m) \neq 0$ for $m \in W$.

2.2.5 Examples

1. Any open subset U of $\mathbb{C}^n$ is a complex manifold of dimension n.

2. The complex projective space $\mathbb{P}^n$.
   We consider the space $\mathbb{C}^{n+1} \setminus \{0\}$ with coordinates $\left(Z^0, \ldots, Z^n\right)$ and we identify two points on the same line, so that

   $$\left(Z^0, \ldots, Z^n\right) \sim \left(Z'^0, \ldots, Z'^n\right)$$

   if and only if

   $$\left(Z'^0, \ldots, Z'^n\right) = \left(\lambda Z^0, \ldots, \lambda Z^n\right)$$

   for some $\lambda \in \mathbb{C}$, $\lambda \neq 0$.

   Then $\mathbb{P}^n = \mathbb{C}^{n+1} \setminus \{0\}/\sim$. We denote by $\left[Z^0, \ldots, Z^n\right]$ a point in $\mathbb{P}^n$, $Z^0, \ldots, Z^n$ are called the "homogeneous coordinates". For $a = 0, \ldots, n$, we define

   $$U_a = \left\{\, \left[Z^0, \ldots, Z^n\right] \in \mathbb{P}^n \;\middle|\; Z^a \neq 0 \right\}.$$

   Then the (U_a) cover $\mathbb{P}^n$ and in U_a, one can define complex coordinates

   $$z_a^j = \frac{Z^j}{Z^a} \text{ for } j \neq a.$$

   Then the mapping $z_a\colon U_a \to z_a(U_a) = \mathbb{C}^n$ and the changes of coordinates are obviously holomorphic.

3. Complex tori.
We consider the space $\mathbb{C}^n$ and $2n$ vectors $(\omega_1, \dots, \omega_{2n})$ which are linearly independent over $\mathbb{R}$. Call

$$L = \left\{ \sum_{k=1}^{2n} p_k \omega_k \,\middle|\, p_k \in \mathbb{Z} \right\}.$$

This is a discrete subgroup of maximal rank. Then $\mathbb{C}^n/L$ is a complex manifold, which is compact.

One can change the description of L, by choosing

$$\omega'_k = \sum a_{kj} \omega_j$$

where (a_{kj}) is an integer valued matrix with determinant 1.

4. Blowing-up of a point.
Let M be a complex manifold, and m a point in M. The blowing-up of M at m is a manifold W equipped with a surjective mapping $\pi\colon W \to M$ such that π induces an isomorphism between $W \setminus \pi^{-1}(m)$ and $M \setminus (m)$, and whose construction can be described locally around m in the following way. Let us identify m with $0 \in \mathbb{C}^n$, and an open neighborhood of m in M with a ball B centered at 0. In $B \times \mathbb{P}^{n-1}$, we use the natural coordinates $(z^1, \dots, z^n)$ in B and the homogeneous coordinates $[Z^1, \dots, Z^n]$ of $\mathbb{P}^{n-1}$ and we consider the submanifold W of $\mathbb{P}^{n-1}$ defined by the equations

$$z^j Z^k - z^k Z^j = 0 \qquad j \neq k.$$

Let us consider the affine open set $U_a \subset \mathbb{P}^{n-1}$ of points such that $Z^a \neq 0$. Then in $B \times U_a$ we find $n-1$ independent equations, equivalent to the previous set:

$$z^j - z^a \frac{Z^j}{Z^a} = 0 \qquad j \neq a.$$

The jacobian matrix of the above system has rank $n-1$, so that W is indeed a submanifold of dimension n. One has a natural projection

$$\begin{aligned} \pi\colon W &\to B \\ (z, [Z]) &\to z. \end{aligned}$$

Then $\pi^{-1}(0) = \{0\} \times \mathbb{P}^{n-1}$, while

$$\pi \big|_{W \setminus (\{0\} \times \mathbb{P}^{n-1})} : W \setminus \left(\{0\} \times \mathbb{P}^{n-1}\right) \to B \setminus \{0\}$$

is a complex isomorphism.

The submanifold $\pi^{-1}(0) = \{0\} \times \mathbb{P}^{n-1}$ is called the exceptional divisor of the blowing-up. In W, the point 0 of B has been replaced by the exceptional divisor, which is the set of complex directions of lines through 0.

5. Blowing-up along a submanifold.
Let M be a complex manifold, and $N \subset M$ a closed submanifold. The blowing-up of M along N is a manifold W equipped with a surjective mapping $\pi \colon W \to M$ such that π induces an isomorphism between $W \setminus \pi^{-1}(N)$ and $M \setminus N$, and whose construction can be described locally. We replace M by an open ball $B \subset \mathbb{C}^n$, and N by the submanifold of B defined by the equations $z^1 = 0, \ldots, z^p = 0$. We can assume that $B = B_p \times B_{n-p}$ where $B_p \subset \mathbb{C}^p$ has coordinates $(z^1, \ldots, z^p)$ and $B_{n-p} \subset \mathbb{C}^{n-p}$ has coordinates $(z^{p+1}, \ldots, z^n)$. Then, we can blow-up 0 in B_p and obtain a manifold
$$\pi \colon M_p \to B_p$$
and the exceptional divisor $\pi^{-1}(0) \simeq \mathbb{P}^{p-1}$ in M_p. The blowing-up of the submanifold $(0) \times B_{n-p} \subset \mathbb{C}^n$ is the manifold
$$\pi \colon M_p \times B_{n-p} \to B_p \times B_{n-p}$$
with the exceptional divisor $\pi^{-1}(\{0\} \times B_{n-p}) \simeq \mathbb{P}^{p-1} \times B_{n-p}$. It is also equivalent to consider the submanifold of $B_n \times \mathbb{P}^{p-1}$ defined by the equations
$$z^j Z^k - z^k Z^j = 0 \qquad j, k = 1, \ldots, p$$
where $(z^1, \ldots, z^n)$ are coordinates in B_n and $[Z^1, \ldots, Z^p]$ are homogeneous coordinates in $\mathbb{P}^{p-1}$.

2.3 Complex vector bundles and divisors

We recall briefly the definitions and the operations on vector bundles.

Definition 2.1. A complex vector bundle of rank d on a C^∞ manifold M, is a C^∞ manifold F, together with a mapping $\pi \colon F \to M$ which is onto and C^∞, such that there exist an open covering $\mathcal{U} = (U_a)$ of M and diffeomorphisms : $\varphi_a \colon \pi^{-1}(U_a) \to \mathbb{C}^d \times U_a$
$$\varphi_a(v) = (v_a(v), \pi(v)) \tag{2.12}$$
where $v_a(v) \in \mathbb{C}^d$, with the property that
$$v_a(v) = \gamma_{ab}(\pi(v))\, v_b(v) \qquad v \in \pi^{-1}(U_a \cap U_b) \tag{2.13}$$
where $\gamma_{ab} \colon U_a \cap U_b \to \mathrm{GL}(d, \mathbb{C})$ is a C^∞ mapping.

The mapping π is called projection, the mappings γ_{ab} are called transition matrices and the mappings φ_a are called trivializations. Hence F is covered by the open sets $\pi^{-1}(U_a) \simeq \mathbb{C}^d \times U_a$. We denote $F_m = \pi^{-1}(m)$ the fiber of a point $m \in M$. It is a complex vector space and thus $F_m \simeq \mathbb{C}^d$, the isomorphism depends on the trivialization $v \xrightarrow{\sim} v_a(v)$. Moreover one has obviously

$$\begin{aligned} \gamma_{ab} &= \gamma_{ba}^{-1} \\ \gamma_{ab}\gamma_{bc}\gamma_{ca} &= \mathrm{id} \qquad \text{(Identity matrix).} \end{aligned} \tag{2.14}$$

A (differentiable) section on an open set $U \subset M$ of the bundle is a C^∞ mapping $s\colon U \to F$ with $\pi(s(m)) = m$ for $m \in U$, or in other words $s(m) \in F_m$. We denote $C^\infty(U, F)$ the vector space of (differentiable) sections of F on U.

If M is a complex manifold, a complex vector bundle $\pi\colon F \to M$ is called holomorphic, if F is a complex manifold, π is a holomorphic mapping, the trivializations φ_a and the transition matrices γ_{ab} are holomorphic on their respecitve domains of definition. In this case, one can speak of holomorphic sections $s\colon U \to F$ and one denotes $\mathcal{O}(U, F)$ the vector space of holomorphic sections of F on U.

One can define a morphim of vector bundles $f\colon F \to F'$ as a C^∞ mapping commuting with the projections $\pi\colon F \to M$, $\pi'\colon F' \to M$:

$$\pi'(f(v)) = \pi(v)$$

so that there exist trivializations φ_a, φ'_a

$$\begin{array}{ccc} \pi^{-1}(U_a) & \xrightarrow{f} & \pi'^{-1}(U_a) \\ \varphi_a \downarrow & & \downarrow \varphi'_a \\ \mathbb{C}^d \times U_a & \longrightarrow & \mathbb{C}^{d'} \times U_a \\ (v, m) & \longrightarrow & (\Phi_a(m)v, m) \end{array}$$

where $\Phi_a\colon U_a \to \mathrm{Mat}\left(\mathbb{C}^d, \mathbb{C}^{d'}\right)$ is a C^∞ mapping whose rank is constant. Thus f is a linear mapping on each fiber. One can define also holomorphic morphisms between holomorphic bundles.

A line bundle is a complex vector bundle of rank 1; hence the fibers $\pi^{-1}(m) \sim \mathbb{C}$ and the transitions matrices are transition functions $\gamma_{ab}\colon U_a \cap U_b \to \mathbb{C}^*$.

One can define also subbundles and quotient bundles.

A complex vector bundle is called C^∞ trivial if it is C^∞ isomorphic to the trivial bundle $\mathbb{C}^d \times M \to M$. A holomorphic vector bundle is called holomorphically trivial if it is holomorphically isomorphic to the trivial bundle.

A complex bundle $\pi\colon F \to M$ of rank d is C^∞ trivial (resp. holomorphically trivial) if and only if one can find d global differentiable (resp. holomorphic) sections $s_1, \ldots, s_d$ such that at each point $m \in M$, $\{s_1(m), \ldots, s_d(m)\}$ is a basis of F_m. Indeed the trivialization is given by

$$v_m = \sum_{k=1}^{d} \zeta^k s_k(m) \to \left(\zeta^1, \ldots, \zeta^d\right) \in \mathbb{C}^d.$$

In particular, a line bundle is C^∞ trivial (resp. holomorphically trivial) if and only if one can find a C^∞ section (resp. a holomorphic section) which vanishes nowhere.

2.3.1 Operations on bundles

Definition 2.2. If $\pi\colon F \to M$ and $\pi'\colon F' \to M$ are bundles one can define the direct sum

$$F \oplus F' \to M$$
$$(F \oplus F')_m = F_m \oplus F'_m.$$

We can also define the tensor product

$$F \otimes F' \to M$$
$$(F \otimes F')_m = F_m \otimes F'_m.$$

with the obvious transition functions.

In particular, if $L \to M$, $L' \to M$ are line bundles, with transition functions $\gamma_{ab}\colon U_a \cap U_b \to \mathbb{C}^*$ and $\gamma'_{ab}\colon U_a \cap U_b \to \mathbb{C}^*$, $L \otimes L' \to M$ is a line bundle with transition functions $\gamma''_{ab} = \gamma_{ab}\gamma'_{ab}$.

If $f\colon M' \to M$ is a mapping and $\pi\colon F \to M$ is a bundle over M, one can define a bundle

$$\pi'\colon f^*F \to M'$$

so that for $m' \in M'$, $(f^*F)_{m'} = F_{f(m)}$. If (U_a, φ_a) is a trivialization of $F \to M$, then a trivialization of $f^*F \to M'$ is $\left(f^{-1}(U_a), \varphi'_a\right)$

$$\varphi'_a(v') = (v_a(v'), \pi'(v'))$$

In particular, if $j\colon N \hookrightarrow M$ is a submanifold, $j^*F \to N$ is denoted $F|_N \to N$ (the restriction of F to N). At each point $m \in N$, the fiber $(F|_N)_m$ is just F_m and the trivialization and transition matrices are the restrictions to N of the corresponding trivializations and transition matrices.

If $\pi\colon F \to M$, one can define the dual bundle $\pi^*\colon F^* \to M$ with $F^*_m = (F_m)^*$. The transition matrices are

$$\gamma^*_{ab}(m) = \gamma^{-1}_{ab}(m).$$

In this way, if $v \in F_m$, $v^* \in F^*_m$, $\langle v, v^*\rangle$ is well defined as

$$\langle v, v^*\rangle = \langle v_a(v), v^*_a(v^*)\rangle$$

and does not depend on the trivialization. One can also define the conjugate bundle $\bar{\pi}\colon \overline{F} \to M$ with transition functions $\bar{\gamma}_{ab}$.

2.3.2 Tangent bundle

Let M be a complex manifold of dimension n and

$$V(M) = \bigcup_{m \in M} V_m(M).$$

This is a complex bundle of rank n over M, *the tangent bundle*. A *vector field* on M is a section of $V(M)$.

Let $\mathcal{U} = (U_a)$ be a coordinate system on M. The trivialization $\varphi_a : \pi^{-1}(U_a) \to \mathbb{C}^{2n} \times U_a$ is given by

$$\varphi_a(v) = (\xi_a^1, \ldots, \xi_a^{2n}, \pi(v))$$

where ξ_a^j are the components of v on the basis $\left(\frac{\partial}{\partial x_a^j}\right)_{j=1,\ldots,2n}$ (see (2.3)). The transition matrices are

$$\gamma_{ba}(m) = \left(\frac{\partial x_b^k}{\partial x_a^j}(m)\right)_{k,j=1,\ldots,2n} \tag{2.15}$$

(see (2.5)).

One has also a subbundle of $V(M)$

$$T(M) = \bigcup_{m \in M} T_m(M) \subset V(M)$$

with trivializations $\varphi'_a : \pi^{-1}(U_a) \to \mathbb{C}^n \times U_a$

$$\varphi'_a(v) = (\zeta_a^1, \ldots, \zeta_a^n, \pi(v))$$

where ζ_a^j are the components of v on the basis $\left(\frac{\partial}{\partial z_a^j}\right)_{j=1,\ldots,n}$ (see (2.8)). The transition matrices are

$$\gamma'_{ba}(m) = \left(\frac{\partial z_b^k}{\partial z_a^j}(m)\right)_{k,j=1,\ldots,n} \tag{2.16}$$

(see (2.7)) and they are holomorphic so that $T(M)$ is a holomorphic bundle. Then, as a bundle, $V(M)$ is the direct sum of $T(M)$ and $\overline{T}(M)$

$$V(M) = T(M) \oplus \overline{T}(M). \tag{2.17}$$

If $W \subset M$ is a real submanifold, one can define the restriction $V(M)|_W \to W$, which contains the tangent bundle to W as a subbundle

$$V(W) \hookrightarrow V(M)|_W.$$

The normal bundle to W is the quotient bundle

$$V^{\perp}(W) = V(M)|_W / V(M).$$

If $W \subset M$ is a complex submanifold, one can also define $T(M)|_W$, $T(W)$, and the normal bundle

$$N(W) = T(M)|_W / T(W)$$

which are holomorphic bundles on W.

2.3.3 Example: complex tori

On a complex torus M of dimension n, the various tangent bundles $V(M)$, $T(M)$ are all trivial. One can find, for example, n global sections of $T(M)$ which form a basis of $T_m(M)$, namely the $\left(\frac{\partial}{\partial z^k}\right)_{k=1,\ldots,n}$ where $\left(z^k\right)$ are the complex coordinates on $\mathbb{C}^n$.

2.3.4 Line bundles and divisors

(i) Let D be a complex irreducible hypersurface in a manifold M. Then one can cover M by a system of coordinates $\mathcal{U} = (U_a)$ so that in each U_a, D is given by an equation $s_a = 0$ where $s_a \in \mathcal{O}\left(U_a\right)$.

Then in $U_a \cap U_b$, one has

$$s_a = \varphi_{ab} s_b$$

where φ_{ab} is a holomorphic function which does not vanish. Obviously

$$\varphi_{ab} = \varphi_{ab}^{-1}, \; \varphi_{ab}\varphi_{bc}\varphi_{ca} = \mathrm{Id}$$

so that the (φ_{ab}) define a system of transition functions for a line bundle $L(D) \to M$ and the collection (s_a) defines a global section $s \in \mathcal{O}\left(M, L\left(D\right)\right)$ whose zero set is D.

(ii) A divisor D on a complex manifold is a finite formal sum

$$D = \sum_{k=1}^{p} n_k D_k \qquad .$$

where D_k are irreducible hypersurfaces of M, n_k are integers. The line bundle associated to D is

$$L(D) = L\left(D_1\right)^{\otimes n_1} \otimes L\left(D_2\right)^{\otimes n_2} \otimes \cdots \otimes L\left(D_p\right)^{\otimes n_p}$$

and if $n < 0$, $L(D)^{\otimes n} = \left(L\left(D\right)^*\right)^{\otimes(-n)}$. In each U_a, D_k has an irreducible equation

$$s_{a,k} = 0.$$

The transition functions of $L(D)$ are

$$\varphi_{ab} = \prod_{k=1}^{p} \left(\varphi_{ab,k}\right)^{n_k}$$

(where $\varphi_{ab,k} = \frac{s_{a,k}}{s_{b,k}}$ are the transition functions of $L(D_k)$).

Moreover if $s_a = \prod_{k=1}^{p} (s_{a,k})^{n_k}$, one has $s_a = \varphi_{ab} s_b$ on $U_a \cap U_b$.

(iii) Two divisors D, D' are linearly equivalent if $D - D'$ is the divisor of a global meromorphic function φ on M.

Recall that a meromorphic function φ on M is defined by an open covering (U_a) of M such that in each U_a, $\varphi = \frac{f_a}{g_a}$ where f_a, g_a are holomorphic functions on U_a and $f_a g_b = f_b g_a$ on $U_a \cap U_b$.

Lemma 2.3. *D is linearly equivalent to D' if and only if $L(D) \simeq L(D')$ as line bundles.*

Indeed, let φ_{ab}, φ'_{ab}, be the transition functions of $L(D)$ and $L(D')$, so that the transition functions of $L(D - D')$ are $\varphi_{ab}\varphi'^{-1}_{ab}$. If there exists a global meromorphic function φ with divisor $D - D'$ this means that $\varphi|_{U_a}$ satisfies $\varphi|_{U_a} \equiv s_a u_a$ where u_a is a non vanishing holomorphic function and $s_a = \prod (s_{ak})^{n_k}$ where the $s_{a,k}$ are local holomorphic equations of the components of $D - D'$. Then because $s_a = \varphi_{ab}\varphi'^{-1}_{ab} s_b$, we have

$$u_b = \varphi_{ab}\varphi'^{-1}_{ab} u_a$$

in $U_a \cap U_b$ away from the poles of φ, but this equation is valid in $U_a \cap U_b$ because both members are holomorphic, so $L(D - D')^*$ has a holomorphic section which vanishes nowhere and it is thus trivial.

Conversely if $L(D - D')^*$ is trivial, it has a holomorphic section u_a which vanishes nowhere. Let s_a be the section of $L(D - D')$ describing $D - D'$ locally, then $\varphi_a \equiv s_a u_a$ defines a global meromorphic function with divisor $D - D'$.

2.3.5 Example: $\mathbb{P}^n$ and its line bundles

Let $[Z^0, \ldots, Z^n]$ the homogeneous coordinates of $\mathbb{P}^n$, U_a, the affine open set where $Z^a \neq 0$. Let H be a hyperplane with equation $\sum_{k=0}^{n} u_k Z^k = 0$. Then $H \cap U_a$ is described by the holomorphic equation

$$s_a \equiv \sum_{k=0}^{n} u_k \frac{Z^k}{Z^a} = 0$$

because $\frac{Z^k}{Z^a}$ are the holomorphic coordinates in U_a. One has in $U_a \cap U_b$

$$s_a = \frac{Z^b}{Z^a} s_b$$

so that $L(H)$ is a line bundle with transition functions $\varphi_{ab} = \frac{Z^b}{Z^a}$. It is called the $\mathcal{O}_{\mathbb{P}^n}(1)$-bundle. It does not depend on the choice of H. Indeed, obviously

$H \sim_L H'$ for any two hyperplanes because $H - H'$ is the divisor of the global rational function $\frac{\sum u_k Z^k}{\sum u'_k Z^k}$.

The holomorphic sections of $\mathcal{O}(1)$ are the homogeneous polynomials of degree 1 in the Z^k. The tensor powers $L(H)^{\otimes d} \equiv \mathcal{O}_{\mathbb{P}^n}(d)$ have transition functions $\left(\frac{Z^b}{Z^a}\right)^d = \varphi^d_{ab}$. Their holomorphic sections are the homogeneous polynomials of degree d in the Z^k. Indeed if $P(Z)$ is a homogeneous polynomial of degree d in the Z^k, it defines a section of $L(H)^{\otimes d}$ by

$$s_a = \frac{P(Z)}{(Z^a)^d}$$

which is holomorphic in U_a, and

$$s_a = \left(\frac{Z^b}{Z^a}\right)^d s_b.$$

2.4 Differential forms on complex manifolds

Let M be a complex manifold of dimension n. We have defined the complex tangent bundle $V(M)$ and the holomorphic tangent bundle $T(M)$, so that

$$V(M) = T(M) \oplus \overline{T}(M).$$

We can consider the dual bundles, so that

$$V^*(M) = T^*(M) \oplus \overline{T}^*(M). \tag{2.18}$$

Definition 2.3. The bundle of k-forms on M is the kth exterior power of $V^*(M)$ and is denoted $\Lambda^k(M)$:

$$\Lambda^k(M) = \Lambda^k V^*(M). \tag{2.19}$$

The bundle of forms of type (p,q) on M is the bundle

$$\Lambda^{p,q}(M) = \Lambda^p T^*(M) \wedge \Lambda^q \overline{T}^*(M). \tag{2.20}$$

A differential form of degree k (resp. a form of type (p,q)) on an open set U of M is a differentiable section on U of the bundle $\Lambda^k(M)$ (resp. $\Lambda^{p,q}(M)$).

It is clear that

$$\Lambda^k(M) = \bigoplus_{p+q=k} \Lambda^{p,q}(M) \tag{2.21}$$

and at each point $m \in M$, a k-form $\varphi(m)$ is decomposed in a unique way as a sum of (p,q)-forms with $p+q=k$.

The bundles $\Lambda^k(M)$ and $\Lambda^{p,q}(M)$ are C^∞-bundles. We denote $\mathcal{E}^k_M(U)$ and $\mathcal{E}^{p,q}_M(U)$ (or $\Gamma(U,\mathcal{E}^k_M)$ and $\Gamma(U,\mathcal{E}^{p,q}_M)$) the space of the C^∞ sections on an open set U.

The bundles $\Lambda^{p,0}(M)$ are holomorphic vector bundles. We denote $\Omega^p_M(U)$ or $\Gamma(U,\Omega^p_M)$ the space of the holomorphic sections on U. An element of $\Omega^p_M(U)$ is called a holomorphic p-form on U.

2.4.1 Expressions in local coordinates

Let U be an open set with complex coordinates $(z^1,\ldots,z^n)$ and real coordinates x^j with $z^j = x^{2j-1} + ix^{2j}$ $j = 1,\ldots,n$.

We can use as a basis of $V(M)$ on U the $\left(\frac{\partial}{\partial x^1}, \frac{\partial}{\partial x^2},\ldots,\frac{\partial}{\partial x^{2n}}\right)$, so a dual basis of $V^*(M)$ on U is $(dx^1, dx^2,\ldots,dx^{2n})$:

$$\langle dx^j, \frac{\partial}{\partial x^k}\rangle = \delta^j_k.$$

A basis of $\Lambda^k(M)$ on M is formed by the exterior products of the dx^j of order k.

Let $I = (i_1,\ldots,i_k)$ be an ordered subset of $\{1,\ldots,2n\}$ with distinct i_l. We denote

$$dx^I = dx^{i_1} \wedge dx^{i_2} \wedge \cdots \wedge dx^{i_k}.$$

A collection I is called a multiindex of length k and we denote its length by $|I| = k$.

A multiindex is increasing if $i_1 < i_2 < \cdots < i_k$.

A basis of $\Lambda^k(M)$ on U is formed by the $\{dx^I\}$ where I are the increasing multiindices of length k and a k-form $\varphi \in \mathcal{E}^k_M(U)$ can be written

$$\varphi = {\sum_{|I|=k}}' \varphi_I \, dx^I \tag{2.22}$$

where $\sum'$ denotes a sum restricted on increasing multiindices I and φ_I are C^∞ functions on U.

Alternatively, one can also write

$$\varphi = \frac{1}{k!} \sum_{|I|=k} \varphi_I \, dx^I$$

where the sum is now on unrestricted multiindices I of length k and φ_I is skew-symmetric with respect to I.

In the same way, we have a basis $\left(\frac{\partial}{\partial z^1},\ldots,\frac{\partial}{\partial z^n}\right)$ of $T(M)$ on U, and a basis $\left(\frac{\partial}{\partial \bar z^1},\ldots,\frac{\partial}{\partial \bar z^n}\right)$ of $\overline{T}(M)$ on U. The dual basis are $(dz^1,\ldots,dz^n)$ and $(d\bar z^1,\ldots,d\bar z^n)$ for $T^*(M)$ and $\overline{T}^*(M)$

$$dz^j = dx^{2j-1} + i\,dx^{2j} \qquad d\bar z^j = dx^{2j-1} - i\,dx^{2j}. \tag{2.23}$$

If $I = (i_1 < \cdots < i_p)$ and $J = (j_1 < \cdots < j_q)$ are increasing multiindices of length p and q respectively, with the indices varying in $\{1, \ldots, n\}$, a basis of $\Lambda^{p,q}(M)$ on U is formed by the

$$dz^I \wedge d\bar{z}^J = dz^{i_1} \wedge \cdots \wedge dz^{i_p} \wedge d\bar{z}^{j_1} \wedge \cdots \wedge d\bar{z}^{j_q}. \tag{2.24}$$

A (p,q)-form $\varphi \in \mathcal{E}_M^{p,q}(U)$ can be written as

$$\varphi = \sum_{\substack{|I|=p \\ |J|=q}}{}' \varphi_{I\bar{J}}\, dz^I \wedge d\bar{z}^J \tag{2.25}$$

where again $\sum'$ denotes a sum restricted on increasing multiindices I, J and $\varphi_{I\bar{J}}$ is $C^\infty(U)$. Alternatively one can write

$$\varphi = \frac{1}{p!q!} \sum_{\substack{|I|=p \\ |J|=q}} \varphi_{I\bar{J}}\, dz^I \wedge d\bar{z}^J \tag{2.26}$$

with the sum on unrestricted multiindices and $\varphi_{I\bar{J}}$ skew-symmetric in I and skew-symmetric in J.

2.4.2 The Hodge filtrations F and $\overline{F}$

If M is a complex manifold, one can write a k-differential form φ on M in local complex coordinates as

$$\varphi = \sum_{|I|+|J|=k}{}' \varphi_{I\bar{J}}\, dz^I \wedge d\bar{z}^J$$

We say that φ has type $\geq p$, if one has $|I| \geq p$ in the above sum (for every system of complex coordinates on M). We define $F^p\mathcal{E}_M^k$ as the subsheaf of $\mathcal{E}_M^k$ of k-forms of type $\geq p$. This defines a decreasing filtration (called *the Hodge filtration*)

$$\cdots F^p\mathcal{E}_M^k \supset F^{p+1}\mathcal{E}_M^k \supset \cdots$$

and d respects this filtration:

$$d(F^p\mathcal{E}_M^k) \subset F^p\mathcal{E}_M^{k+1}$$

The *conjugate Hodge filtration* is defined by

$$\overline{F}^q\mathcal{E}_M^k = \overline{F^q\mathcal{E}_M^k}.$$

2.4.3 Pullback

If $f\colon N \to M$ is a C^∞ mapping between complex manifolds and φ is a k-form on M, one can define $f^*\varphi$ as a k-form in N. Locally, if (V, x') and

(U, x) are coordinate open sets in N and M such that $f(V) \subset U$, and φ is defined by (2.22), we define on V

$$f^*\varphi = \sum_{|I|=k}' (\varphi_I \circ f)\, f^*\left(dx^I\right)$$

$$f^*\left(dx^I\right) = f^*\left(dx^{i_1}\right) \wedge f^*\left(dx^{i_2}\right) \wedge \cdots \wedge f^*\left(dx^{i_k}\right)$$

and

$$f^*\, dx^i = d\left(x^i \circ f\right)$$

is the differential of the function $x^i \circ f$ on V and one verifies that the definition is intrinsic.

In particular, if $W \subset M$ is a real submanifold, φ a k-form, the restriction $\varphi|_W$ is a k-form on W which is the pullback of φ by the injection $j\colon W \hookrightarrow M$.

If W is defined locally by equations

$$x^r = 0 \qquad r \in R$$

(R is a subset of $\{1, \ldots, 2n\}$), then $\varphi|_W$ is obtained from (2.22) by killing all the dx^I where $I \cap R \neq \emptyset$ and restricting the φ_I with $I \cap R = \emptyset$ to W

$$\varphi|_W = \sum_{\substack{|I|=k \\ I\cap R=\emptyset}}' \varphi_I|_W\, dx^I.$$

If $f\colon N \to M$ is a holomorphic mapping, and φ a (p, q)-form on M, one can define in a similar way $f^*\varphi$ which is also a (p, q)-form on N. If φ is a $(p, 0)$-form holomorphic on M, $f^*\varphi$ is a $(p, 0)$-form holomorphic in N.

If $W \subset M$ is a complex submanifold of M, one can define $\varphi|_W$ as a (p, q)-form in W if φ is a (p, q)-form in M.

If W is defined locally by holomorphic equations

$$z^{r+1} = \cdots = z^n = 0$$

so that $\dim W = r$, $\varphi|_W$ is obtained as

$$\varphi = \sum_{\substack{1\leq i_l\leq r \\ 1\leq j_l\leq r}}' \left(\varphi_{I\bar{J}}|_W\right) dz^I \wedge d\bar{z}^J.$$

2.4.4 Exterior differentials

For a function $f\colon M \to \mathbb{C}$, one defines the differentials

$$
\begin{aligned}
df &= \sum_{j=1}^{2n} \frac{\partial f}{\partial x^j}\, dx^j \\
\partial f &= \sum_{j=1}^{n} \frac{\partial f}{\partial z^j}\, dz^j \\
\bar\partial f &= \sum_{j=1}^{n} \frac{\partial f}{\partial \bar z^j}\, d\bar z^j
\end{aligned}
\tag{2.27}
$$

so that $df = \partial f + \bar\partial f$.

For a k-form $\varphi \in \mathcal{E}^k_M(U)$, we write:

$$\varphi = \frac{1}{k!} \sum_{|I|=k} \varphi_I\, dx^I \qquad (\varphi_I \in C^\infty(U))$$

and we define in U

$$d\varphi = \frac{1}{k!} \sum_{\substack{|I|=k\\ j}} \frac{\partial \varphi_I}{\partial x^j}\, dx^j \wedge dx^I \tag{2.28}$$

Then we show easily that

$$d \circ d = 0 \tag{2.29}$$

In the sum of (2.28), $dx^j \wedge dx^I \equiv 0$ if $j \in I$ hence $\frac{\partial \varphi_I}{\partial x^j}$ never appears if $j \in I$.

For a (p,q)-form $\varphi \in \mathcal{E}^{p,q}_M(U)$, which is written as

$$\varphi = \frac{1}{p!q!} \sum_{\substack{|I|=p\\ |J|=q}} \varphi_{I\bar J}\, dz^I \wedge d\bar z^J$$

one defines

$$\partial\varphi = \frac{1}{p!q!} \sum_{\substack{|I|=p\\ |J|=q\\ l}} \frac{\partial \varphi_{I\bar J}}{\partial z^l}\, dz^l \wedge dz^I \wedge d\bar z^J \tag{2.30}$$

$$\bar\partial\varphi = \frac{(-1)^p}{p!q!} \sum_{\substack{|I|=p\\ |J|=q\\ l}} \frac{\partial \varphi_{I\bar J}}{\partial \bar z^l}\, dz^I \wedge d\bar z^l \wedge d\bar z^J \tag{2.31}$$

These definitions are extended to any k-form by decomposing the form in pure types.

One checks easily that

$$\partial \circ \partial = 0 \qquad \bar{\partial} \circ \bar{\partial} = 0 \tag{2.32}$$

$$d = \partial + \bar{\partial} \tag{2.33}$$

and as a consequence of $d^2 = 0$

$$\partial\bar{\partial} + \bar{\partial}\partial = 0 \tag{2.34}$$

One has by definition

$$\begin{aligned} \partial \mathcal{E}_M^{p,q}(U) &\subset \mathcal{E}_M^{p+1,q}(U) \\ \bar{\partial} \mathcal{E}_M^{p,q}(U)) &\subset \mathcal{E}_M^{p,q+1}(U) \end{aligned} \tag{2.35}$$

2.4.5 Exterior differentials and pullback

The main theorem, which can be checked by local calculations is that the differentials commute with pullback.

Theorem 2.1. *If $f\colon N \to M$ is a C^∞ mapping between manifolds and φ is a k-form on M*

$$df^*\varphi = f^*\, d\varphi. \tag{2.36}$$

If N, M are complex manifolds and f is holomorphic, then

$$\begin{aligned} \partial f^*\varphi &= f^*\partial\varphi \\ \bar{\partial} f^*\varphi &= f^*\bar{\partial}\varphi. \end{aligned} \tag{2.37}$$

2.4.6 Differentials and exterior products

If φ is a k-form, ψ is a l-form, $\varphi \wedge \psi$ is a $(k + l)$-form, and one verifies

$$d(\varphi \wedge \psi) = d\varphi \wedge \psi + (-1)^k \varphi \wedge d\psi. \tag{2.38}$$

If φ is a (p, q)-form and ψ a (r, s)-form, $\varphi \wedge \psi$ is a $(p + r, q + s)$-form and

$$\begin{aligned} \partial(\varphi \wedge \psi) &= \partial\varphi \wedge \psi + (-1)^{p+q} \varphi \wedge \partial\psi \\ \bar{\partial}(\varphi \wedge \psi) &= \bar{\partial}\varphi \wedge \psi + (-1)^{p+q} \varphi \wedge \bar{\partial}\psi. \end{aligned} \tag{2.39}$$

2.4.7 Forms with coefficients in a vector bundle

Let $\pi\colon F \to M$ a complex bundle. A k-form with coefficients in F is a section of the bundle $\Lambda^k(M) \otimes F$. Locally, we can use a trivialization of F on U_a, so that

$$\pi^{-1}(U_a) \xrightarrow{\sim} \mathbb{C}^r \times U_a.$$

Then a k-form with coefficients in F can be expressed on U_a as a r-uple

$$\left(\omega_a^1, \ldots, \omega_a^r\right)$$

where ω_a^α are k-forms on U_a.

Moreover if we have another trivialization of F

$$\pi^{-1}(U_b) \simeq \mathbb{C}^r \times U_b$$

and transitions $\gamma_{ab}\colon U_a \cap U_b \to \mathrm{GL}(r, \mathbb{C})$, the element $v \in \pi^{-1}(U_a \cap U_b)$ is expressed either as $v_a(v) \in \mathbb{C}^r$ or $v_b(v) \in \mathbb{C}^r$ in each of the trivializations, with

$$v_a(v) = \gamma_{ab}(\pi(v))\, v_b(v) \tag{2.40}$$

So the k-form with coefficient in F will be expressed on U_b as a r-uple

$$\left(\omega_b^1, \ldots, \omega_b^r\right)$$

where ω_b^β are k-forms on U_b and (2.40) means that :

$$\begin{pmatrix} \omega_a^1(m) \\ \vdots \\ \omega_a^r(m) \end{pmatrix} = \gamma_{ab}(m) \begin{pmatrix} \omega_b^1(m) \\ \vdots \\ \omega_b^r(m) \end{pmatrix}$$

for $m \in U_a \cap U_b$, or in other words

$$\omega_a^\alpha(m) = \sum_{\beta=1}^{r} \gamma_{ab,\beta}^\alpha(m)\omega_b^\beta(m). \tag{2.41}$$

In general, it is not possible to define the exterior differentiation of a form with coefficients in F as a form with coefficients in F. Indeed, one can define for each trivialization the differentials $\left(d\omega_a^1, \ldots, d\omega_a^r\right)$ and $\left(d\omega_b^1, \ldots, d\omega_b^r\right)$ but these differentials do not satisfy (2.41) because in general

$$d\gamma_{ab} \neq 0.$$

It is only for locally flat bundles, namely when one can choose constant transition matrices γ_{ab}, that one can define intrinsically the differential of a k-form with coefficient in F.

On the other hand, let $\pi\colon F \to M$ be a holomorphic vector bundle. One can define also the bundle of (p, q)-forms with coefficients in F, as $\Lambda^{p,q}(M) \otimes F$. *If ω is a (p, q)-form with coefficients in F, one can define $\bar{\partial}\omega$ intrinsically.*

In a trivialization $\pi^{-1}(U_a) \simeq \mathbb{C} \times U_a$, where ω is given as a r-uple

$$\left(\omega_a^1, \ldots, \omega_a^r\right)$$

one defines $\left(\bar{\partial}\omega_a^1, \ldots, \bar{\partial}\omega_a^r\right)$: this is indeed a $(p, q+1)$-form with coefficients in F because:

$$\bar{\partial}\omega_a^\alpha = \sum_{\beta=1}^{r} \gamma_{ab,\beta}^\alpha(m)\bar{\partial}\omega_b^\beta \tag{2.42}$$

as one can see immediately by taking the $\bar{\partial}$ of (2.41) and using the fact that the $\gamma_{ab,\beta}^\alpha$ are holomorphic in $U_a \cap U_b$.

2.5 Local solutions of d- and $\bar{\partial}$-equations

2.5.1 Poincaré lemma

Theorem 2.2. *Let U be a star-shaped open set in $\mathbb{R}^n$ (there exists a point $0 \in U$ so that any point x can be joined to 0 by a linear segment). Let φ be a C^∞ p-form in U such that*

$$d\varphi = 0. \tag{2.43}$$

There exists a C^∞ $(p-1)$-form ψ in U, with

$$d\psi = \varphi. \tag{2.44}$$

Proof. Obviously $d\varphi = 0$ is a necessary condition to be able to solve $d\psi = \varphi$ because $d^2 = 0$. We prove theorem 2.2 by recursion on dimension $n \geq p$.
1) First, suppose that $n = p$, so that

$$\varphi = \varphi(x)\, dx^1 \wedge \cdots \wedge dx^p \qquad x = \left(x^1, \ldots, x^p\right)$$

where $\varphi(x)$ is a function. Define

$$\psi = \psi_{1,\ldots,p-1}\, dx^1 \wedge \cdots \wedge dx^{p-1}$$

with

$$\psi_{1,\ldots,p-1}(x) = (-1)^{p-1} \int_0^{x^p} \varphi\left(x^1, \ldots, x^{p-1}, t\right) dt. \tag{2.45}$$

Then $d\psi = \varphi$.
2) Suppose now $n > p$ and assume that Poincaré lemma is correct for $n-1 \geq p$. Now write

$$\varphi = \frac{1}{p!} \sum_{\substack{1 \leq i \leq n-1 \\ |I|=p}} \varphi_I\, dx^I + \frac{1}{(p-1)!}\, dx^n \wedge \sum_{\substack{|I|=p-1 \\ 1 \leq i \leq n-1}} \varphi_{nI}\, dx^I.$$

Define

$$\psi_I\left(x^1, \ldots, x^n\right) = \int_0^{x^n} \varphi_{nI}\left(x^1, \ldots, x^{n-1}, t\right) dt$$

$$\psi = \frac{1}{(p-1)!} \sum_{\substack{|I|=p-1 \\ 1 \leq i \leq n-1}} \psi_I\, dx^I.$$

Then, it is easy to see that $\varphi - d\psi$ is a p-form involving only the dx^j for $j \leq n-1$. Moreover $d(\varphi - d\psi) = 0$, so that the coefficients of $\varphi - d\psi$ do not depend on x^n. Hence $\varphi - d\psi$ is a closed form of $(n-1)$ variables so by induction $\varphi - d\psi = d\omega$, and finally

$$\varphi = d(\psi + \omega).$$

□

2.5.2 Dolbeault lemma

Theorem 2.3. *Let U_R be a polydisk in $\mathbb{C}^n$*

$$U_R = \left\{ (z_1, \ldots, z_n) \in \mathbb{C}^n \;\middle|\; |z_j| < R_j \right\}$$

and φ be a (p,q)-form in $\mathcal{E}^{p,q}_{\mathbb{C}^n}(U_R)$. Suppose

$$\bar\partial\varphi = 0 \text{ in } U_R \tag{2.46}$$

Then there exists $\psi \in \mathcal{E}^{p,q-1}_{\mathbb{C}^n}(U_R)$ with

$$\bar\partial\psi = \varphi. \tag{2.47}$$

Proof. We can assume $p = 0$, so we deal with $(0,q)$-forms. If one wants to solve (2.46) in a smaller polydisk $U'_{R'}$ $\left(R'_j < R_j\right)$, one can take a C^∞ cut-off function χ which is 1 on $U''_{R''}$ with $R' < R''_j < R_j$ and with compact support in U_R so that $\chi\varphi \equiv \varphi$ in $U''_{R''}$ and is $\bar\partial$ closed there. Then the logic of the proof is exactly the same as the logic of the proof of Poincaré lemma, except one has to invert $\frac{\partial}{\partial \bar z}$ instead of $\frac{\partial}{\partial x}$. To invert $\frac{\partial}{\partial x}$, one had to take a primitive along a straight line segment. To invert $\frac{\partial}{\partial \bar z}$ one uses the fact that $\frac{1}{\pi z}$ satisfies

$$\frac{\partial}{\partial \bar z}\frac{1}{\pi z} = \delta_0.$$

So instead of (2.45), we use the formula

$$\psi(z) = (-1)^{q-1} \iint \frac{\varphi\left(z^1, \ldots, z^{q-1}, \zeta\right)\chi\left(z^1, \ldots, z^{q-1}, \zeta\right)}{z^q - \zeta}\,\frac{d\zeta\, d\bar\zeta}{\pi}$$

which gives

$$\frac{\partial\psi}{\partial \bar z^q} = \varphi\chi$$

so we have solved in $U'_{R'} \subset \mathbb{C}^q$ the equation

$$\bar\partial\psi = \varphi\, d\bar z^1 \wedge \cdots \wedge d\bar z^q.$$

The rest of the proof is exactly the same as in theorem 2.2. □

2.5.3 Poincaré lemma for holomorphic forms

Theorem 2.4. *Let U be a star-shaped open set in $\mathbb{C}^n$ and φ a d-closed holomorphic $(p,0)$-form on U. Then there exists a holomorphic $(p-1,0)$-form ψ on U with*

$$d\psi = \varphi.$$

Proof. The proof is just identical to the proof of Poincaré lemma, except the dx^j becomes dz^j. The main point is that if $g(z)$ is holomorphic, then by choosing a straight line γ_z from 0 to z the integral

$$f(z) = \int_{\gamma_z} g(z')\, dz'$$

is a holomorphic function of z and $\frac{\partial f}{\partial z} = g$. □

Chapter 3

Sheaves and cohomology

3.1 Sheaves

We refer to [Br], [G], [GR] for more details on the contents of the present chapter.

A *presheaf* $\mathcal{F}$ of abelian groups (or vector spaces, rings...) on a topological space X associates to each open set $U \subset X$ an abelian group (or vector space, ring...) $\mathcal{F}(U)$, and to each pair $U \subset V$ of open sets a homomorphism $r_{V,U} \colon \mathcal{F}(V) \to \mathcal{F}(U)$, called the restriction map, such that:

(1) for any triple $U \subset V \subset W$ of open sets

$$r_{W,U} = r_{V,U} \circ r_{W,V}$$

so that we can write $s|_U$ instead of $r_{V,U}(s)$.

The elements of $\mathcal{F}(U)$ are called the *sections* of $\mathcal{F}$ over U. The presheaf is a *sheaf* if moreover

(2) let $U \subset X$ be an open set, $s \in \mathcal{F}(U)$ a section, $U = \bigcup_i U_i$ an open covering of U such that $s|_{U_i} = 0$ for all i; then $s = 0$.

(3) for any open set $U \subset X$, any open covering $U = \bigcup_i U_i$ of U and sections $s_i \in \mathcal{F}(U_i)$ with $s_i|_{U_i \cap U_j} = s_j|_{U_i \cap U_j}$ there exists $s \in \mathcal{F}(U)$ such that $s|_U = s_i$. Such s is unique by (2).

A sub(pre)-sheaf of $\mathcal{F}$ is a (pre)-sheaf $\mathcal{G}$ such that $\mathcal{G}(U) \subset \mathcal{F}(U)$ is a subgroup (a vector subspace...) and the restrictions in $\mathcal{G}$ are induced by those in $\mathcal{F}$.

If a presheaf $\mathcal{F}$ is not a sheaf, one can construct the *associated sheaf* $\tilde{\mathcal{F}}$ simply by adding to each $\mathcal{F}(U)$ the missing sections: an element of $\tilde{\mathcal{F}}(U)$ is a family $(s_i \in \mathcal{F}(U_i))$ with $s_i|_{U_i \cap U_j} = s_j|_{U_i \cap U_j}$, $U = \bigcup_i U_i$ being an open covering (two such families must be identified if their restrictions to some refinement of the respective coverings coincide).

There is a canonical morphism $\mathcal{F} \to \tilde{\mathcal{F}}$; if the presheaf $\mathcal{F}$ already satisfies the property (2), which is true in most interesting cases, $\mathcal{F}$ is a subpresheaf of $\tilde{\mathcal{F}}$.

Examples. The following are examples of sheaves (the restriction maps are the usual ones).

(i) The constant sheaf $\mathbb{C}_X$ on a topological space X, whose sections on an open set U are the locally constant $\mathbb{C}$-valued functions on U. More generally, the constant sheaf K_X, where K is any abelian group.

(ii) The sheaf $\mathcal{E}^k_M$ of complex differential k-forms on a differentiable manifold M.

(iii) The sheaf $\mathcal{O}_M$ of holomorphic functions, and the sheaf Ω^p_M of holomorphic p-forms on a complex manifold M.

(iv) The sheaf $\mathcal{F}$ of differentiable (resp. holomorphic) sections of a differentiable (resp. holomorphic) vector bundle F on a differentiable (resp. complex) manifold M: the elements of $\mathcal{F}(U)$ are the differentiable (resp. holomorphic) sections of F on U.

A *global section* of a (pre)sheaf $\mathcal{F}$ on X is an element of $\mathcal{F}(X)$. Quite often $\mathcal{F}(X)$ is denoted by $\Gamma(X, \mathcal{F})$.

A (pre)sheaf $\mathcal{F}$ on X restricts to a (pre)sheaf $\mathcal{F}|_U$ on any open set $U \subset X$: if $V \subset U$ is open, $\mathcal{F}|_U(V) = \mathcal{F}(V)$.

A *morphism of presheaves* $f\colon \mathcal{F} \to \mathcal{G}$ is a collection of morphisms $f_U\colon \mathcal{F}(U) \to \mathcal{G}(U)$ (U open in X), such that if $U \subset V$, f_U and f_V commute with the restriction maps. A *morphism of sheaves* is a morphism of the underlying presheaves. If $f\colon \mathcal{F} \to \mathcal{G}$ is a morphism of sheaves, the presheaf defined by the assignment $U \mapsto \ker f_U$ is easily seen to be a sheaf, which is called *the kernel* of f and denoted $\ker f$; hence $(\ker f)(U) = \ker f_U$ for any open set U. On the contrary, the assignment $U \mapsto \operatorname{im} f_U$ defines a presheaf, which is not a sheaf in general. Hence we must define the *image* sheaf $\operatorname{im} f$ as the sheaf associated to the presheaf $U \mapsto \operatorname{im} f_U$. Thus in general $(\operatorname{im} f)(U) \supset \operatorname{im} f_U$ and $(\operatorname{im} f)(U) \neq \operatorname{im} f_U$. In the same way we define the sheaf $\operatorname{coker} f$ as the sheaf associated to the presheaf $U \mapsto \operatorname{coker} f_U$.

If $\mathcal{F} \subset \mathcal{G}$ is a subsheaf, the quotient $\mathcal{G}/\mathcal{F}$ is the sheaf associated to the presheaf $U \mapsto \mathcal{G}(U)/\mathcal{F}(U)$, and in general $(\mathcal{G}/\mathcal{F})(U) \neq \mathcal{G}(U)/\mathcal{F}(U)$.

Let $\mathcal{F}$ be a (pre)-sheaf on X; if A is a subset of X, the set $\mathcal{F}(A)$ of the sections of $\mathcal{F}$ on A can be defined. If A has a fundamental system of paracompact neighborhoods in X, $\mathcal{F}(A)$ is the quotient of the set

$$\{ s \in \mathcal{F}(U),\ U \text{ an open neighborhood of } A \text{ in } X \}$$

under the equivalence relation: if $s \in \mathcal{F}(U)$, $t \in \mathcal{F}(V)$, $s \equiv t$: there exists a neighborhood $W \subset U \cap V$ of A with $s|_W = t|_W$. In other words, an element of $\mathcal{F}(A)$ is a section of $\mathcal{F}$ in an open neighborhood of A, provided one identifies two such sections when they coincide in a smaller neighborhood. $\mathcal{F}(A)$ carries the same structure (abelian group, vector space...) as the $\mathcal{F}(U)$.

If x is a point of X the *stalk*, or *fiber*, $\mathcal{F}_x$ of $\mathcal{F}$ at x is $\mathcal{F}(A)$, $A = \{x\}$. An element of $\mathcal{F}_x$ is a section of $\mathcal{F}$ in an open neighborhood of x, and two such sections must be identified when they coincide in a smaller neighborhood of x. A section s of $\mathcal{F}$ defined in a open neighborhood of x detects an element

$s_x \in \mathcal{F}_x$ called *the germ* of s at x. Also the stalks $\mathcal{F}_x$ carry the same structure (abelian group, vector space...) as the $\mathcal{F}(U)$.

Let s be a section of $\mathcal{F}$ on an open set U of X. If $s_x = 0$ at a point $x \in U$, there exists an open neighborhood V of x in U such that $s|_V = 0$ so that $s_y = 0$ for $y \in V$. Hence the *support of* s, defined as the subset

$$\operatorname{supp}(s) = \{\, x \in U \mid s_x \neq 0 \,\}$$

is closed in U.

The *restriction* $\mathcal{F}|_A$ of the sheaf $\mathcal{F}$ on X to a subset A of X is defined by $\mathcal{F}|_A(V) = \mathcal{F}(V)$, V being an open subset of A. Let us remark that at $x \in A$, the stalks are the same: $(\mathcal{F}|_A)_x = \mathcal{F}_x$.

A morphism of (pre)-sheaves $f\colon \mathcal{F} \to \mathcal{G}$ induces for $x \in X$ a morphism $f_x\colon \mathcal{F}_x \to \mathcal{G}_x$.

If $\mathcal{F}$ is a presheaf, $\tilde{\mathcal{F}}$ is the associated sheaf and $\mathcal{F} \subset \tilde{\mathcal{F}}$, for all $x \in X$ we have $\mathcal{F}_x = \tilde{\mathcal{F}}_x$. In particular let us remark that if $f\colon \mathcal{F} \to \mathcal{G}$ is a morphism of sheaves, $(\ker f)_x = \ker f_x$, $(\operatorname{im} f)_x = \operatorname{im} f_x$, $(\operatorname{coker} f)_x = \operatorname{coker} f_x$. If $\mathcal{F} \subset \mathcal{G}$ is a subsheaf, $(\mathcal{G}/\mathcal{F})_x = \mathcal{G}_x/\mathcal{F}_x$.

Note also that if $\mathcal{F}_1 \subset \mathcal{F}_2$ are sheaves and for all $x \in X$, $\mathcal{F}_{1,x} = \mathcal{F}_{2,x}$, it follows that $\mathcal{F}_1 = \mathcal{F}_2$.

We say that a sequence of sheaf morphisms

$$\mathcal{E} \xrightarrow{\;f\;} \mathcal{F} \xrightarrow{\;g\;} \mathcal{G} \tag{3.1}$$

is *a complex* if $g \circ f = 0$, and is *exact* if it is a complex and the following equality of sheaves holds: $\ker g = \operatorname{im} f$. This is equivalent to $\ker g_x = \operatorname{im} f_x$ for all x, and thus to the exactness of the sequence of abelian groups (or vector spaces...)

$$\mathcal{E}_x \xrightarrow{\;f_x\;} \mathcal{F}_x \xrightarrow{\;g_x\;} \mathcal{G}_x \tag{3.2}$$

for all $x \in X$. So finally the sequence (3.1) is a complex (resp. is exact) if and only if (3.2) is a complex (resp. is exact) for all $x \in X$.

More generally we say that a sequence of maps of sheaves

$$\cdots \longrightarrow \mathcal{F}_{n-1} \xrightarrow{\;f_{n-1}\;} \mathcal{F}_n \xrightarrow{\;f_n\;} \mathcal{F}_{n+1} \longrightarrow \cdots \tag{3.3}$$

is a complex (resp. is exact) if the sequence

$$\cdots \longrightarrow \mathcal{F}_{n-1,x} \xrightarrow{\;f_{n-1,x}\;} \mathcal{F}_{n,x} \xrightarrow{\;f_{n,x}\;} \mathcal{F}_{n+1,x} \longrightarrow \cdots \tag{3.4}$$

is a complex (resp. is exact) for all $x \in X$. If (3.3) is a complex, it is easy to see that for an open set $U \subset X$ the induced sequence

$$\cdots \longrightarrow \mathcal{F}_{n-1}(U) \xrightarrow{\;f_{n-1,U}\;} \mathcal{F}_n(U) \xrightarrow{\;f_{n,U}\;} \mathcal{F}_{n+1}(U) \longrightarrow \cdots \tag{3.5}$$

is a complex. But if (3.3) is exact, (3.5) is not exact in general. *The lack of exactness in* (3.5) *gives rise in fact to the cohomology theory of sheaves.*

Nevertheless: if

$$0 \longrightarrow \mathcal{E} \longrightarrow \mathcal{F} \longrightarrow \mathcal{G}$$

is an exact sequence of sheaves, for each open set $U \subset X$ the sequence

$$0 \longrightarrow \mathcal{E}(U) \longrightarrow \mathcal{F}(U) \longrightarrow \mathcal{G}(U)$$

is exact.

Let us note also that there is no useful definition of exact sequence for presheaves. Nevertheless the following remark will be helpful.

Remark 3.1. Let

$$\cdots \longrightarrow \mathcal{F}_{n-1} \xrightarrow{f_{n-1}} \mathcal{F}_n \xrightarrow{f_n} \mathcal{F}_{n+1} \longrightarrow \cdots$$

be a sequence of morphisms of presheaves such that for any open set $U \subset X$ the induced sequence

$$\cdots \longrightarrow \mathcal{F}_{n-1}(U) \xrightarrow{f_{n-1,U}} \mathcal{F}_n(U) \xrightarrow{f_{n,U}} \mathcal{F}_{n+1}(U) \longrightarrow \cdots$$

is exact. Then the sequence of associated sheaves

$$\cdots \longrightarrow \tilde{\mathcal{F}}_{n-1} \xrightarrow{\tilde{f}_{n-1}} \tilde{\mathcal{F}}_n \xrightarrow{\tilde{f}_n} \tilde{\mathcal{F}}_{n+1} \longrightarrow \cdots$$

is exact.

Let $Y \subset X$ be a locally closed subset. A sheaf $\mathcal{G}$ on Y can be "extended by zero" to a sheaf on X, i.e. to a sheaf $\tilde{\mathcal{G}}$ whose restriction to Y is $\mathcal{G}$, and whose restriction to $X \setminus Y$ is identically zero, by formula

$$\tilde{\mathcal{G}}(U) = \{\, s \in \mathcal{G}(U \cap Y) \mid \operatorname{supp}(s) \text{ is closed in } U \,\} \tag{3.6}$$

(Note that $\operatorname{supp}(s)$ is apriori closed in $U \cap Y$). If $x \notin Y$, let $s \in \tilde{\mathcal{G}}(U)$ (U being an open neighborhood of x in X); then $S = \operatorname{supp}(s)$ is closed in U, and $x \notin S$, so there is an open neighborhood V of x in U such that $s|_V = 0$; thus $s_x = 0$, which implies $\tilde{\mathcal{G}}_x = 0$. If $x \in Y$, there is a fundamental system of open neighborhoods U of x in X such that $U \cap Y$ is closed in U (here we use the assumption that Y is locally closed). Then for every $s \in \mathcal{G}(U \cap Y)$, $\operatorname{supp}(s)$ is closed in U, or $\tilde{\mathcal{G}}(U) = \mathcal{G}(U \cap Y)$; it follows that $\tilde{\mathcal{G}}_x = \mathcal{G}_x$. We have thus proved that $\tilde{\mathcal{G}}|_Y = \mathcal{G}$ and $\tilde{\mathcal{G}}|_{X \setminus Y} = 0$.

The extension of $\mathcal{G}$ to X is still denoted $\mathcal{G}$ when no confusion arises.

Let $Y \subset X$ be a locally closed subset and $\mathcal{L}$ be a sheaf on X. We can restrict $\mathcal{L}$ to Y, obtaining $\mathcal{L}|_Y$, and then extend $\mathcal{L}|_Y$ by zero to X. We denote by $\mathcal{L}^Y$ such extension, so that $(\mathcal{L}^Y)\,|_Y = \mathcal{L}|_Y$ and $(\mathcal{L}^Y)\,|_{X \setminus Y} = 0$.

If Y is open, $\mathcal{L}^Y$ is a subsheaf of $\mathcal{L}$; in fact for any open set U of X, by definition (3.6) we have $\mathcal{L}^Y(U) = \{\, s \in \mathcal{L}|_Y(U \cap Y) \mid \operatorname{supp}(s) \text{ is closed in } U\}$; but $\mathcal{L}|_Y(U \cap Y) = \mathcal{L}(U \cap Y)$ because $U \cap Y$ is open in X; finally $\mathcal{L}^Y(U) \subset \mathcal{L}(U)$ because every $s \in \mathcal{L}(U \cap Y)$ whose support is closed in U can be uniquely extended by zero to a section of $\mathcal{L}(U)$.

If Y is closed , $\mathcal{L}^Y$ is a quotient of $\mathcal{L}$; in this case the relation $\mathcal{L}^Y(U) = \mathcal{L}(U \cap Y)$ shows that there is a canonical surjective morphism $\mathcal{L} \to \mathcal{L}^Y$, sending a section of $\mathcal{L}$ on U to its restriction to $U \cap Y$. In conclusion:

Proposition 3.1. *Let Y be a closed subset of X. For any sheaf $\mathcal{L}$ on X there is an exact sequence of sheaves on X*

$$0 \longrightarrow \mathcal{L}^{X \setminus Y} \longrightarrow \mathcal{L} \longrightarrow \mathcal{L}^Y \longrightarrow 0 \tag{3.7}$$

3.2 The cohomology of sheaves

Let Φ be a family of closed sets of a topological space X verifying the conditions

(1) The union of two sets of Φ belongs to Φ.

(2) If $S \in \Phi$, every closed subset of S belongs to Φ.

We say that Φ is a *family of supports* in X. If $\mathcal{F}$ is a sheaf (of abelian groups) on X, we denote by $\Gamma_\Phi(X, \mathcal{F})$, or simply $\Gamma_\Phi(\mathcal{F})$, the subgroup of $\Gamma(X, \mathcal{F})$ of the sections s such that $\operatorname{supp}(s)$ belongs to Φ. For an open set $U \subset X$ we write $\Gamma_\Phi(U, \mathcal{F})$ for $\Gamma_\Phi(U, \mathcal{F}|_U)$. When Φ is the family of all the compact subsets of X, we denote $\Gamma_\Phi(X, \mathcal{F})$ by $\Gamma_c(X, \mathcal{F})$, the set of sections of $\mathcal{F}$ with compact support. When Φ is the family of all the closed subsets of X, we simply write $\Gamma(X, \mathcal{F})$.

A family Φ of supports is called a *paracompactifying family of supports* if every $S \in \Phi$ is paracompact, and possesses a neighborhood belonging to Φ. If X is paracompact, the family of all the closed subsets, and the family of all the compact subsets of X, are paracompactifying.

We say that a sheaf (of abelian groups) $\mathcal{F}$ on X is *flabby* if for each open set $U \subset X$ the restriction morphism $\mathcal{F}(X) \to \mathcal{F}(U)$ is surjective.

The restriction of a flabby sheaf to an open set is flabby. Note also that a sheaf $\mathcal{G}$ on a locally closed subset Y of X is flabby if and only if its extension by zero to X is flabby.

It is easy to see that if $\mathcal{F}$ is a flabby sheaf on X, for any family Φ of supports in X, and any open set U, the restriction $\Gamma_\Phi(X, \mathcal{F}) \to \Gamma_\Phi(U, \mathcal{F})$ remains surjective.

3.2.1 The canonical flabby sheaf $\mathcal{C}^0\mathcal{F}$ associated to a given sheaf $\mathcal{F}$

A section $s \in \mathcal{C}^0\mathcal{F}$ on an open set U is a family $\left(s_x^{(x)} \in \mathcal{F}_x\right)_{x\in U}$ (with no gluing conditions); the lack of gluing conditions implies that the restriction $\left(\mathcal{C}^0\mathcal{F}\right)(X) \to \left(\mathcal{C}^0\mathcal{F}\right)(U)$ is surjective.

The sheaf $\mathcal{F}$ injects as a subsheaf of $\mathcal{C}^0\mathcal{F}$: $s \in \mathcal{F}(U)$ goes to $(s_x \in \mathcal{F}_x)_{x\in U}$. Moreover $\mathcal{C}^0\mathcal{F}$ carries the same structure as $\mathcal{F}$.

When we need to be precise about the space X on which we construct $\mathcal{C}^0\mathcal{F}$, we write $\mathcal{C}^0(X, \mathcal{F})$ instead of $\mathcal{C}^0\mathcal{F}$.

The construction of $\mathcal{C}^0\mathcal{F}$ is functorial: a morphism $\mathcal{F} \to \mathcal{G}$ extends to a natural morphism $\mathcal{C}^0\mathcal{F} \to \mathcal{C}^0\mathcal{G}$. More precisely it can be proved that

Proposition 3.2. *If*

$$0 \longrightarrow \mathcal{E}_1 \longrightarrow \mathcal{E}_2 \longrightarrow \mathcal{E}_3 \longrightarrow 0$$

is an exact sequence of sheaves, the sequence

$$0 \longrightarrow \mathcal{C}^0\mathcal{E}_1 \longrightarrow \mathcal{C}^0\mathcal{E}_2 \longrightarrow \mathcal{C}^0\mathcal{E}_3 \longrightarrow 0$$

is also exact.

Lemma 3.1. *Let*

$$0 \longrightarrow \mathcal{E}_1 \longrightarrow \mathcal{E}_2 \longrightarrow \mathcal{E}_3 \longrightarrow 0$$

be an exact sequence of sheaves.

(1) If $\mathcal{E}_1$ is flabby, for each open set $U \subset X$ the sequence

$$0 \longrightarrow \mathcal{E}_1(U) \longrightarrow \mathcal{E}_2(U) \longrightarrow \mathcal{E}_3(U) \longrightarrow 0$$

is exact. If moreover Φ is a family of supports in X, the sequence

$$0 \longrightarrow \Gamma_\Phi(X, \mathcal{E}_1) \longrightarrow \Gamma_\Phi(X, \mathcal{E}_2) \longrightarrow \Gamma_\Phi(X, \mathcal{E}_3) \longrightarrow 0$$

is exact.

(2) If $\mathcal{E}_1$ and $\mathcal{E}_2$ are flabby, $\mathcal{E}_3$ is flabby.

Corollary 3.1. *Let*

$$0 \longrightarrow \mathcal{E}_1 \longrightarrow \mathcal{E}_2 \longrightarrow \mathcal{E}_3 \longrightarrow \cdots$$

be an exact sequence of flabby sheaves on X. For each open subset $U \subset X$ the sequence

$$0 \longrightarrow \mathcal{E}_1(U) \longrightarrow \mathcal{E}_2(U) \longrightarrow \mathcal{E}_3(U) \longrightarrow \cdots$$

is exact. If moreover Φ is a family of supports in X, the sequence

$$0 \longrightarrow \Gamma_\Phi(X, \mathcal{E}_1) \longrightarrow \Gamma_\Phi(X, \mathcal{E}_2) \longrightarrow \Gamma_\Phi(X, \mathcal{E}_3) \longrightarrow \cdots$$

is exact.

3.2.2 Resolutions of sheaves

A complex of sheaves $(\mathcal{E}^k, f^k)$, is a sequence of sheaves $\mathcal{E}^k$ $(k \in \mathbb{N})$ and morphisms of sheaves $f^k \colon \mathcal{E}^k \to \mathcal{E}^{k+1}$ such that $f^{k+1} \circ f^k = 0$. The complex is said to be exact if the sequence of sheaves

$$\cdots \longrightarrow \mathcal{E}^{k-1} \xrightarrow{f^{k-1}} \mathcal{E}^k \xrightarrow{f^k} \mathcal{E}^{k+1} \longrightarrow \cdots$$

is exact.

A morphism of sheaves $\mathcal{E} \to \mathcal{F}$ is by definition injective (resp. surjective) if the sequence $0 \to \mathcal{E} \to \mathcal{F}$ (resp. $\mathcal{E} \to \mathcal{F} \to 0$) is exact, or in other words, if for every $x \in X$ the morphism of stalks $\mathcal{E}_x \to \mathcal{F}_x$ is injective (resp. surjective).

A *resolution* of a sheaf $\mathcal{E}$ is an exact sequence of sheaves

$$0 \longrightarrow \mathcal{E} \xrightarrow{\epsilon} \mathcal{E}^0 \xrightarrow{f^0} \mathcal{E}^1 \xrightarrow{f^1} \mathcal{E}^2 \xrightarrow{f^2} \mathcal{E}^3 \longrightarrow \cdots$$

where the morphism ϵ is called the *augmentation*. So a resolution of $\mathcal{E}$ is an exact complex $(\mathcal{E}^k, f^k)_{k\geq 0}$ together with an augmentation morphism $\epsilon \colon \mathcal{E} \to \mathcal{E}^0$ inducing an isomorphism $\mathcal{E} \simeq \ker f^0$.

There is an obvious definition of a morphism of complexes and morphism of resolutions.

A resolution is called flabby if all the sheaves $\mathcal{E}^k$ are flabby for $k \geq 0$.

Lemma 3.2. *Every sheaf $\mathcal{E}$ has a canonical flabby resolution $(\mathcal{C}^k\mathcal{E}, d^k)_{k\geq 0}$. Every morphism $\mathcal{E}_1 \to \mathcal{E}_2$ extends to a morphism of the canonical resolutions $(\mathcal{C}^k\mathcal{E}_1, d_1^k) \to (\mathcal{C}^k\mathcal{E}_2, d_2^k)$. When we need to be precise about the space X on which we construct the resolution, we write $\mathcal{C}^k(X, \mathcal{E})$ instead of $\mathcal{C}^k\mathcal{E}$.*

The proof is very easy. Let $\epsilon \colon \mathcal{E} \to \mathcal{C}^0\mathcal{E}$ be the injection into the canonical flabby sheaf of $\mathcal{E}$. Let $\mathcal{C}^1\mathcal{E} = \mathcal{C}^0(\mathcal{Z}_1)$, where $\mathcal{Z}_1 = \mathcal{C}^0\mathcal{E}/\mathcal{E}$; hence $\mathcal{Z}_1$ is a subsheaf of $\mathcal{C}^1\mathcal{E}$ and we obtain an exact sequence

$$0 \longrightarrow \mathcal{E} \xrightarrow{\epsilon} \mathcal{C}^0\mathcal{E} \xrightarrow{d^0} \mathcal{C}^1\mathcal{E} \tag{3.8}$$

Then we define inductively $\mathcal{C}^2\mathcal{E} = \mathcal{C}^0(\mathcal{Z}_2)$, where $\mathcal{Z}_2 = \mathcal{C}^1\mathcal{E}/\mathcal{Z}_1 \dots$. The construction shows that a morphism $\mathcal{E}_1 \to \mathcal{E}_2$ extends to the resolutions.

3.2.3 Cohomology of sheaves

Definition 3.1. Let $\mathcal{E}$ be a sheaf on a topological space X and Φ be a family of supports in X. The cohomology groups $H^k_\Phi(X, \mathcal{E})$ of X with supports in Φ and coefficients in $\mathcal{E}$ are the cohomology groups of the following complex of groups

$$0 \longrightarrow \Gamma_\Phi(X, \mathcal{C}^0\mathcal{E}) \xrightarrow{d^0} \Gamma_\Phi(X, \mathcal{C}^1\mathcal{E}) \xrightarrow{d^1} \Gamma_\Phi(X, \mathcal{C}^2\mathcal{E}) \xrightarrow{d^2} \cdots \tag{3.9}$$

that is

$$H^k_\Phi(X,\mathcal{E}) = \frac{\ker\{\, d^k : \Gamma_\Phi(X,\mathcal{C}^k\mathcal{E}) \to \Gamma_\Phi(X,\mathcal{C}^{k+1}\mathcal{E})\,\}}{d^{k-1}\Gamma_\Phi(X,\mathcal{C}^{k-1}\mathcal{E})} \tag{3.10}$$

$H^k_\Phi(X,\mathcal{E})$ is called the k-th cohomology group, or the cohomology group of degree k, of $\mathcal{E}$ on X, with supports in Φ. If Φ is the family of all the closed subsets of X, we denote by $H^k(X,\mathcal{E})$ the corresponding cohomology group, and we simply call it the k-th cohomology group of $\mathcal{E}$ on X. If Φ is the family of all the compact subsets of X, we denote by $H^k_c(X,\mathcal{E})$ the corresponding cohomology group, and we call it the k-th cohomology group of $\mathcal{E}$ on X with compact supports. If $U \subset X$ is open, we define $H^k(U,\mathcal{E}) = H^k(U,\mathcal{E}|_U)$ and call it the cohomology group of degree k of $\mathcal{E}$ on U.

We deduce the following properties of the cohomology groups.

- The $H^k_\Phi(X,\mathcal{E})$ carry the same structure as the $\Gamma_\Phi(X,\mathcal{E})$.
- $H^k_\Phi(X,\mathcal{E}) = 0$ for $k < 0$.
- $H^0_\Phi(X,\mathcal{E}) = \Gamma_\Phi(X,\mathcal{E})$.
  In fact the sequence (3.8) is exact, so that the sequence

  $$0 \longrightarrow \Gamma_\Phi(X,\mathcal{E}) \xrightarrow{\ \epsilon\ } \Gamma_\Phi(X,\mathcal{C}^0\mathcal{E}) \xrightarrow{\ d^0\ } \Gamma_\Phi(X,\mathcal{C}^1\mathcal{E})$$

  is exact, which implies, by (3.10), $H^0_\Phi(X,\mathcal{E}) = \Gamma_\Phi(X,\mathcal{E})$.
- If $\mathcal{E}$ is a flabby sheaf, $H^k_\Phi(X,\mathcal{E}) = 0$ for $k > 0$.
  If $\mathcal{E}$ is flabby, the canonical flabby resolution

  $$0 \longrightarrow \mathcal{E} \xrightarrow{\ \epsilon\ } \mathcal{C}^0\mathcal{E} \xrightarrow{\ d^0\ } \mathcal{C}^1\mathcal{E} \xrightarrow{\ d^1\ } \cdots \tag{3.11}$$

  is an exact sequence of flabby sheaves, hence by corollary 3.1 the sequence

  $$\begin{aligned} 0 \longrightarrow \Gamma_\Phi(X,\mathcal{E}) &\xrightarrow{\ \epsilon\ } \Gamma_\Phi(X,\mathcal{C}^0\mathcal{E}) \\ &\xrightarrow{\ d^0\ } \Gamma_\Phi(X,\mathcal{C}^1\mathcal{E}) \xrightarrow{\ d^1\ } \cdots \end{aligned} \tag{3.12}$$

  remains exact, and coincides in positive degrees with the sequence (3.9), hence the cohomology groups vanish in positive degrees.
- A morphism of sheaves $\mathcal{E}_1 \to \mathcal{E}_2$ induces natural homomorphisms of cohomology groups $H^k_\Phi(X,\mathcal{E}_1) \to H^k_\Phi(X,\mathcal{E}_2)$.
- A (short) exact sequence of sheaves

  $$0 \longrightarrow \mathcal{E}_1 \longrightarrow \mathcal{E}_2 \longrightarrow \mathcal{E}_3 \longrightarrow 0$$

gives rise to a long exact sequence in cohomology:

$$\begin{aligned} 0 \to H^0_\Phi(X, \mathcal{E}_1) \to H^0_\Phi(X, \mathcal{E}_2) \to H^0_\Phi(X, \mathcal{E}_3) \to H^1_\Phi(X, \mathcal{E}_1) \cdots \\ \cdots H^k_\Phi(X, \mathcal{E}_1) \to H^k_\Phi(X, \mathcal{E}_2) \to H^k_\Phi(X, \mathcal{E}_3) \to H^{k+1}_\Phi(X, \mathcal{E}_1) \cdots \end{aligned} \tag{3.13}$$

where the morphisms $H^k_\Phi(X, \mathcal{E}_3) \to H^{k+1}_\Phi(X, \mathcal{E}_1)$ are the so called *connecting homomorphisms*, and the others are natural homomorphisms as above.

The construction of the long exact sequence starts with a diagram

$$\begin{array}{ccccccccc}
 & & 0 & & 0 & & 0 & & \\
 & & \downarrow & & \downarrow & & \downarrow & & \\
0 & \longrightarrow & \mathcal{E}_1 & \longrightarrow & \mathcal{E}_2 & \longrightarrow & \mathcal{E}_3 & \longrightarrow & 0 \\
 & & \downarrow{\scriptstyle \epsilon_1} & & \downarrow{\scriptstyle \epsilon_2} & & \downarrow{\scriptstyle \epsilon_3} & & \\
0 & \longrightarrow & \mathcal{C}^0\mathcal{E}_1 & \longrightarrow & \mathcal{C}^0\mathcal{E}_2 & \longrightarrow & \mathcal{C}^0\mathcal{E}_3 & \longrightarrow & 0 \\
 & & \downarrow{\scriptstyle d^0_1} & & \downarrow{\scriptstyle d^0_2} & & \downarrow{\scriptstyle d^0_3} & & \\
0 & \longrightarrow & \mathcal{C}^1\mathcal{E}_1 & \longrightarrow & \mathcal{C}^1\mathcal{E}_2 & \longrightarrow & \mathcal{C}^1\mathcal{E}_3 & \longrightarrow & 0 \\
 & & \downarrow{\scriptstyle d^1_1} & & \downarrow{\scriptstyle d^1_2} & & \downarrow{\scriptstyle d^1_3} & & \\
 & & \vdots & & \vdots & & \vdots & &
\end{array}$$

where the columns are the canonical flabby resolutions of the three sheaves, and the morphisms in the rows are the extensions of the morphisms in the first row to the canonical resolutions. The rows are exact (as it follows from proposition 3.2). Passing to global sections in the above diagram, and replacing the first row by the zero row, we obtain a diagram

$$\begin{array}{ccccccccc}
 & & 0 & & 0 & & 0 & & \\
 & & \downarrow & & \downarrow & & \downarrow & & \\
0 & \longrightarrow & \Gamma_\Phi(X, \mathcal{C}^0\mathcal{E}_1) & \longrightarrow & \Gamma_\Phi(X, \mathcal{C}^0\mathcal{E}_2) & \longrightarrow & \Gamma_\Phi(X, \mathcal{C}^0\mathcal{E}_3) & \longrightarrow & 0 \\
 & & \downarrow{\scriptstyle d^0_1} & & \downarrow{\scriptstyle d^0_2} & & \downarrow{\scriptstyle d^0_3} & & \\
0 & \longrightarrow & \Gamma_\Phi(X, \mathcal{C}^1\mathcal{E}_1) & \longrightarrow & \Gamma_\Phi(X, \mathcal{C}^1\mathcal{E}_2) & \longrightarrow & \Gamma_\Phi(X, \mathcal{C}^1\mathcal{E}_3) & \longrightarrow & 0 \\
 & & \downarrow{\scriptstyle d^1_1} & & \downarrow{\scriptstyle d^1_2} & & \downarrow{\scriptstyle d^1_3} & & \\
 & & \vdots & & \vdots & & \vdots & &
\end{array}$$

whose rows are exact by lemma 3.1 and whose columns are the complexes whose cohomologies are the $H^k_\Phi(X,\mathcal{E}_j)$. Hence we obtain the long cohomology sequence by general results (see chapter 1, section 1.15).

The cohomology groups of a sheaf $\mathcal{E}$ have been defined by means of the flabby canonical resolution, which is not much suitable for the computations (the sheaves $\mathcal{C}^k\mathcal{E}$ are too big). Fortunately, many other resolutions can be used to compute the cohomology of $\mathcal{E}$.

Theorem 3.1. *Let* $\left(\mathcal{L}^k, d^k\right)_{k\geq 0}$ *be a resolution of* $\mathcal{E}$ *such that* $H^p_\Phi\left(X,\mathcal{L}^k\right)=0$ *for every* $p>0$ *and every* $k\geq 0$*. Then the cohomology groups* $H^p_\Phi(X,\mathcal{E})$ *are isomorphic to the cohomology groups of the complex of the global sections* $\left(\Gamma_\Phi\left(X,\mathcal{L}^{\cdot}\right),d\right)$*:*

$$0\longrightarrow\Gamma_\Phi\left(X,\mathcal{L}^0\right)\longrightarrow\Gamma_\Phi\left(X,\mathcal{L}^1\right)\longrightarrow\Gamma_\Phi\left(X,\mathcal{L}^2\right)\longrightarrow\cdots \tag{3.14}$$

In particular, any flabby resolution of a sheaf is suitable for the computation of its cohomology.

Proof. We limit ourselves to the case of the cohomology groups $H^p(X,\mathcal{E})$, that is, Φ is the family of all the closed subsets of X. We consider the exact sequence

$$0\longrightarrow\mathcal{E}\longrightarrow\mathcal{L}^0\xrightarrow{d^0}\mathcal{L}^1\xrightarrow{d^1}\mathcal{L}^2\xrightarrow{d^2}\cdots$$

and we remark that the sequence

$$0\longrightarrow\mathcal{E}(X)\longrightarrow\mathcal{L}^0(X)\xrightarrow{d^0}\mathcal{L}^1(X)$$

remains exact, hence $H^0(X,\mathcal{E})=\mathcal{E}(X)=\ker\left\{d_0\colon\mathcal{L}^0(X)\to\mathcal{L}^1(X)\right\}$.

For $i\geq 1$ let $\mathcal{Z}^i=\ker d^i=\operatorname{im} d^{i-1}$. The exact sequence

$$0\longrightarrow\mathcal{Z}^i\longrightarrow\mathcal{L}^i\longrightarrow\mathcal{Z}^{i+1}\longrightarrow 0$$

gives rise to the long exact sequence

$$0\longrightarrow H^0\left(X,\mathcal{Z}^i\right)\longrightarrow H^0\left(X,\mathcal{L}^i\right)\longrightarrow H^0\left(X,\mathcal{Z}^{i+1}\right)\longrightarrow H^1\left(X,\mathcal{Z}^i\right)\cdots$$
$$\cdots H^p\left(X,\mathcal{Z}^i\right)\longrightarrow H^p\left(X,\mathcal{L}^i\right)\longrightarrow H^p\left(X,\mathcal{Z}^{i+1}\right)\longrightarrow H^{p+1}\left(X,\mathcal{Z}^i\right)\cdots$$

From the assumptions $H^p\left(X,\mathcal{L}^i\right)=0$ $(p>0)$ we obtain

$$H^1\left(X,\mathcal{Z}^i\right)\simeq\frac{H^0\left(X,\mathcal{Z}^{i+1}\right)}{d^i\left(H^0\left(X,\mathcal{L}^i\right)\right)}\simeq\frac{\mathcal{Z}^{i+1}(X)}{d^i\left(\mathcal{L}^i(X)\right)}$$
$$H^p\left(X,\mathcal{Z}^i\right)\simeq H^{p+1}\left(X,\mathcal{Z}^{i-1}\right)\simeq\cdots\simeq H^{p+i}(X,\mathcal{E})$$

and in particular $H^i(X,\mathcal{E})\simeq H^1\left(X,\mathcal{Z}^{i-1}\right)$.

On the other hand we compute the cohomology of the complex (3.14)

$$H^i\left(\mathcal{L}^{\cdot}(X)\right) = \frac{\mathcal{Z}^i(X)}{d^{i-1}\left(\mathcal{L}^{i-1}(X)\right)} \simeq H^1\left(X, \mathcal{Z}^{i-1}\right) \simeq H^i\left(X, \mathcal{E}\right)$$

which proves the theorem. □

Let $\left(\mathcal{L}^{\cdot}, d\right)$ be a complex of sheaves (for example, of vector spaces) on X. For each open set U of X, $\left(\mathcal{L}^{\cdot}(U), d_U\right)$ is a complex of vector spaces, whose cohomology groups are $H^q\left(\mathcal{L}^{\cdot}(U)\right)$. We denote them also by $H^q\left(U, \mathcal{L}^{\cdot}\right)$:

$$H^q\left(U, \mathcal{L}^{\cdot}\right) = H^q\left(\mathcal{L}^{\cdot}(U)\right)$$

We construct a presheaf (which is not a sheaf in general) by the assignment

$$U \mapsto H^q\left(\mathcal{L}^{\cdot}(U)\right)$$

and denote by $\mathcal{H}^q\left(\mathcal{L}^{\cdot}\right)$ the associated sheaf, which we call the *q-th cohomology sheaf* of the complex $\left(\mathcal{L}^{\cdot}, d\right)$. If $x \in X$ the stalk $\mathcal{H}^q\left(\mathcal{L}^{\cdot}\right)_x$ is the cohomology $H^q\left(\mathcal{L}^{\cdot}_x\right)$ of the complex $\left(\mathcal{L}^{\cdot}_x, d_x\right)$. In particular, $\mathcal{H}^q\left(\mathcal{L}^{\cdot}\right) = 0$ if and only if $\left(\mathcal{L}^{\cdot}, d\right)$ is exact in degree q.

Of course $\mathcal{H}^q\left(\mathcal{L}\right)(X)$ it is not equal to $H^q\left(\mathcal{L}^{\cdot}(X)\right)$, but there is a spectral sequence, whose second term is (in our notations of chapter 1)

$$E_2^{-p,p+q} = H^p\left(X, \mathcal{H}^q\left(\mathcal{L}^{\cdot}\right)\right) \implies H^{p+q}\left(X, \mathcal{L}^{\cdot}\right)$$

There are other interesting spectral sequences related to the above situation, but we will not need them.

3.3 The cohomology sequence associated to a closed subspace

Let X be a paracompact topological space, $\mathcal{E}$ a sheaf on X, A a locally closed subset of X. There are natural morphisms of canonical resolutions

$$\mathcal{C}^{\cdot}\left(X, \mathcal{E}\right) \longrightarrow \mathcal{C}^{\cdot}\left(X, \mathcal{E}\right)|_A \longrightarrow \mathcal{C}^{\cdot}\left(A, \mathcal{E}\right)$$

where the first is obtained by restriction of sections from X to A, and the second is obtained in a natural way by following the construction of the canonical flabby resolutions. If Φ is a family of supports in X, let us denote by $\Phi \cap A$ the family of the $S \cap A$, $S \in \Phi$; $\Phi \cap A$ is a family of supports on A. From the composition of the above morphisms we get a morphism of complexes

$$\Gamma_{\Phi}\left(X, \mathcal{C}^{\cdot}\mathcal{E}\right) \rightarrow \Gamma_{\Phi \cap A}\left(A, \mathcal{C}^{\cdot}\mathcal{E}\right)$$

and passing to the cohomology, homomorphisms

$$H^k_\Phi(X,\mathcal{E}) \to H^k_{\Phi\cap A}(A,\mathcal{E}) \tag{3.15}$$

We quote, without proofs, the following results.

Proposition 3.3. *Let us suppose that A is closed in X and $\mathcal{E}$ is concentrated on A (that is, $\mathcal{E}|_{X\setminus A} = 0$). Then the homomorphisms (3.15) are isomorphisms. Moreover if $\mathcal{F}$ is a sheaf on A and $\mathcal{E}$ is the extension of $\mathcal{F}$ by zero to X, there are isomorphisms*

$$H^k(X,\mathcal{E}) \to H^k(A,\mathcal{F}) \tag{3.16}$$

Let us remark that since A is closed, if Φ is the family of all the closed sets of X, $\Phi\cap A$ is the family of all the closed sets of A.

The above proposition is not true if we suppose that A is only locally closed, or even open.

Proposition 3.4. *Let us suppose that A is open in X, and let Φ be the family of closed sets of X contained in A. Then there are natural isomorphisms*

$$H^k_\Phi(A,\mathcal{E}) \to H^k(X,\mathcal{E}^A) \tag{3.17}$$

where $\mathcal{E}^A$ is the extension of $\mathcal{E}|_A$ by zero to X.

Taking the cohomology long sequence associated to the exact sequence (3.7), by the above propositions 3.3 and 3.4, we obtain:

Proposition 3.5. *Let $\mathcal{E}$ be a sheaf on X, A a closed subset of X, Φ the family of closed sets of X contained in $X\setminus A$. Then there are long exact sequences of cohomology*

$$\begin{aligned} \cdots &\longrightarrow H^k_\Phi(X\setminus A,\mathcal{E}^{X\setminus A}) \longrightarrow H^k(X,\mathcal{E}) \longrightarrow \\ &\longrightarrow H^k(A,\mathcal{E}^A) \longrightarrow H^{k+1}_\Phi(X\setminus A,\mathcal{E}^{X\setminus A}) \longrightarrow \cdots \end{aligned} \tag{3.18}$$

$$\begin{aligned} \cdots &\longrightarrow H^k_c(X\setminus A,\mathcal{E}^{X\setminus A}) \longrightarrow H^k_c(X,\mathcal{E}) \longrightarrow \\ &\longrightarrow H^k_c(A,\mathcal{E}^A) \longrightarrow H^{k+1}_c(X\setminus A,\mathcal{E}^{X\setminus A}) \longrightarrow \cdots \end{aligned} \tag{3.19}$$

We are especially interested in the case $\mathcal{E}$ = a constant sheaf K_X on X (where K is an abelian group). We denote by $H^k(X,K)$ the cohomology groups of X with coefficients in K_X. Then $\mathcal{E}^{X\setminus A} = K_{X\setminus A}$, $\mathcal{E}^A = K_A$, so that the exact sequences (3.18) and (3.19) become

$$\begin{aligned} \cdots &\longrightarrow H^k_\Phi(X\setminus A,K) \longrightarrow H^k(X,K) \longrightarrow \\ &\longrightarrow H^k(A,K) \longrightarrow H^{k+1}_\Phi(X\setminus A,K) \longrightarrow \cdots \end{aligned} \tag{3.20}$$

$$\begin{aligned} \cdots &\longrightarrow H^k_c(X\setminus A,K) \longrightarrow H^k_c(X,K) \longrightarrow \\ &\longrightarrow H^k_c(A,K) \longrightarrow H^{k+1}_c(X\setminus A,K) \longrightarrow \cdots \end{aligned} \tag{3.21}$$

3.4 Soft and fine sheaves

Let X be a paracompact space.

Definition 3.2. A sheaf $\mathcal{E}$ on X is soft if for every closed subset A of X the restriction map $\mathcal{E}(X) \to \mathcal{E}(A)$ is surjective, that is for every $t \in \mathcal{E}(U)$, U an open neighborhood of A, there is $s \in \mathcal{E}(X)$ whose restriction to a possibly smaller open neighborhood of A is t.

A flabby sheaf is clearly soft. Other examples of soft sheaves are: the sheaf $\mathcal{C}_X$ of continuous functions on X, the sheaf $\mathcal{E}_X^0$ of differentiable functions or the sheaf $\mathcal{E}_X^k$ of k-differential forms on a differentiable manifold X.

Lemma 3.3. *Let*

$$0 \longrightarrow \mathcal{E}_1 \longrightarrow \mathcal{E}_2 \longrightarrow \mathcal{E}_3 \longrightarrow 0$$

be an exact sequence of sheaves on X.

(1) If $\mathcal{E}_1$ is soft, the sequence

$$0 \longrightarrow \mathcal{E}_1(X) \longrightarrow \mathcal{E}_2(X) \longrightarrow \mathcal{E}_3(X) \longrightarrow 0$$

is exact.

(2) If $\mathcal{E}_1$ and $\mathcal{E}_2$ are soft, $\mathcal{E}_3$ is soft.

(Let us remember that we are supposing that X is paracompact).

Proof.

(1) Let t be a section of $\mathcal{E}_3$ on X. We must extend it to a section of $\mathcal{E}_2$. Since it is possible to extend it locally, there is an open, locally finite covering $X = \bigcup_{i\in I} U_i$ and $s_i \in \mathcal{E}_2(U_i)$ which extend t on U_i. Let $(V_i)_{i\in I}$ be a covering of X such that $F_i = \overline{V}_i \subset U_i$. Let us consider the set E of the pairs (s, J) where $J \subset I$ and s is a section of $\mathcal{E}_2$ on $F_J = \bigcup_{i\in J} F_i$ which extends t (let us note that F_J is closed, because the covering (V_i) is locally finite). The (non empty) set E is partially ordered by extension, and every ascending chain of E has a maximal element. Then there exists $(s, J) \in E$ which is maximal. We show that $J = I$, and this will implies that s is an extension of t defined on all of X. Now, if there is $i \in I \setminus J$, let s_i be an extension of t to U_i; in $F_J \cap F_i$, s and s_i differ only by a section of $\mathcal{E}_1$, which extends to a section s' on U_i, because $\mathcal{E}_1$ is soft. Then $s - s'$ and s coincide on $F_J \cap F_i$, so that they define an extension of t on $F_J \cup F_i$, which is a contradiction.

(2) is an easy corollary of (1). □

Corollary 3.2. *Let*

$$0 \longrightarrow \mathcal{E}_1 \longrightarrow \mathcal{E}_2 \longrightarrow \mathcal{E}_3 \longrightarrow \cdots$$

be an exact sequence of soft sheaves on X. The sequence

$$0 \longrightarrow \mathcal{E}_1(X) \longrightarrow \mathcal{E}_2(X) \longrightarrow \mathcal{E}_3(X) \longrightarrow \cdots$$

is exact.

Theorem 3.2. *Let $\mathcal{E}$ be a soft sheaf. Then the cohomology groups $H^p(X, \mathcal{E}) = 0$ for $p > 0$.*

We consider the canonical flabby resolution of $\mathcal{E}$, so that the sequence of sheaves (3.11) is exact. Since a flabby sheaf is soft, by corollary 3.2 the sequence (3.12) remains exact. The cohomology groups of $\mathcal{E}$ are the cohomology groups of the complex (3.9) (without supports), so that they are zero in positive degree.

Definition 3.3. A sheaf $\mathcal{E}$ on X is called fine if for any pair of closed subset A, B of X with $A \cap B = \emptyset$, there exists an endomorphism $\phi\colon \mathcal{E} \to \mathcal{E}$ such that $\phi|_A = 0$ and $\phi|_B$ is the identity.

Lemma 3.4. *A fine sheaf is soft, in particular its cohomology groups vanish in positive degree.*

Proof. Let $\mathcal{E}$ be a fine sheaf on X, $A \subset X$ a closed set, and $s \in \mathcal{E}(U)$ a section of $\mathcal{E}$ on an open neighborhood U of A. Let V be an open set such that $A \subset V \subset \overline{V} \subset U$. There is an endomorphism $\phi\colon \mathcal{E} \to \mathcal{E}$ such that $\phi|_{(X \setminus V)} = 0$ and $\phi|_A$ is the identity. The section $\phi(s)$ of $\mathcal{E}(U)$ can be extended by zero on all of X, obtaining a global section t which extends s. □

By theorem 3.2 in order to compute the cohomology groups of a sheaf, it is possible to make use of soft, or fine, resolutions.

The direct sum of soft (or fine) sheaves, is soft (fine).

Remark. If Φ is a paracompactifying family of supports on X, it is possible to define the notion of Φ-soft and Φ-fine sheaf, which are useful in order to compute the cohomologies with supports in Φ.

3.5 Direct images of sheaves

Let $f\colon X \to Y$ be a continuous map, $\mathcal{E}$ a sheaf on X. The presheaf: $V \mapsto \mathcal{E}\left(f^{-1}(V)\right)$ is a sheaf on Y, denoted $f_*\mathcal{E}$ and called the *direct image* of $\mathcal{E}$ through f:

$$(f_*\mathcal{E})(V) = \mathcal{E}\left(f^{-1}(V)\right)$$

The sheaf $f_*\mathcal{E}$ carries the same structure as $\mathcal{E}$.

Lemma 3.5. *The direct image of a flabby (resp. soft, resp. fine) sheaf on X is a flabby (resp. soft, resp. fine) sheaf on Y.*

Let $f\colon X \to Y$ be a continuous map, $\mathcal{E}$ a sheaf on X. We define the *higher direct image sheaves, or derived sheaves* of $\mathcal{E}$, as follows. Let p be an integer. The assignment $V \mapsto H^p\left(f^{-1}(V), \mathcal{E}\right)$ defines a presheaf on Y, which is not a sheaf in general. The associated sheaf is denoted by $R^p f_*\mathcal{E}$ and is called the p-th direct image sheaf, or p-th derived sheaf, of $\mathcal{E}$ through f. By definition, the 0-th direct image sheaf is $R^0 f_*\mathcal{E} = f_*\mathcal{E}$.

The sheaves $R^p f_*\mathcal{E}$ carry the same structure as $\mathcal{E}$.

Proposition 3.6. *Let $f\colon X \to Y$ be a continuous map, and*

$$0 \longrightarrow \mathcal{E}_1 \longrightarrow \mathcal{E}_2 \longrightarrow \mathcal{E}_3 \longrightarrow 0$$

be a short exact sequence of sheaves on X. Then there is a long exact sequence of direct images sheaves on Y

$$\begin{aligned} 0 \longrightarrow f_*\mathcal{E}_1 \longrightarrow f_*\mathcal{E}_2 \longrightarrow f_*\mathcal{E}_3 \longrightarrow R^1 f_*\mathcal{E}_1 \cdots \\ \cdots R^k f_*\mathcal{E}_1 \longrightarrow R^k f_*\mathcal{E}_2 \longrightarrow R^k f_*\mathcal{E}_3 \longrightarrow R^{k+1} f_*\mathcal{E}_1 \cdots \end{aligned} \tag{3.22}$$

Proof. For any open set U of X there is the long exact sequence of the cohomologies

$$\begin{aligned} 0 \longrightarrow H^0(U, \mathcal{E}_1) \longrightarrow H^0(U, \mathcal{E}_2) \longrightarrow H^0(U, \mathcal{E}_3) \longrightarrow H^1(U, \mathcal{E}_1) \cdots \\ \cdots H^k(U, \mathcal{E}_1) \longrightarrow H^k(U, \mathcal{E}_2) \longrightarrow H^k(U, \mathcal{E}_3) \longrightarrow H^{k+1}(U, \mathcal{E}_1) \cdots \end{aligned}$$

The conclusion follows from the remark 3.1 in section 3.1 □

Remark. If X is a complex space, we can consider the trivial morphism $f\colon X \to \{*\}$, where $\{*\}$ is a point. Then the direct images of the sheaf $\mathcal{E}$ on X are the cohomology groups: $R^p f_*\mathcal{E} = H^p(X, \mathcal{E})$.

3.6 $\mathbb{C}$-ringed spaces

A *ringed space* is a pair $(X, \mathcal{O}_X)$ where X is a topological space and $\mathcal{O}_X$ is a sheaf of commutative rings with units on X *(the structure sheaf)*. To simplify the notations we will say that X is a ringed space. An open subset U of X is a ringed space $(U, \mathcal{O}_U)$ where $\mathcal{O}_U = \mathcal{O}_X|_U$. A *sheaf of $\mathcal{O}_X$-modules*, or an *$\mathcal{O}_X$-module*, or simply a *module over X*, is a sheaf $\mathcal{E}$ on X such that for each open set U of X, $\mathcal{E}(U)$ is an $\mathcal{O}_X(U)$-module, and the restriction maps $\mathcal{E}(U) \to \mathcal{E}(V)$ are compatible with the respective module structure (we leave the explicit details to the reader). In particular, for $x \in X$, $\mathcal{E}_x$ is an $\mathcal{O}_{X,x}$-module.

Examples (of ringed spaces).

(1) $(X, \mathcal{C}_X)$, where $\mathcal{C}_X$ is the sheaf of complex valued continuous functions on X.

(2) $(X, \mathcal{E}_X^0)$, where X is a differentiable manifold, and $\mathcal{E}_X^0$ is the sheaf of complex valued differentiable functions on X.

(3) $(M, \mathcal{O}_M)$, where M is a complex manifold and $\mathcal{O}_M$ is the sheaf of holomorphic functions on M.

(4) In the above examples, $\mathcal{O}_X$ is a sheaf of functions. Though this is not always the case, in the present book we will use only ringed spaces whose structure sheaf is a subsheaf of $\mathcal{C}_X$. Hence the reader can think of sections of $\mathcal{O}_X$ as particular continuous functions.

The sheaves $\mathcal{C}_X$, $\mathcal{E}_X^0$, $\mathcal{O}_M$ in the above examples are not just sheaves of rings, in addition all the rings $\mathcal{O}_X(U)$ are $\mathbb{C}$-algebras and all the stalks $\mathcal{O}_{X,x}$ are local $\mathbb{C}$-algebras: the maximal ideal $m_x \subset \mathcal{O}_{X,x}$ consists of the germs represented, in a neighborhood of x, by functions which vanish at x.

If X is a topological space, a *sheaf of* $\mathbb{C}$*-algebras* on X is a sheaf of rings $\mathcal{A}$ which contains the constant sheaf $\mathbb{C}_X$ as a subsheaf of rings, and the $\mathcal{A}(U)$ are $\mathbb{C}_X(U)$-algebras for each open set U. It follows from the definition that $\mathcal{A}(X)$ contains the constant functions on X, in particular the unit in $\mathcal{A}(X)$ can be identified with $1 \in \mathbb{C}$.

A sheaf of $\mathbb{C}$-algebras $\mathcal{A}$ is called a sheaf of *local* $\mathbb{C}$*-algebras* if every stalk $\mathcal{A}_x$ is a local ring with (unique) maximal ideal m_x so that the quotient morphism $\mathbb{C} \to \mathcal{A}_x/m_x$ is an isomorphism.

A sheaf mapping $\phi\colon \mathcal{A} \to \mathcal{B}$ of $\mathbb{C}$-algebras is called a $\mathbb{C}$*-morphism* if every stalk map $\phi_x : \mathcal{A}_x \to \mathcal{B}_x$ is a $\mathbb{C}$-algebra homomorphism; if $\mathcal{A}$ and $\mathcal{B}$ are sheaves of local $\mathbb{C}$-algebras, ϕ_x is automatically local, i.e. sends the maximal ideal of $\mathcal{A}$ to the maximal ideal of $\mathcal{B}$.

A ringed space $(X, \mathcal{O}_X)$ is called a $\mathbb{C}$*-ringed space* if $\mathcal{O}_X$ is a sheaf of local $\mathbb{C}$-algebras.

All the above examples of ringed spaces are $\mathbb{C}$-ringed spaces.

Given a $\mathbb{C}$-ringed space X and a family of $\mathcal{O}_X$-modules $\mathcal{E}_i$ we define the direct sum $\mathcal{E} = \bigoplus_i \mathcal{E}_i$ by $\mathcal{E}(U) = \bigoplus_i \mathcal{E}_i(U)$; it is a sheaf of $\mathcal{O}_X$-modules. The tensor product (over $\mathcal{O}_X$) of two $\mathcal{O}_X$-modules $\mathcal{E}$ and $\mathcal{F}$, denoted by $\mathcal{E} \otimes_{\mathcal{O}_X} \mathcal{F}$, is defined as the sheaf associated to the presheaf $U \mapsto \mathcal{E}(U) \otimes_{\mathcal{O}_X(U)} \mathcal{F}(U)$; this is not a sheaf in general, so $\Gamma(X, \mathcal{E} \otimes_{\mathcal{O}_X} \mathcal{F})$ *is not equal* to $\mathcal{E}(X) \otimes_{\mathcal{O}_X(X)} \mathcal{F}(X)$; but the stalk at a point $x \in X$ is given by the tensor product of the stalks:

$$(\mathcal{E} \otimes_{\mathcal{O}_X} \mathcal{F})_x = \mathcal{E}_x \otimes_{\mathcal{O}_{X,x}} \mathcal{F}_x$$

Proposition 3.7. *Let $(X, \mathcal{O}_X)$ be a ringed space.*

(1) If $\mathcal{O}_X$ is a soft sheaf, any $\mathcal{O}_X$-module is fine. In particular, a sheaf of rings is soft if and only if it is fine.

(2) The direct sum of flabby (resp. soft, fine) sheaves, is flabby (resp. soft, fine).

(3) If $\mathcal{E}$ and $\mathcal{F}$ are $\mathcal{O}_X$-modules and $\mathcal{E}$ is fine, the tensor product $\mathcal{E} \otimes_{\mathcal{O}_X} \mathcal{F}$ is fine.

Proof.

1. Let A, B two closed subsets of X with $A \cap B = \emptyset$; there exists a section $f \in \mathcal{O}_X(X)$ such that $f \equiv 0$ on A and $f \equiv 1$ on B. Let $\phi \colon \mathcal{E} \to \mathcal{E}$ be the multiplication by f. Then $\phi|_A = 0$ and $\phi|_B$ is the identity.

2. is obvious.

3. Let A, B as above. If $\phi \colon \mathcal{E} \to \mathcal{E}$ is an endomorphism such that $\phi|_A = 0$ and $\phi|_B$ is the identity, the endomorphism $\psi = \phi \oplus \mathrm{id}_{\mathcal{F}}$ of $\mathcal{E} \otimes_{\mathcal{O}_X} \mathcal{F}$ will satisfy $\psi|_A = 0$ and $\psi|_B =$ identity. □

Let X be a differentiable manifold. The sheaf $\mathcal{E}^0_X$ of the complex valued differentiable functions is soft, hence it is fine. It follows that the sheaves $\mathcal{E}^k_X$ of complex k-differential forms, which are $\mathcal{E}^0_X$-modules, are fine.

Hence we can recall the following fundamental theorem

Theorem 3.3 (De Rham theorem). *Let X be a differentiable manifold. The De Rham complex $(\mathcal{E}^{\cdot}_X, d)$, provided with the natural augmentation $\mathbb{C}_X \to \mathcal{E}^0_X$, is a fine resolution of the constant sheaf $\mathbb{C}_X$. In particular*

$$H^k(X, \mathbb{C}) = \frac{\ker\{d \colon \Gamma(X, \mathcal{E}^k_X) \to \Gamma(X, \mathcal{E}^{k+1}_X)\}}{d\Gamma(X, \mathcal{E}^{k-1}_X)}$$

The fact that the De Rham complex is a resolution of $\mathbb{C}_X$ is a restatement of Poincaré lemma.

In the same way, Dolbeault lemma (theorem 2.3 of chapter 2) can be restated as

Theorem 3.4 (Dolbeault theorem). *Let X be a complex manifold. The Dolbeault complex $(\mathcal{E}^{p,\cdot}_X, \bar\partial)$, provided with the natural augmentation $\Omega^p_X \to \mathcal{E}^{p,0}_X$, is a fine resolution of the sheaf Ω^p_X of holomorphic p-forms on X. In particular*

$$H^q(X, \Omega^p_X) = \frac{\ker\left\{\bar\partial \colon \Gamma(X, \mathcal{E}^{p,q}_X) \to \Gamma\left(X, \mathcal{E}^{p,q+1}_X\right)\right\}}{\bar\partial\Gamma\left(X, \mathcal{E}^{p,q-1}_X\right)}$$

The cohomology groups

$$H^{p,q}(X) = H^q(X, \Omega_X^p)$$

are called the Dolbeault groups of X.

Let $(X, \mathcal{O}_X)$, $(Y, \mathcal{O}_Y)$ be $\mathbb{C}$-ringed spaces. A *morphism of* $\mathbb{C}$*-ringed spaces* is a pair (f, ϕ) where $f\colon X \to Y$ is a continuous map and $\phi\colon \mathcal{O}_Y \to f_*\mathcal{O}_X$ is a $\mathbb{C}$-algebra morphism of sheaves of rings on Y. Note that $f_*\mathcal{O}_X$ is not, in general, a sheaf of $\mathbb{C}$-algebras on Y. But for $x \in X$ the morphism ϕ canonically determines stalk maps

$$\phi_x\colon \mathcal{O}_{Y,f(x)} \to \mathcal{O}_{X,x}$$

which are automatically local $\mathbb{C}$-algebra homomorphisms.

For each open set $V \subset Y$ the ring homomorphism $\phi_V\colon \mathcal{O}_Y(V) \to \mathcal{O}_X\left(f^{-1}(V)\right)$ allows to "pullback" to $\mathcal{O}_X$ the sections of $\mathcal{O}_Y$, so the morphism ϕ replaces in our context the composition of functions $g \mapsto g \circ f$. In general ϕ is not uniquely determined by f, except when $\mathcal{O}_X$ and $\mathcal{O}_Y$ are subsheaves of sheaves of continuous functions:

Proposition 3.8. *Let* $(f, \phi) : (X, \mathcal{C}_X) \to (Y, \mathcal{C}_Y)$ *be a morphism of* $\mathbb{C}$*-ringed spaces. Then* ϕ *is the composition* $g \mapsto g \circ f$.

Hence, if $\mathcal{O}_X$ and $\mathcal{O}_Y$ are subsheaves of the respective sheaves of continuous functions, a morphism of ringed spaces $(X, \mathcal{O}_X) \to (Y, \mathcal{O}_Y)$ is nothing else that a continuous function $f\colon X \to Y$ such that for any open set $V \subset Y$ and $g \in \mathcal{O}_Y(V)$, $g \circ f$ belongs to $\mathcal{O}_X\left(f^{-1}(V)\right)$. For the purposes of the present book, this can be taken as the definition of morphism of ringed spaces.

A morphism between $\mathbb{C}$-ringed spaces will often be denoted $f\colon X \to Y$.

We can deal also with $\mathbb{R}$*-ringed spaces*: they are ringed spaces $\left(X, \mathcal{O}_X^{\mathbb{R}}\right)$, where $\mathcal{O}_X^{\mathbb{R}}$ is a ring of local $\mathbb{R}$-algebras. Important examples are the real analytic manifolds M: there $\mathcal{O}_M^{\mathbb{R}}$ is the sheaf of real analytic functions on M.

3.7 Coherent sheaves

Let $(X, \mathcal{O}_X)$ be a ringed space (not necessarily a $\mathbb{C}$-ringed space). We are going to introduce the notion of coherent sheaf $\mathcal{E}$ of $\mathcal{O}_X$-modules.

We denote by $\mathcal{O}_X^{\oplus p}$ ($\mathcal{O}_X^p$ if no confusion arises) the direct sum of p copies of $\mathcal{O}_X$. Let $s_1, \ldots, s_p \in \mathcal{E}(U)$ a finite set of sections of $\mathcal{E}$ on an open set U of X. They define a morphism of sheaves

$$\mathcal{O}_U^{\oplus p} \to \mathcal{E}|_U \tag{3.23}$$

by the assignment

$$(a_1, \ldots, a_p) \mapsto a_1 s_1 + \cdots + a_p s_p$$

Conversely, any morphism (3.23) defines sections $s_1, \ldots, s_p \in \mathcal{E}(U)$ which are the images of the sections $(1, 0, \ldots, 0), \ldots, (0, \ldots, 0, 1)$ of $\mathcal{O}_X^{\oplus p}(U)$.

Let $x \in X$. A *finite system of generators* of $\mathcal{E}$ on an open set $U \subset X$ is a family of sections $s_1, \ldots, s_p \in \mathcal{E}(U)$ such that for every $x \in U$, every germ $s_x \in \mathcal{E}_x$ is a combination, with coefficients in $\mathcal{O}_{X,x}$, of the germs $s_{1,x}, \ldots, s_{p,x}$:

$$s_x = a_{1,x} s_{1,x} + \cdots + a_{p,x} s_{p,x}$$

where $a_1, \ldots, a_p \in \mathcal{O}_X(V)$, V is an open neighborhood of x. (It is important to remark that the s_j are defined on U, but the a_j usually live in a smaller neighborhood of x).

The sections $(s_1, \ldots, s_p)$ define the morphism of sheaves (3.23) and they are a system of generators of $\mathcal{E}$ on U if and only if the morphism (3.23) is surjective.

An $\mathcal{O}_X$-module $\mathcal{E}$ is called *finitely generated, or of finite type, at a point* $x \in X$, if there exists an open neighborhood U of x and a system of generators of $\mathcal{E}$ on U. We say that $\mathcal{E}$ is *finitely generated, or of finite type*, if it is finitely generated at any point of X. We note that this implies that all the stalks $\mathcal{E}_x$ are finitely generated $\mathcal{O}_{X,x}$-modules, but being of finite type is a much stronger property. Let us note also that a more appropriate expression would be "locally finitely generated," or "locally of finite type."

Examples.

(1) $\mathcal{O}_X^{\oplus p}$ is of finite type: the sections $(1, 0, \ldots, 0), \ldots, (0, \ldots, 0, 1)$ are a system of generators on every open set U.

(2) If $\mathcal{E} \to \mathcal{F}$ is a surjective morphism and $\mathcal{E}$ is of finite type, also $\mathcal{F}$ is of finite type. In particular a quotient of a module of finite type is of finite type.

(3) Subsheaves of modules of finite type are not necessarily of finite type.

Lemma 3.6. *Let $\mathcal{E}$ be an $\mathcal{O}_X$-module of finite type at a point $x \in X$. Let $(t_1, \ldots, t_m)$ be a family sections of $\mathcal{E}$ on an open neighborhood U of x, such that $(t_{1,x}, \ldots, t_{m,x})$ generate the $\mathcal{O}_{X,x}$-module $\mathcal{E}_x$; then there exists an open neighborhood $V \subset U$ of x, such that $(t_1, \ldots, t_m)$ are a system of generators of $\mathcal{E}$ on V.*

Proof. There exist sections $s_1, \ldots, s_p$ in a neighborhood W of x whose germs at z generate $\mathcal{E}_z$ over $\mathcal{O}_{X,z}$ for $z \in W$. Since $(t_{1,x}, \ldots, t_{m,x})$ generate $\mathcal{E}_x$ over $\mathcal{O}_{X,x}$, we can write

$$s_{i,x} = \sum_j a_{ij,x} t_{j,x}$$

$(a_{ij,x} \in \mathcal{O}_{X,x})$. The above equality extends to an open neighborhood V:

$$s_i|_V = \sum_j a_{ij} t_j|_V$$

which implies that $(t_1, \ldots, t_m)$ are a system of generators of $\mathcal{E}$ on V. □

The following lemma (whose proof is an exercise) summarizes some properties of modules of finite type.

Lemma 3.7.

(i) *Let $\mathcal{E}$ be an $\mathcal{O}_X$-module, and $\mathcal{G} \subset \mathcal{E}$, $\mathcal{F} \subset \mathcal{E}$ two subsheaves. If $\mathcal{G}$ is of finite type at a point x and $\mathcal{G}_x \subset \mathcal{F}_x$, then $\mathcal{G}|_V \subset \mathcal{F}|_V$ in a suitable open neighborhood V of x. In particular if $\mathcal{G}$ is a sheaf of finite type and $\mathcal{G}_x = 0$, then $\mathcal{G}|_V = 0$ in an open neighborhood V of x.*

(ii) *Let $\phi\colon \mathcal{G} \to \mathcal{E}$ be a morphism of $\mathcal{O}_X$-modules. If $\phi\colon \mathcal{G}_x \to \mathcal{E}_x$ is surjective at a point x, then $\phi|_V\colon \mathcal{G}|_V \to \mathcal{E}|_V$ is surjective in a suitable open neighborhood V of x.*

Let $\mathcal{E}$ be an $\mathcal{O}_X$-module. Let $s_1, \ldots, s_p \in \mathcal{E}(U)$ be sections on an open set U and $\psi\colon \mathcal{O}_U^{\oplus p} \to \mathcal{E}|_U$ the morphism defined by them. The kernel of ψ is an $\mathcal{O}_U$-module called the *sheaf of relations* of $s_1, \ldots, s_p$ and denoted by $\mathcal{R}(s_1, \ldots, s_p)$. In fact for an open set $V \subset U$:

$$\mathcal{R}(s_1, \ldots, s_p)(V) = \left\{ (g_1, \ldots, g_p) \in \mathcal{O}_U^{\oplus p}(V) \;\middle|\; g_1 s_1|_V + \cdots + g_p s_p|_V = 0 \right\}.$$

Definition 3.4. A coherent $\mathcal{O}_X$-module is an $\mathcal{O}_X$-module $\mathcal{E}$ with the following properties:

(1) $\mathcal{E}$ is of finite type.

(2) For every open set U and every finite set of sections $s_1, \ldots, s_p \in \mathcal{E}(U)$, the sheaf of relations $\mathcal{R}(s_1, \ldots, s_p)$ is of finite type on U.

The coherence of a sheaf depends heavily on the structure sheaf $\mathcal{O}_X$, but if no confusion arises, we will talk about coherent sheaves, meaning $\mathcal{O}_X$-coherent sheaves.

The most important examples of coherent sheaves appear in the theory of complex spaces and complex algebraic varieties. The structure sheaf of a complex space, or of a complex algebraic variety, is coherent (see chapter 7).

The coherence is a local property: a sheaf $\mathcal{E}$ on X is coherent if and only if every point $x \in X$ there is an open neighborhood U of x such that $\mathcal{E}|_U$ is coherent.

A subsheaf of a coherent sheaf is coherent if and only if it is of finite type.

The main technical result on coherent sheaves is the following theorem (stated without proof).

Theorem 3.5. *Let*

$$0 \longrightarrow \mathcal{E}_1 \longrightarrow \mathcal{E}_2 \longrightarrow \mathcal{E}_3 \longrightarrow 0$$

be an exact sequence of $\mathcal{O}_X$-modules. If two of them are coherent, so it is the third.

Proposition 3.9.

(1) The direct sum of a finite number of coherent sheaves is coherent.

(2) Let $\phi\colon \mathcal{E} \to \mathcal{F}$ be a morphism of coherent $\mathcal{O}_X$-modules. Then $\ker\phi$, $\operatorname{im}\phi$, $\operatorname{coker}\phi$ are coherent.

(3) Let $\mathcal{E}_1 \xrightarrow{f} \mathcal{E}_2 \xrightarrow{g} \mathcal{E}_3$ be a sequence of coherent $\mathcal{C}_X$-modules which is exact at a point $x \in X$. There exists an open neighborhood U of x such that the sequence $\mathcal{E}_1|_U \xrightarrow{f|_U} \mathcal{E}_2|_U \xrightarrow{g|_U} \mathcal{E}_3|_U$ is exact.

Proof.

(1) The result can be obtained by repeatedly applying the above theorem.

(2) $\operatorname{im}\phi$, as a quotient of $\mathcal{E}$, is of finite type, and as a subsheaf of finite type of the coherent sheaf $\mathcal{F}$ is coherent. Then the exact sequences

$$0 \longrightarrow \ker\phi \longrightarrow \mathcal{E} \longrightarrow \operatorname{im}\phi \longrightarrow 0$$

$$0 \longrightarrow \operatorname{im}\phi \longrightarrow \mathcal{F} \longrightarrow \operatorname{coker}\phi \longrightarrow 0$$

and the theorem 3.5 imply that $\ker\phi$ and $\operatorname{coker}\phi$ are coherent.

(3) The sheaf $\mathcal{E}_1/\ker(g\circ f)$ is coherent and its stalk at x is zero, hence it is zero on a neighborhood V of x. This means $g \circ f = 0$ on V. Then $\operatorname{im} f \subset \ker g$ on V, and $\ker g/\operatorname{im} f$ is a coherent sheaf whose stalk at x is zero. Again we find $\ker g/\operatorname{im} f = 0$ on a smaller neighborhood U of x. □

The *support* of a coherent sheaf $\mathcal{E}$ is the set

$$Y = \operatorname{supp}\mathcal{E} = \{\, x \in X \mid \mathcal{E}_x \neq 0 \,\}$$

Y is a closed subset of X (the complement is open: $\mathcal{E}_x = 0$ implies $\mathcal{E}_z = 0$ for z in a neighborhood of x).

A subsheaf of ideals $\mathcal{I} \subset \mathcal{O}_X$ will be shortly called an ideal of $\mathcal{O}_X$.

Proposition 3.10. *Let us suppose that the structure sheaf $\mathcal{O}_X$ is coherent, and let $\mathcal{I} \subset \mathcal{O}_X$ be an ideal of finite type (hence coherent). The quotient $\mathcal{O}_X/\mathcal{I}$ is a coherent sheaf of $\mathcal{O}_X$-modules and is a sheaf of rings. Let Y be its support, and $\mathcal{O}_Y = (\mathcal{O}_X/\mathcal{I})\,|_Y$. Then $\mathcal{O}_Y$ is a coherent $\mathcal{O}_Y$-module. Moreover an $\mathcal{O}_Y$-module $\mathcal{F}$ is coherent if and only if its extension by zero to X is a coherent $\mathcal{O}_X$-module.*

The proof is based on the fact that the sheaf $\mathcal{O}_X/\mathcal{I}$ is the extension of $\mathcal{O}_Y$ to X by zero.

The ringed space $(Y, \mathcal{O}_Y)$ of the above theorem can be given the name of *ringed subspace* of $(Y, \mathcal{O}_X)$.

Definition 3.5. Let $(X, \mathcal{O}_X)$ be a ringed space. A locally free sheaf of rank r on X is an $\mathcal{O}_X$-module locally isomorphic to $\mathcal{O}_X^r$; an invertible sheaf is a locally free sheaf of rank 1.

If $\mathcal{O}_X$ is coherent, a locally free sheaf is clearly coherent.

For example, if F is a rank r holomorphic vector bundle on a complex manifold M, the sheaf of holomorphic sections of F is a locally free $\mathcal{O}_M$-module.

Chapter 4

Harmonic forms on hermitian manifolds

4.1 Introduction

We have seen that the cohomology $H^k(M, \mathbb{C})$ on a manifold M is the quotient space

$$H^k(M, \mathbb{C}) = \frac{\ker \{d\colon \Gamma(M, \mathcal{E}_M^k) \to \Gamma(M, \mathcal{E}_M^{k+1})\}}{d\Gamma(M, \mathcal{E}_M^{k-1})}$$

of the space of d-closed forms modulo the subspace of exact forms. So, a cohomology class is a set of forms $\{\omega + d\rho\}$ for a fixed closed form ω and varying forms ρ. The purpose of the classical theory of De Rham-Hodge is to find, in a given cohomology class $\{\omega + d\rho\}$, a canonical representative ω_0 which will be d-closed, and which will be specified by other equations. Moreover the cohomology class is 0, if and only if this canonical representation of the cohomology class is identically 0 as a form.

To specify the other equations satisfied by the canonical representatives of each cohomology class, one has to fix an extra structure on M, namely the data of a riemannian metric. Once this is fixed, the space of forms carries a natural scalar product. The extra equation to be satisfied by their representative ω_0 is $\delta\omega_0 = 0$ where δ is the adjoint of d with respect to this scalar product. It turns out that on a compact manifold, the set of equations $d\omega_0 = 0$, $\delta\omega_0 = 0$ is equivalent to a second order equation $\Delta\omega_0 = 0$ where Δ is the so called De Rham Laplacian on forms. Then ω_0 is called the harmonic representative of the given cohomology class.

The choice of a canonical representative presents several advantages. For example, it becomes easier to prove that a cohomology group is 0 (vanishing theorems of Bochner, Kodaira etc...) because one has now to prove that harmonic forms are 0.

This means that one has to prove that the Laplace operator has a kernel reduced to 0, and, this is reduced, by integration by parts, to prove that the zero order coefficients of the operator are positive in a suitable sense.

A very important consequence is the Hodge theory of compact kählerian manifolds. Essentially, this theory says that harmonic forms decompose in sums of harmonic forms of pure type, which in addition to the equations $d\omega_0 = 0$, $\delta\omega_0 = 0$ are also $\bar{\partial}$-closed, so that they induce cohomology classes of the sheaves of holomorphic forms.

4.2 Hermitian metrics on an exterior algebra

4.2.1 Hermitian forms on a complex vector space

Let $V_{\mathbb{R}}$ be an even dimensional real vector space, $\dim V_{\mathbb{R}} = 2n$. We fix a basis $(\underline{e}_1, \ldots, \underline{e}_{2n})$ of $V_{\mathbb{R}}$. We denote by V the complexification of $V_{\mathbb{R}}$ so that a vector $\underline{v} \in V_{\mathbb{C}}$ can be written

$$\underline{v} = \sum_{j=1}^{2n} \xi^j \underline{e}_j \qquad \xi \in \mathbb{C}.$$

We define

$$\begin{aligned} \underline{u}_j &= \frac{1}{2}\left(\underline{e}_{2j-1} - i\underline{e}_{2j}\right) \\ \underline{\bar{u}}_j &= \frac{1}{2}\left(\underline{e}_{2j-1} + i\underline{e}_{2j}\right) \end{aligned} \tag{4.1}$$

so that

$$\underline{v} = \sum_{j=1}^{n} \zeta'^j \underline{u}_j + \sum_{j=1}^{n} \zeta''^j \underline{\bar{u}}_j$$

where $\zeta'^j = \xi^{2j-1} + i\xi^{2j}$, $\zeta''^j = \xi^{2j-1} - i\xi^{2j}$. Call T the complex subspace of V generated by the vectors $\underline{u}_j$ $j = 1, \ldots, n$ and $\overline{T}$ the complex subspace generated by the $\underline{\bar{u}}_j$, so that

$$V = T \oplus \overline{T}.$$

Let h be a hermitian form on T, so that $h(v_1, v_2)$ is $\mathbb{C}$-linear in $v_1 \in T$ and $\mathbb{C}$-antilinear in $v_2 \in T$ and

$$h(v_1, v_2) = \overline{h(v_2, v_1)}. \tag{4.2}$$

In the basis of the $\underline{\bar{u}}_j$, h has a hermitian matrix $(h_{i\bar{j}})$ so that

$$h(v_1, v_2) = \sum_{i,j} h_{i\bar{j}} \zeta_1^i \bar{\zeta}_2^j \tag{4.3}$$

for $v_1 = \sum_j \zeta_1^j \underline{u}_j$, $v_2 = \sum_j \zeta_2^j \underline{u}_j$. One can split h in its real part and imaginary part:

$$h(v_1, v_2) = g(v_1, v_2) + i\omega(v_1, v_2) \tag{4.4}$$

where g and ω are real. Now, from (4.2) it is clear that $g(v_1, v_2)$ is symmetric and $\omega(v_1, v_2)$ is skew-symmetric in v_1, v_2.

Moreover one can identify $V_{\mathbb{R}}$ to T, using $\zeta^j = \xi^{2j-1} + i\xi^{2j}$ for ξ^l real, so that $g(v_1, v_2)$ induces a symmetric form on $V_{\mathbb{R}}$.

Now, we can find a unitary matrix $U \in \mathrm{U}(n)$ which diagonalizes the matrix $(h_{i\bar{j}})$, so, if we define $\zeta = Uz$, we obtain

$$\sum_{i,j} h_{i\bar{j}}\zeta_1^i\bar{\zeta}_2^j = {}^t\zeta_1\underline{h}\bar{\zeta}_2 = {}^tz_1\left({}^tU\underline{h}\bar{U}\right)\bar{z}_2 = \sum_{j=1}^n \lambda_j z_1^j\bar{z}_2^j$$

where the λ_j are the (real) eigenvalues of h.

In particular

$$g(v_1, v_2) = \mathrm{Re}\left(\sum_{i,j} h_{i\bar{j}}\zeta_1^j\bar{\zeta}_2^j\right) = \sum_{j=1}^n \lambda_j\left(x_1^{2j-1}x_2^{2j-1} + x_1^{2j}x_2^{2j}\right) \tag{4.5}$$

and

$$\omega(v_1, v_2) = \mathrm{Im}\left(\sum_{i,j} h_{i\bar{j}}\zeta_1^i\bar{\zeta}_2^j\right) = \sum_{j=1}^n \lambda_j\left(x_1^{2j}x_2^{2j-1} - x_1^{2j-1}x_2^{2j}\right). \tag{4.6}$$

where we have defined

$$z^j = x^{2j-1} + ix^{2j}$$

So, the linear unitary transformation U induces on the $2n$-dimensional space $V_{\mathbb{R}}$ a linear transformation Ω which diagonalizes the symmetric form g and reduces the skew-symmetric form ω to a skew-symmetric canonical form with 2×2 blocks of the type

$$\begin{pmatrix} 0 & -\lambda_j \\ \lambda_j & 0 \end{pmatrix}.$$

Moreover writing

$$\zeta^j = \xi^j + i\eta^j \qquad z^j = a^j + ib^j \qquad j = 1, \ldots, n$$

the transformation Ω can be read from the equation $\zeta = Uz$ as

$$\begin{pmatrix} \xi \\ \eta \end{pmatrix} = \begin{pmatrix} \mathrm{Re}\,U & -\mathrm{Im}\,U \\ \mathrm{Im}\,U & \mathrm{Re}\,U \end{pmatrix}\begin{pmatrix} a \\ b \end{pmatrix} \equiv \Omega\begin{pmatrix} a \\ b \end{pmatrix}$$

and the conditions of unitarity for U

$$\sum_{i=1}^n \bar{u}_{ij}u_{ik} = \delta_{jk}$$

imply that Ω is orthogonal.

As a consequence its determinant is ± 1, and from (4.5) one deduces

$$\det g = \left(\prod_{j=1}^{n} \lambda_j \right)^2 = (\det h)^2 \tag{4.7}$$

If h is positive definite, i.e. $h(v,v) > 0$ for $v \neq 0$, then g is also positive definite and induces a euclidean metrics on $V_{\mathbb{R}}$.

4.2.2 The exterior algebra of V^*

We call $\underline{e}^j$ the dual basis of the $\underline{e}_j$, so that the dual basis of the $\underline{u}_j$ or $\underline{\bar{u}}_j$ is

$$\underline{u}^j = \underline{e}^{2j-1} + i\underline{e}^{2j}, \quad \underline{\bar{u}}^j = \underline{e}^{2j-1} - i\underline{e}^{2j}. \tag{4.8}$$

The hermitian form h can be considered as an element of $T^* \otimes \overline{T}^*$

$$h = \sum_{i,\bar{j}} h_{i\bar{j}}\, \underline{u}^i \otimes \underline{\bar{u}}^j. \tag{4.9}$$

The alternating form ω is thus

$$\omega = i \sum_{i,\bar{j}} h_{i\bar{j}}\, \underline{u}^i \wedge \underline{\bar{u}}^j. \tag{4.10}$$

and $\omega = \bar{\omega}$.

As in chapter 2, formula (2.21) one has

$$\Lambda^k V^* = \bigoplus_{p+q=k} (\Lambda^p T^*) \wedge \left(\Lambda^q \overline{T}^*\right). \tag{4.11}$$

We denote

$$\Lambda^{p,q} = (\Lambda^p T^*) \wedge \left(\Lambda^q \overline{T}^*\right).$$

Any element $\varphi \in \Lambda^k V^*$ has a unique decomposition:

$$\varphi = \sum_{p+q=k} Q^{p,q}(\varphi) \tag{4.12}$$

where $Q^{p,q}(\varphi)$ is the (p,q)-part of φ.

4.2.3 Volume form

One has

$$\underline{u}^j \wedge \underline{\bar{u}}^j = -2i\, \underline{e}^{2j-1} \wedge \underline{e}^{2j} \tag{4.13}$$

In particular

$$\underline{e}^1 \wedge \underline{e}^2 \wedge \cdots \wedge \underline{e}^{2n-1} \wedge \underline{e}^{2n} = \left(\frac{i}{2}\right)^n \underline{u}^1 \wedge \underline{\bar{u}}^1 \wedge \cdots \wedge \underline{u}^n \wedge \underline{\bar{u}}^n \tag{4.14}$$

Moreover using (4.10) one has

$$\omega^n = n!\, i^n\, (\det h)\, \underline{u}^1 \wedge \underline{\bar{u}}^1 \wedge \cdots \wedge \underline{u}^n \wedge \underline{\bar{u}}^n. \tag{4.15}$$

In particular, the volume element for the euclidean metrics s induced by h on the real space $V_{\mathbb{R}}$ is

$$dV = (\det s)^{1/2}\, \underline{e}^1 \wedge \underline{e}^2 \wedge \cdots \wedge \underline{e}^{2n-1} \wedge \underline{e}^{2n}$$

so it is

$$dV = \frac{\omega^n}{2^n n!} \tag{4.16}$$

4.2.4 Metrics on $\Lambda^{p,q}$

A hermitian metrics h on T induces a dual hermitian metrics on T^* by taking the inverse $\left(h^{\bar{j}i}\right)$ of the matrix $(h_{i\bar{j}})$ so that

$$\sum_{i=1}^{n} h^{\bar{j}i} h_{i\bar{k}} = \delta^{\bar{j}}_{\bar{k}}.$$

Then for two elements $\varphi, \psi \in T^* \equiv \Lambda^{1,0}$

$$\varphi = \sum_{j=1}^{n} \varphi_j \underline{u}^j, \quad \psi = \sum_{j=1}^{n} \psi_j \underline{u}^j$$

one has

$$(\,\varphi \mid \psi\,) = \sum_{i,j} h^{\bar{j}i} \varphi_i \overline{\psi}_j. \tag{4.17}$$

This metrics can be extended as a hermitian metrics on $\Lambda^{p,q}$. Recall that an element $\varphi \in \Lambda^{p,q}$ can be written

$$\varphi = \frac{1}{p!q!} \sum_{\substack{|I|=p \\ |J|=q}} \varphi_{I\bar{J}}\, \underline{u}^I \wedge \underline{\bar{u}}^J \tag{4.18}$$

where I (resp. J) are multiindices $I = (i_1, \ldots, i_p)$ (resp. $J = (j_1, \ldots, j_q)$) and

$$\begin{aligned} \underline{u}^I &= \underline{u}^{i_1} \wedge \underline{u}^{i_2} \wedge \cdots \wedge \underline{u}^{i_p} \\ \underline{\bar{u}}^J &= \underline{\bar{u}}^{j_1} \wedge \underline{\bar{u}}^{j_2} \wedge \cdots \wedge \underline{\bar{u}}^{j_q} \end{aligned}$$

and the $\varphi_{I\bar{J}}$ are skew-symmetric with respect to I and with respect to J. In (4.18) the summation is extended over all the multiindices. Then one defines the hermitian product

$$(\,\varphi \mid \psi\,) = \frac{1}{p!q!} \sum_{\substack{I,J \\ K,L}} h^{\bar{j}_1 l_1} \ldots h^{\bar{j}_q l_q} h^{\bar{k}_1 i_1} \ldots h^{\bar{k}_p i_p} \varphi_{I\bar{J}} \overline{\psi}_{K\bar{L}} \tag{4.19}$$

which is positive definite.

4.2.5 The $*$-operator

The $*$-operator of De Rham-Hodge is an isomorphism

$$* : \Lambda^{p,q} \to \Lambda^{n-q,n-p} \tag{4.20}$$

with the properties that for $\varphi \in \Lambda^{p,q}$, $\psi \in \Lambda^{p,q}$

$$\begin{aligned} \varphi \wedge *\overline{\psi} &= (\varphi \mid \psi) \frac{\omega^n}{n!} \\ \overline{*\psi} &= *\overline{\psi} \\ **\psi &= (-1)^{p+q}\psi \quad \text{for } \psi \in \Lambda^{p,q}. \end{aligned} \tag{4.21}$$

In particular, $*$ is a real operator.

The exact formula is as follows: write

$$\psi = \sum_{\substack{|I|=p \\ |J|=q}}{}' \psi_{I\bar{J}} \, \underline{u}^I \wedge \underline{\bar{u}}^J .$$

Then for any increasing multiindex I, denote by $\hat{I}$ the complementary multiindex, that is

(i) $\hat{I}$ is the complementary set of I in $\{1, \dots, n\}$

(ii) $\hat{I}$ is ordered by increasing order.

Then $I.\hat{I}$, in this order, is, as an ordered n-index, equal to $\{1, \dots, n\}$ up to a shuffle. Call

$$\begin{aligned} I.\hat{I} &= (k_1, \dots, k_n) \\ J.\hat{J} &= (l_1, \dots, l_n) . \end{aligned}$$

Let σ_I be the signature of the permutation reordering $I.\hat{I}$ as $(1, \dots, n)$ and

$$h_{I\hat{I}\bar{J}\hat{\bar{J}}} = \det\left(h_{k_i \bar{l}_j}\right) = (\operatorname{sgn} \sigma_I)(\operatorname{sgn} \sigma_J)(\det h)$$

Then

$$*\psi = i^n (-1)^{\frac{(n-1)n}{2} + pn} \sum_{\substack{|I|=q \\ |J|=p}}{}' h_{I\hat{I}\bar{J}\hat{\bar{J}}} \psi^{\bar{J}I} \, \underline{u}^{\hat{I}} \wedge \underline{\bar{u}}^{\hat{J}} . \tag{4.22}$$

Here in $\psi^{\bar{J}I}$ the multiindices have been lifted by $h^{\bar{j}l}$ as usual:

$$\psi^{\bar{J}I} = \sum_{\substack{|L|=p \\ |K|=q}} h^{\bar{j}_1 l_1} \dots h^{\bar{j}_p l_p} h^{\bar{k}_1 i_1} \dots h^{\bar{k}_q i_q} \psi_{L\bar{K}} . \tag{4.23}$$

The proof that the operator $*$, defined by (4.22) satisfies the properties of (4.21) is given in [MK] ch.3 §2. In fact, it is sufficient to do this proof in a basis $\underline{u}_j$ where $\left(h^{\bar{j}l}\right)$ is diagonalized.

4.2.6 Determination of $*$ in an orthonormal basis

Choose for $\underline{u}^j$ an orthonormal basis for the dual hermitian metrics on T^*, so that

$$\left(\underline{u}^j \mid \underline{u}^k \right) = \delta^{jk}.$$

In this basis we have

$$h^{\bar{j}k} = \delta^{\bar{j}k}.$$

Using (4.19), we deduce that an orthonormal basis for $\Lambda^{p,q}$ is formed by the $\underline{u}^I \wedge \underline{\bar{u}}^J$, with increasing multiindices I, J of length p, q respectively.

Nevertheless it is more convenient to use another basis for Λ^k, namely we define for $M \subset \{1, \ldots, n\}$

$$\underline{w}^M = \bigwedge_{m \in M} \left(\underline{u}^m \wedge \underline{\bar{u}}^m \right) \tag{4.24}$$

so that this does not depend of the ordering of M and then we define the orthonormal basis

$$\underline{u}^I \wedge \underline{\bar{u}}^J \wedge \underline{w}^M \qquad I, J, M \text{ disjoint subsets of } \{1, \ldots, n\}. \tag{4.25}$$

Now

$$\frac{\omega^n}{n!} = i^n \underline{w}^{\{1,\ldots,n\}}$$

and the first equation (4.21) can be written as

$$\varphi \wedge \overline{* \left(\underline{u}^I \wedge \underline{\bar{u}}^J \wedge \underline{w}^M \right)} = \left(\underline{u}^j \mid \underline{u}^I \wedge \underline{\bar{u}}^J \wedge \underline{w}^M \right) i^n \underline{w}^{\{1,\ldots,n\}} \tag{4.26}$$

To have a non-zero term on the right-hand side, we need to take $\varphi = \underline{u}^I \wedge \underline{\bar{u}}^J \wedge \underline{w}^M$, this implies that

$$* \left(\underline{u}^I \wedge \underline{\bar{u}}^J \wedge \underline{w}^M \right) = i^n (-1)^{\beta(I,J,M)} \, \underline{u}^I \wedge \underline{\bar{u}}^J \wedge \underline{w}^{\widehat{M \cup I \cup J}} \tag{4.27}$$

where as usual $\widehat{M \cup I \cup J}$ is the complementary subset of $M \cup I \cup J$ in $\{1, \ldots, n\}$. To determine β, we use (4.26)

$$\begin{aligned} \underline{u}^I \wedge \underline{\bar{u}}^J \wedge \underline{w}^M \wedge \overline{* \left(\underline{u}^I \wedge \underline{\bar{u}}^J \wedge \underline{w}^M \right)} = i^n (-1)^n (-1)^{\beta(I,J,M)} & \left(\underline{u}^I \wedge \underline{\bar{u}}^J \wedge \underline{w}^M \right) \\ & \wedge \left(\underline{\bar{u}}^I \wedge \underline{u}^J \wedge \overline{\underline{w}}^{\widehat{M \cup I \cup J}} \right) \end{aligned}$$

and this should be $i^n \underline{w}^{\{1,\ldots,n\}}$ from which we deduce

$$\beta(I, J, M) = |M| + \frac{1}{2} \left(|I| + |J| \right)^2 + \frac{1}{2} \left(|I| - |J| \right). \tag{4.28}$$

Then one checks that $*$ is a real operator

$$\overline{* \left(\underline{u}^I \wedge \underline{\bar{u}}^J \wedge \underline{w}^M \right)} = * \left(\underline{\bar{u}}^I \wedge \underline{u}^J \wedge \overline{\underline{w}}^M \right).$$

4.3 Hermitian metrics on a complex manifold

Definition 4.1. Let M be a complex manifold. A hermitian metrics h on M is the data for each point m of a hermitian form on the complex space $T_m(M)$ which is positive definite and C^∞ with respect to m.

Thus in local complex coordinates $\left(z^j\right)$ one has

$$ds^2 = \sum_{i,j} h_{i\bar{j}}(z)\, dz^i \otimes d\bar{z}^j \tag{4.29}$$

where $(h_{i\bar{j}})$ is an hermitian matrix which is positive definite and depends C^∞ on z.

If $v = \sum \zeta^j \frac{\partial}{\partial z^j}$ is a vector in $T_m(M)$, with m having the coordinates z, one has for the square length

$$\|v\|^2 = \sum_{i,j} h_{i\bar{j}}(z)\, \zeta^i \bar{\zeta}^j. \tag{4.30}$$

If one changes coordinates from a system $\left(z_a^i\right)$ to a system $\left(z_b^j\right)$, the vector v is

$$v = \sum \zeta_a^i \frac{\partial}{\partial z_a^i} = \sum \zeta_b^k \frac{\partial}{\partial z_b^k}$$

with

$$\zeta_b^k = \sum \zeta_a^i \frac{\partial z_b^k}{\partial z_a^i}$$

so that the expressions of h in these coordinate systems are related by

$$h_{a,i\bar{j}} = \sum_{k,l} h_{b,k\bar{l}} \frac{\partial z_b^k}{\partial z_a^i} \overline{\left(\frac{\partial z_b^l}{\partial z_a^j}\right)} \tag{4.31}$$

indicating that h is a global C^∞ section of the bundle $T^*(M) \otimes \overline{T}^*(M)$, which justifies the notation of (4.29)

It is easy to construct hermitian metrics on a complex paracompact manifold. For such a manifold, one can find an open covering $\mathcal{U} = (U_a)$ with coordinates $\left(z_a^j\right)$. Let p_a be a partition of unity such that $\sum_a p_a = 1$ and the support of p_a is in U_a, then one defines

$$ds^2 = \sum_a p_a \sum_i dz_a^i \otimes d\bar{z}_a^i.$$

If $W \subset M$ is a complex submanifold, a hermitian metrics on M induces a hermitian metrics on W (by restriction to the tangent vectors to W).

4.3.1 Application of the results of section 4.2

All the results of section 4.2 apply pointwise in M.

(i) A hermitian metrics induces a riemannian metrics $(g_{ij})_{i,j=1,\dots,2n}$ on the real tangent bundle of M, $V_{\mathbb{R}}(M)$. Moreover by (4.7)

$$\det(g_{ij}) = (\det h)^2 \tag{4.32}$$

(ii) A hermitian metrics induces a C^∞ $(1,1)$-form

$$\omega = i \sum_{i,j} h_{i\bar{j}}(z)\, dz^i \wedge d\bar{z}^j \tag{4.33}$$

which is real: $\omega = \overline{\omega}$.

(iii) The riemannian volume element dV is given by (4.16)

$$dV = \frac{\omega^n}{2^n n!} \qquad \omega^n = \omega \wedge \cdots \wedge \omega \ (n \text{ times}) \tag{4.34}$$

(iv) The hermitian metrics induces a hermitian scalar product on $\Lambda_m^{p,q}(M)$ at each point $m \in M$.
If

$$\varphi = \frac{1}{p!q!} \sum_{\substack{|I|=p\\|J|=q}} \varphi_{I\bar{J}}(z)\, dz^I \wedge d\bar{z}^J$$

$$\psi = \frac{1}{p!q!} \sum_{\substack{|I|=p\\|J|=q}} \psi_{I\bar{J}}(z)\, dz^I \wedge d\bar{z}^J$$

the hermitian scalar product at z is given by

$$(\varphi \mid \psi)(z) = \frac{1}{p!q!} \sum_{\substack{I,J\\K,L}} h^{\bar{j}_1 l_1}(z) \cdots h^{\bar{j}_q l_q}(z) h^{\bar{k}_1 i_1}(z) \cdots h^{\bar{k}_p i_p}(z) \varphi_{I\bar{J}} \overline{\psi_{K\bar{L}}(z)} \tag{4.35}$$

(v) One can define at each point z, the $*$-operator by

$$\varphi(z) \wedge *\overline{\psi}(z) = (\varphi \mid \psi)(z) \frac{\omega^n}{n!} \tag{4.36}$$

and the precise formula is given by (4.22). One has the properties of (4.21) pointwise. Evidently

$$*\colon \Lambda_m^{p,q}(M) \to \Lambda_m^{n-q,n-p}(M).$$

(vi) Lifting of indices.
This is given by (4.23) which we rewrite

$$\psi^{\bar{J}I} = \sum_{\substack{|K|=q \\ |L|=p}} h^{\bar{J}L} h^{\overline{K}I} \psi_{L\overline{K}} \tag{4.37}$$

with the abbreviation

$$h^{\bar{J}L} = h^{\bar{j}_1 l_1} h^{\bar{j}_2 l_2} \dots h^{\bar{j}_p l_p}. \tag{4.38}$$

Then, one rewrites (4.35) as

$$(\varphi \mid \psi)(z) = \frac{1}{p!q!} \sum_{\substack{|K|=p \\ |L|=q}} \varphi^{\overline{K}L}(z) \overline{\psi_{K\bar{L}}(z)} \tag{4.39}$$

$$(\varphi \mid \psi)(z) = \frac{1}{p!q!} \sum_{\substack{|I|=p \\ |J|=q}} \varphi_{I\bar{J}}(z) \overline{\psi^{\bar{I}J}(z)}. \tag{4.40}$$

4.4 Adjoints of d, ∂, $\bar{\partial}$. De Rham-Hodge Laplacian

Let M be a compact hermitian manifold, and φ, $\psi \in \mathcal{E}_M^{p,q}(M)$. One can define the L^2-scalar product on M

$$(\varphi \mid \psi) = \int_M (\varphi \mid \psi)(z) \frac{\omega^n}{n!} = \int_M (\varphi \mid \psi)(z) 2^n \, dV(z). \tag{4.41}$$

Lemma 4.1. *Let M be a compact hermitian manifold; one can define first order differential operators δ, ϑ, $\bar{\vartheta}$:*

$$\begin{aligned} (d\varphi \mid \psi) &= (\varphi \mid \delta\psi) \quad \varphi \in \mathcal{E}_M^k(M), \psi \in \mathcal{E}_M^{k+1}(M) \\ (\bar{\partial}\varphi \mid \psi) &= (\varphi \mid \vartheta\psi) \quad \varphi \in \mathcal{E}_M^{p,q}(M), \psi \in \mathcal{E}_M^{p,q+1}(M) \\ (\partial\varphi \mid \psi) &= (\varphi \mid \bar{\vartheta}\psi) \quad \varphi \in \mathcal{E}_M^{p,q}(M), \psi \in \mathcal{E}_M^{p+1,q}(M) \end{aligned} \tag{4.42}$$

so that

$$\begin{aligned} \delta &\colon \mathcal{E}_M^k(M) \to \mathcal{E}_M^{k-1}(M) \\ \vartheta &\colon \mathcal{E}_M^{p,q}(M) \to \mathcal{E}_M^{p,q-1}(M) \\ \bar{\vartheta} &\colon \mathcal{E}_M^{p,q}(M) \to \mathcal{E}_M^{p-1,q}(M) \end{aligned} \tag{4.43}$$

The formulas are

$$\begin{aligned} \delta &= -{*}d{*} \\ \vartheta &= -{*}\partial{*} \\ \bar{\vartheta} &= -{*}\bar{\partial}{*}\,. \end{aligned}$$

Definition 4.2. δ, ϑ, $\bar{\vartheta}$ are respectively the adjoints of d, $\bar{\partial}$, ∂ for the hermitian scalar product.

Proof. Let us calculate $(\bar{\partial}\varphi \mid \psi)$, $\varphi \in \mathcal{E}_M^{p,q-1}(M)$, $\psi \in \mathcal{E}_M^{p,q}(M)$.

$$(\bar{\partial}\varphi \mid \psi) = \int_M (\bar{\partial}\varphi \mid \psi)(z)\frac{\omega^n}{n!} = \int_M \bar{\partial}\varphi \wedge *\overline{\psi}(z)$$

where we have used the definition of the scalar product and the main property of (4.36) of the $*$-operator.

Now

$$\begin{aligned}\bar{\partial}\varphi \wedge *\overline{\psi} &= \bar{\partial}\left(\varphi \wedge *\overline{\psi}\right) + (-1)^{p+q}\varphi \wedge \bar{\partial}\left(*\overline{\psi}\right)\\ &= d\left(\varphi \wedge *\overline{\psi}\right) + (-1)^{p+q}\varphi \wedge \overline{\partial *\psi}\end{aligned}$$

because $\varphi \wedge *\overline{\psi}$ is of type $(n, n-1)$ and thus is trivially ∂-closed. Now by Stokes formula

$$\int_M d\left(\varphi \wedge *\overline{\psi}\right) = 0$$

because M is compact without boundary. Moreover $\overline{\partial *\psi}$ has degree $2n-(p+q)+1$, so

$$(-1)^{p+q}\varphi \wedge \overline{\partial *\psi} = -\varphi \wedge *\overline{(*\partial *\psi)}$$

using (4.21). So finally

$$\int_M \bar{\partial}\varphi \wedge *\overline{\psi} = \int_M \varphi \wedge *\overline{(-*\partial *\psi)} = \int_M (\varphi \mid \vartheta\psi)(z)\frac{\omega^n}{n!}$$

(using once more (4.36) for $*$). □

Lemma 4.2. *For* $\psi \in \mathcal{E}_M^{p,q+1}(M)$*, one has*

$$(\vartheta\psi)^{\bar{I}J} = (-1)^{p+1}\frac{1}{\det h}\sum_{j=1}^{n}\frac{\partial}{\partial z^j}\left(\psi^{\bar{I}jJ}\det h\right) \tag{4.44}$$

where $|I| = p$, $|J| = q$.

Proof. One writes $(\bar{\partial}\varphi \mid \psi) = (\varphi \mid \vartheta\psi)$ in local coordinates. Because $\bar{\partial}$ does not act on the p-type we can forget about the I. We use (4.40) with no I, $q \mapsto q+1$, $|J| = q+1$ and $\bar{\partial}\varphi$ instead of φ. We know that for φ for type $(0, q)$:

$$\left(\bar{\partial}\varphi\right)_{\bar{\jmath}_0\cdots\bar{\jmath}_q} = \sum_{l=0}^{q}\frac{\partial}{\partial\bar{z}^{j_l}}\left(\varphi_{\bar{\jmath}_0\cdots\hat{\bar{\jmath}}_l\cdots\bar{\jmath}_q}\right) \tag{4.45}$$

where the index $\hat{\bar{\jmath}}_l$ is forgotten. From (4.40) we deduce

$$\begin{aligned}(\bar{\partial}\varphi \mid \psi)(z) &= \frac{1}{(q+1)!}\sum_{|J|=q+1}\left(\bar{\partial}\varphi\right)_{\bar{J}}\overline{\psi^J}\\ &= \frac{1}{q!}\sum_{|J|=q}\sum_{j=1}^{n}\frac{\partial\varphi_{\bar{J}}}{\partial\bar{z}^j}\overline{\psi^{jJ}}.\end{aligned}$$

Then

$$\begin{aligned}
(\bar\partial\varphi \mid \psi) &= \int_M (\bar\partial\varphi \mid \psi)(z) 2^n (\det h)\, dx^1 \wedge dx^2 \wedge \cdots \wedge dx^{2n} \\
&= -\frac{1}{q!}\int_M \sum_{|J|=q} \varphi_{\bar J} \overline{\sum_j \frac{\partial}{\partial z^j}(\psi^{jJ}(\det h))} 2^n\, dx^1 \wedge \cdots \wedge dx^{2n} \\
&= -\frac{1}{q!}\int_M \sum_{|J|=q} \varphi_{\bar J} \overline{\left(\frac{1}{\det h}\sum_j \frac{\partial}{\partial z^j}(\psi^{jJ}\det h)\right)} 2^n dV \\
&\equiv \int_M (\varphi \mid \vartheta\psi)(z)\, 2^n dV
\end{aligned}$$

hence

$$(\vartheta\psi)^J = -\frac{1}{\det h}\sum_{j=1}^n \frac{\partial}{\partial z^j}(\psi^{jJ}\det h).$$

The extra $(-1)^p$ of (4.44) comes from the fact that if φ has dz^I, one has to include a $(-1)^p$ for calculating the $\bar\partial$ (see formula (2.31) of chapter 2), so that (4.45) must contain a $(-1)^p$ for a (p,q)-form φ. □

Remark. The equation (4.44) is a typical "adjoint formula".

4.4.1 De Rham-Hodge operators

We define three operators

$$\begin{aligned}
\Delta = d\delta + \delta d &\colon \mathcal{E}^k_M(M) \to \mathcal{E}^k_M(M) \\
\Box = \bar\partial\vartheta + \vartheta\bar\partial &\colon \mathcal{E}^{p,q}_M(M) \to \mathcal{E}^{p,q}_M(M) \\
\overline{\Box} = \partial\bar\vartheta + \bar\vartheta\partial &\colon \mathcal{E}^{p,q}_M(M) \to \mathcal{E}^{p,q}_M(M)
\end{aligned}$$

Lemma 4.3. *The operators* $\Box$, $\overline{\Box}$ *and* Δ *are second-order operators. For any form* φ*, one has*

$$(\Box\varphi)_{I\bar J} = -2\sum_{i,j} h^{\bar j i}\frac{\partial^2 \varphi_{I\bar J}}{\partial z^i \partial \bar z^j} + \textit{first and zero order terms} \tag{4.46}$$

$$(\Delta\varphi)_L = -4\sum_{i,j} h^{\bar j i}\frac{\partial^2 \varphi_L}{\partial z^i \partial \bar z^j} + \textit{first and zero order terms.} \tag{4.47}$$

So the operators $\Box$ and Δ acting on a form φ define matrices with coefficients second order operators. Their second order terms appear only on the diagonal and are identical for $\Box$, $\overline{\Box}$ and Δ. Out of the diagonal, only first and zero order terms appear.

Remark. In other words, $\Box$, $\overline{\Box}$ and Δ are diagonal in their principal symbols.

Proof of lemma 4.3. In the formulas for $\square$, Δ and $\overline{\square}$, one has to identify the second order terms. This means that we must identify only the first order terms of δ and ϑ. So, for example, on (4.44) we need not take derivatives of the $h^{\bar{\jmath}l}$ or $\det h$ but only first derivatives of $\psi_{I\bar{J}}$. We fix a point of M, and we have to calculate at that point. Then, we can choose the coordinate system in such a way that at that point $h_{i\bar{\jmath}} = \delta_{i\bar{\jmath}}$ and (4.44) reduces at that point to

$$(\vartheta\psi)_{I\bar{J}} = (-1)^{p+1} \sum_{j=1}^{n} \frac{\partial \psi_{I\bar{\jmath}\bar{J}}}{\partial z^j} + 0^{\text{th}}\text{-order}$$

Now we apply this to $\psi = \bar{\partial}\varphi$, so that

$$\psi_{I,\overline{K}} = (-1)^p \sum_{i=0}^{q} (-1)^i \frac{\partial}{\partial \bar{z}^{k_i}} \varphi_{I,\bar{k}_0 \cdots \hat{\bar{k}}_i \cdots \bar{k}_q}$$

and

$$(\vartheta\bar{\partial}\varphi)_{I\bar{J}} = -\sum_{j=1}^{n} \left(\frac{\partial^2 \varphi_{I\bar{J}}}{\partial z^j \partial \bar{z}^j} + \sum_{l=1}^{q} (-1)^l \frac{\partial^2}{\partial z^j \partial \bar{z}^{j_l}} \varphi_{I,\bar{\jmath}\bar{\jmath}_1 \cdots \hat{\bar{\jmath}}_l \cdots \bar{\jmath}_q} \right). \tag{4.48}$$

Moreover

$$(\vartheta\psi)_{I\overline{K}} = (-1)^{p+1} \sum_{j=1}^{n} \frac{\partial \varphi_{I\bar{\jmath}\overline{K}}}{\partial z^j} + 0^{\text{th}}\text{-order}.$$

Then

$$\begin{aligned} (\bar{\partial}\vartheta\varphi)_{I\bar{J}} &= (-1)^p \sum_{l=1}^{q} (-1)^{l+1} \frac{\partial}{\partial \bar{z}^{j_l}} (\vartheta\varphi)_{I,\bar{\jmath}_1 \cdots \hat{\bar{\jmath}}_l \cdots \bar{\jmath}_q} + \cdots \\ (\bar{\partial}\vartheta\varphi)_{I\bar{J}} &= -\sum_{l=1}^{q} (-1)^{l+1} \frac{\partial}{\partial \bar{z}^{j_l}} \sum_{j=1}^{n} \frac{\partial}{\partial z^j} \left(\varphi_{I,\bar{\jmath}\bar{\jmath}_1 \cdots \hat{\bar{\jmath}}_l \cdots \bar{\jmath}_q} \right) + \cdots \end{aligned} \tag{4.49}$$

We obtain the result by adding (4.48) and (4.49). $\square$

4.5 Hermitian metrics and Laplacian for holomorphic bundles

Definition 4.3. Let $\pi\colon F \to M$ be a complex bundle of rank r. A hermitian metrics on F is the data for every $m \in M$, of a positive hermitian form $\eta(m)$ over the vector space F_m, depending C^∞ on m.

Let U_a be an open set with a trivialization

$$\pi^{-1}(U_a) \simeq \mathbb{C}^r \times U_a$$

so that a vector $v \in F_m$, $m \in U_a$, has components $\zeta_a = (\zeta_a^1, \ldots, \zeta_a^r) \in \mathbb{C}^r$. Then the hermitian form is given by

$$\|v\|^2 = \sum_{\alpha,\beta} \eta_{a,\alpha\bar{\beta}}(m)\zeta_a^\alpha \overline{\zeta_a^\beta}. \tag{4.50}$$

In another trivialization

$$\pi^{-1}(U_b) \simeq \mathbb{C}^r \times U_b$$

the vector v has now components ζ_b and

$$\zeta_a = \gamma_{ab}(m)\zeta_b \qquad \gamma_{ab} \in \mathrm{GL}(r, \mathbb{C}) \tag{4.51}$$

so that

$$\eta_{b,\alpha\bar{\beta}}(m) = \sum_{\delta,\epsilon} \eta_{a,\delta\bar{\epsilon}}(m)\, (\gamma_{ab}(m))^\delta_\alpha\, \overline{(\gamma_{ab}(m))^\epsilon_\beta}. \tag{4.52}$$

4.5.1 Metrics on forms with coefficients in a bundle

Now, we consider (p, q)-forms with coefficients in F, $\varphi \in C^\infty(M, \Lambda^{p,q} \otimes F)$. In the trivialization U_a, such a form φ has components $\varphi_a \equiv (\varphi_a^\alpha)_{\alpha=1,\ldots,r}$ where φ_a^α are (p, q)-forms in U_a.

Let ψ another (p, q)-form with coefficients in F. We can define a pointwise hermitian scalar product at m.

$$(\varphi \mid \psi)(m) = \sum_{\alpha,\bar{\beta}} \eta_{a,\alpha\bar{\beta}}(m) \left(\varphi_a^\alpha \mid \psi_a^\beta\right)(m). \tag{4.53}$$

In this formula, $\left(\varphi_a^\alpha \mid \psi_a^\beta\right)(m)$ is the hermitian scalar product at m, of the (p, q)-forms φ_a^α and ψ_a^β which are the components of φ and ψ respectively in the trivialization U_a. This scalar product $\left(\varphi_a^\alpha \mid \psi_a^\beta\right)$ is well defined when a hermitian metrics is given on M.

Moreover, the expression $(\varphi \mid \psi)(m)$ given by (4.53) is independent of the trivialization chosen because

$$\varphi_a(m) = \gamma_{ab}(m)\varphi_b(m)$$
$$\psi_a(m) = \gamma_{ab}(m)\psi_b(m)$$

and because of (4.52).

4.5.2 The adjoint of $\bar{\partial}$

Now, we consider a holomorphic vector bundle $\pi\colon F \to M$, so that the $\bar{\partial}$ operator is well defined on (p, q)-forms with coefficients in M. In particular, if φ is a (p, q)-form with coefficients in F, and $\varphi_a = (\varphi_a^\alpha)_{\alpha=1,\ldots,r}$ are its components in a trivialization U_a, we define

$$\left(\bar{\partial}\varphi\right)_a = \left(\bar{\partial}\varphi_a^\alpha\right)_{\alpha=1,\ldots,r} \tag{4.54}$$

component by component.

Moreover, we can define the L^2-scalar product for (p,q)-forms with coefficients in F

$$(\varphi \mid \psi) = \int_M (\varphi \mid \psi)(m)\frac{\omega^m}{n!} = \int_M (\varphi \mid \psi)(m)2^n \, dV. \tag{4.55}$$

Then we have the analogue of lemma 4.1.

Lemma 4.4. *Let M be a compact hermitian manifold and $\pi\colon F \to M$ a holomorphic bundle with an hermitian metrics. One can define a first order operator ϑ_η such that*

$$(\bar\partial\varphi \mid \psi) = (\varphi \mid \vartheta_\eta \psi) \tag{4.56}$$

where

$$\varphi \in C^\infty(M, \Lambda^{p,q} \otimes F)$$
$$\psi \in C^\infty\left(M, \Lambda^{p,q+1} \otimes F\right)$$

and

$$\vartheta_\eta \colon C^\infty(M, \Lambda^{p,q} \otimes F) \to C^\infty\left(M, \Lambda^{p,q-1} \otimes F\right). \tag{4.57}$$

Moreover in a trivialization U_a of $F \to M$, one has for the components of $\vartheta_\eta\psi$ in this trivialization

$$(\vartheta_\eta\psi)_a^\alpha = -\sum_{\beta=1}^{r} \eta_a^{\bar\beta\alpha} * \partial\left(\sum_{\delta=1}^{r} \eta_{a,\delta\bar\beta}\left(*\psi_a^\delta\right)\right) \tag{4.58}$$

where $\eta_a^{\bar\beta\alpha}$ is the inverse matrix of $\eta_{a,\alpha\bar\beta}$. Finally,

$$(\vartheta_\eta\psi)_a^\alpha = \vartheta\psi_a^\alpha + 0^{th}\text{-order term}. \tag{4.59}$$

Definition 4.4. ϑ_η is the adjoint of $\bar\partial$ for the hermitian scalar product defined by (4.55).

The proof of (4.58) is entirely analogous to the proof of lemma 4.1. The formula (4.59) is an immediate consequence of (4.58) and the fact that

$$\sum_{\beta=1}^{r} \eta_a^{\bar\beta\alpha}\eta_{a,\delta\bar\beta} = \delta_\delta^\alpha$$

and the expression of $\vartheta = -*\partial*$.

4.5.3 De Rham-Hodge Laplace operator for holomorphic bundles

We now define the Laplace operator

$$\square_\eta = \bar\partial\vartheta_\eta + \vartheta_\eta\bar\partial. \tag{4.60}$$

Lemma 4.5. *$\Box_\eta$ is a second order operator which is diagonal in its principal symbol. If U_a is a trivialization of F and $\varphi \in C^\infty(U_a, \Lambda^{p,q} \otimes F)$, then the components of $\Box_\eta \varphi$ in this trivialization are*

$$(\Box_\eta \varphi)_a^\alpha = \Box \varphi_a^\alpha + 1^{st} or\ 0^{th}\text{-order terms} \tag{4.61}$$

so that

$$\left((\Box_\eta \varphi)_a^\alpha\right)_{I\bar{J}} = -\sum_{i,j=1}^{n} h^{i\bar{j}} \frac{\partial^2 \varphi_{a,I\bar{J}}^\alpha}{\partial z^i \partial \bar{z}^j} + 1^{st} or\ 0^{th}\text{-order terms.} \tag{4.62}$$

Proof. (4.61) comes from (4.59) and the definition of $\Box_\eta$. Then (4.62) is a consequence of lemma 4.3. □

4.6 Harmonic forms and cohomology

4.6.1 Harmonic forms on compact hermitian manifold

First, let M be a compact hermitian manifold. Then, one can define d, δ and the De Rham operator $\Delta = \Delta_k$ on k-forms. The main result is the following.

Theorem 4.1. *One can define two operators G_k (the Green operator) and H_k (the harmonic projector)*

$$G_k, H_k : \mathcal{E}_M^k(M) \to \mathcal{E}_M^k(M)$$

such that any k-form $\varphi \in \mathcal{E}_M^k(M)$ can be uniquely decomposed as

$$\varphi = \Delta G_k \varphi + H_k \varphi = G_k \Delta \varphi + H_k \varphi \tag{4.63}$$

with the following properties:

(i) the decomposition of (4.63) is orthogonal.

(ii) $dG_k = G_{k+1} d$

(iii) $H_k \varphi$ is harmonic, namely

$$\Delta H_k \varphi = 0.$$

(iv) The kernel of Δ_k is a finite dimensional space, as well as all eigenspaces, and the spectrum is discrete and tends to $+\infty$.

(v) $H_k \circ H_k = H_k$

(vi) If $\varphi \in \mathcal{E}_M^k(M)$, φ is harmonic if and only if $d\varphi = 0$, $\delta\varphi = 0$. Moreover harmonic forms are orthogonal to $d\mathcal{E}_M^{k-1}(M)$ and to $\delta\mathcal{E}_M^{k+1}(M)$.

We refer to [DR] for the proof of these facts.

We denote

$$\mathcal{H}^k(M) = \ker \text{of } \Delta \text{ in } \mathcal{E}_M^k(M).$$

From theorem 4.1, we deduce immediately the following consequence.

Theorem 4.2. *One has*

$$H^k(M, \mathbb{C}) \simeq \mathcal{H}^k(M). \tag{4.64}$$

This means that in any cohomology class one can find a unique harmonic form. Moreover, a harmonic form that represents the cohomology class 0 is identically 0.

Proof. Let $[\varphi] \in H^k(M, \mathbb{C})$ and φ a representative of $[\varphi]$, so φ is d-closed. Then

$$\varphi = (d\delta + \delta d)\, G_k\varphi + H_k\varphi.$$

Moreover $dG_k\varphi = G_{k+1}d\varphi = 0$ because φ is closed, so that

$$\varphi = d\delta G_k\varphi + H_k\varphi$$

and thus $[\varphi]$ contains $H_k\varphi$. Moreover if ψ is harmonic, it is d-closed and thus represents a cohomology class. But if $\psi = d\theta$, one would have

$$\|\psi\|^2 = (\, d\theta \mid \psi\,) = 0$$

because ψ being harmonic is orthogonal to $d\mathcal{E}_M^{k-1}(M)$, so $\psi \equiv 0$. Thus we deduce the isomorphism (4.64) □

4.6.2 The case of holomorphic bundles

Let M be a compact hermitian manifold and $\pi\colon F \to M$ a holomorphic vector bundle with a hermitian metrics η.

The analogue of theorem 4.1 is now

Theorem 4.3. *There exists operators G_q, H_q*

$$G_q, H_q\colon C^\infty\left(M, \Lambda^{0,q} \otimes F\right) \to C^\infty\left(M, \Lambda^{0,q} \otimes F\right)$$

such that any $(0,q)$-form φ with values in F can be decomposed as

$$\varphi = \Box_\eta G_q\varphi + H_q\varphi = G_q\Box_\eta\varphi + H_q\varphi \tag{4.65}$$

with the properties

(i) The decomposition (4.65) is orthogonal

(ii) $\bar{\partial} G_q = G_{q+1} \bar{\partial}$

(iii) $H_q \varphi$ *is harmonic namely*

$$\square_\eta H_q \varphi = 0$$

(iv) The kernel of $\square_\eta$ *is finite dimensional*

(v) $H_q \circ H_q = H_q$

(vi) If $\varphi \in C^\infty\left(M, \Lambda^{0,q} \otimes F\right)$, φ *is harmonic if and only if* $\bar{\partial}\varphi = 0$ *and* $\vartheta_\eta \varphi = 0$. *Harmonic forms are orthogonal to* $\bar{\partial} C^\infty\left(M, \Lambda^{0,q-1} \otimes F\right)$ *and* $\vartheta_\eta C^\infty\left(M, \Lambda^{0,q+1} \otimes F\right)$.

As a consequence, we deduce the analogue of theorem 4.2:

Theorem 4.4. *One has an isomorphism*

$$H^q\left(M, F\right) \simeq \mathcal{H}^q(M, F) \tag{4.66}$$

where $\mathcal{H}^q(M, F)$ *is the space of harmonic* $(0, q)$*-forms with coefficients in* F.

Proof. We can represent $H^q(M, F)$ as

$$H^q(M, F) = \frac{\ker\left\{\bar{\partial} \colon C^\infty\left(M, \Lambda^{0,q} \otimes F\right) \to C^\infty\left(M, \Lambda^{0,q+1} \otimes F\right)\right\}}{\bar{\partial} C^\infty\left(M, \Lambda^{0,q-1} \otimes F\right)}$$

Let $[\varphi]$ be a cohomology class, φ a representative which is $\bar{\partial}$-closed. We use (4.65)

$$\varphi = \bar{\partial} \vartheta_\eta G_q \varphi + H_q \varphi$$

(where we have used $\bar{\partial} G_q \varphi = G_{q+1} \bar{\partial} \varphi = 0$) and thus $H_q \varphi \in [\varphi]$. □

Let $\mathcal{H}^{(p,q)}(M, F)$ be the finite dimensional space of harmonic (p, q)-forms with values in F, namely the kernel of $\square_\eta$ on $C^\infty\left(M, \Lambda^{p,q} \otimes F\right)$. When F is the trivial bundle of rank 1, $\mathcal{H}^{(p,q)}(M, F)$ will be denoted $\mathcal{H}^{(p,q)}(M)$ (the space of $\square$-harmonic (p, q)-forms on M).

As a consequence of theorem 4.4, because

$$\Lambda^{p,q}(M) \otimes F = \Lambda^{0,q}(M) \otimes \left(\Lambda^{p,0}(M) \otimes F\right)$$

and because $\Lambda^{p,0}(M) \otimes F$ is a holomorphic bundle, we deduce:

Theorem 4.5. *One has an isomorphism*

$$H^q\left(M, \Omega^p_M \otimes F\right) \simeq \mathcal{H}^{(p,q)}(M, F)$$

In particular, when F *is the trivial bundle of rank* 1, *one obtains isomorphisms of the Dolbeault groups* $H^{p,q}(M) = H^q\left(M, \Omega^p_M\right)$:

$$H^{p,q}(M) \simeq \mathcal{H}^{(p,q)}(M)$$

4.7 Duality

4.7.1 Poincaré duality

Lemma 4.6. *Let M be a compact hermitian manifold and $\varphi \in \mathcal{H}^k(M)$ a harmonic k-form. Then $*\varphi \in \mathcal{H}^{2n-k}(M)$ and $*$ is an isomorphim*

$$*\colon \mathcal{H}^k(M) \to \mathcal{H}^{2n-k}(M). \tag{4.67}$$

Proof. To say that φ is a harmonic form, is equivalent to saying that $d\varphi = 0$, $\delta\varphi = 0$ or

$$d\varphi = 0 \qquad d{*}\varphi = 0$$

(because $\delta = - * d*$ and $*$ is an isomorphism on forms). Then $*\varphi$ is closed and moreover $\delta{*}\varphi = -{*}d{*}{*}\varphi = (-1)^{k+1} * d\varphi = 0$ so

$$d\,(*\varphi) = 0 \qquad \delta\,(*\varphi) = 0$$

and $*\varphi$ is harmonic. As $** = (-1)^k$ on k-forms, $*$ is an isomorphism as in (4.67). □

Theorem 4.6. *The isomorphism $P_k\varphi = *\bar{\varphi}$*

$$P_k\colon \mathcal{H}^k(M) \to \mathcal{H}^{2n-k}(M)$$

is a realization at the level of harmonic forms, of the Poincaré duality.

Proof. Indeed the Poincaré duality P is given by

$$[\varphi] \cap P[\varphi] = \frac{\omega^n}{n!}.$$

□

4.7.2 Serre duality

Let $\pi\colon F \to M$ be a holomorphic vector bundle with a hermitian metrics η. Let $\pi^*\colon F^* \to M$ be the dual bundle. We construct the dual hermitian metrics in F^*: if U_a is a trivialization for F (and thus for F^*), we define

$$\eta^*_{a,\alpha\bar{\beta}} = \eta_a^{\bar{\beta}\alpha} \qquad \text{(inverse metrics)} \tag{4.68}$$

Then we define a morphism of bundles

$$\sigma\colon F \to F^*.$$

If $v \in F_m$ has components $(\zeta_a^\alpha)_{\alpha=1,\dots,r}$ in the trivialization U_a, $\sigma(v)$ has components

$$\sigma\,(v)_a^\beta = \sum \eta_{a,\beta\bar{\alpha}}\overline{\zeta_a^\alpha} \tag{4.69}$$

(always in U_a), then one can see that this defines an element $\sigma(v) \in F_m^*$ and σ preserves the hermitian metrics.

If F is the trivial line bundle with the trivial metrics, then σ just the conjugation

$$\sigma(v) = \bar{v}.$$

Lemma 4.7.

(1) The morphism $P\varphi = *\bar{\varphi}$ *is an isomorphism*

$$P\colon \mathcal{E}_M^{p,q}(M) \to \mathcal{E}_M^{n-p,n-q}(M)$$

(2) the mapping

$$P_F\varphi = \sigma\,(*\varphi) \tag{4.70}$$

is an isomorphism

$$P_F\colon C^\infty\left(M, \Lambda^{p,q} \otimes F\right) \to C^\infty\left(M, \Lambda^{n-p,n-q} \otimes F^*\right).$$

Proof. Recall that $*\colon \Lambda^{p,q}(M) \to \Lambda^{n-q,n-p}(M)$ and the conjugation reestablishes $\Lambda^{n-p,n-q}(M)$. Moreover

$$P_F^2 = (-1)^{p+q} \text{ on } \Lambda^{p,q} \otimes F \tag{4.71}$$

□

Lemma 4.8. P_F *induces an isomorphism of harmonic forms*

$$P_F\colon \mathcal{H}^{p,q}(F) \to \mathcal{H}^{n-p,n-q}(F^*). \tag{4.72}$$

Proof. We start with a harmonic form $\varphi \in \mathcal{H}^{p,q}(F)$ so that $\bar{\partial}\varphi = 0$, $\vartheta_\eta\varphi = 0$. So one has, using (4.58) for ϑ_η in a trivialization,

$$\partial\left(\sum_{\delta=1}^{n} \eta_{a,\delta\bar{\beta}}\left(*\varphi_a^\delta\right)\right) = 0$$

Taking the conjugate and using the definition of σ we see that

$$\bar{\partial}P_F\varphi = 0.$$

Moreover writing

$$\varphi = (-1)^{p+q}P_F\left(P_F\varphi\right) \quad \text{(see (4.71))}$$

we obtain

$$\bar{\partial}P_F\left(P_F\varphi\right) = 0.$$

But this equation is exactly

$$\overline{\vartheta_{\eta^*}P_F\varphi}$$

where ϑ_{η^*} is the adjoint of $\bar{\partial}$ for the hermitian metrics η^* on F^* defined by (4.68) and where we have used (4.58) for ϑ_{η^*} as well as the definition of P_F. So $P_F\varphi$ is harmonic. □

As a consequence we deduce

Theorem 4.7. *One has an isomorphism*

$$[H^q(M, \Omega^p \otimes F)]^* \simeq H^{n-q}(M, \Omega^{n-p} \otimes F^*) \tag{4.73}$$

where $[\cdots]^*$ *denotes the dual space. In particular*

$$[H^q(M, \Omega^p)]^* \simeq H^{n-q}(M, \Omega^{n-p})$$

4.7.3 Application to modifications

Theorem 4.8. *Let* $f\colon M_1 \to M_2$ *be a modification between two compact complex manifolds. Then the induced morphisms*

$$f^*\colon H^q(M_2, \Omega^p_{M_2}) \to H^q(M_1, \Omega^p_{M_1}) \tag{4.74}$$

are all injective.

Proof. We shall use Serre duality i.e. the previous theorem 4.7. Let $[\varphi_2] \in H^q(M_2, \Omega^p_{M_2})$, with $f^*[\varphi_2] = 0$. To prove that $[\varphi_2] = 0$, it is sufficient to prove that for any $[\omega_2] \in H^{n-q}(M_2, \Omega^{n-p}_{M_2})$

$$(\varphi_2 \mid \omega_2) = \int_{M_2} \varphi_2 \wedge \omega_2 = 0.$$

By definition of a modification, there exists proper subvarieties $E_j \subset M_j$, $E_1 = f^{-1}(E_2)$ such that f induces an isomorphism $f\colon M_1 \setminus E_1 \xrightarrow{\sim} M_2 \setminus E_2$. Let U_ε be the set of points of M_2 at distance $< \varepsilon$ of E_2.

Then $f\colon f^{-1}(M_2 \setminus U_\varepsilon) \xrightarrow{\sim} M_1 \setminus U_\varepsilon$ is an isomorphism. Then:

$$\begin{aligned}\int_{M_2} \varphi_2 \wedge \omega_2 &= \lim_{\varepsilon \to 0} \int_{M_2 \setminus U_\varepsilon} \varphi_2 \wedge \omega_2 = \\ &= \lim_{\varepsilon \to 0} \int_{f^{-1}(M_2 \setminus U_\varepsilon)} f^*\varphi_2 \wedge f^*\omega_2 = \\ &= \int_{M_1} f^*\varphi_2 \wedge f^*\omega_2 = 0\end{aligned} .$$

because $f^*\varphi_2$ is exact and ω_2 is closed. ∎

Chapter 5

Hodge theory on compact kählerian manifolds

5.1 Introduction

This chapter explains the central results concerning the cohomology of a compact kählerian manifold M. We shall prove that there is a direct sum decomposition of the cohomology of M as

$$H^k(M, \mathbb{C}) = \bigoplus_{p+q=k} H^{p,q}(M)$$

where $H^{p,q}(M) = H^q(M, \Omega_M^p)$ is the $\bar{\partial}$-cohomology of $(p,0)$-forms. This decomposition goes much beyond mere formalities. It means that if $[\varphi]$ is a d-cohomology class of degree k, one can find in this class a representative $\hat{\varphi}$ such that its decomposition in types $\hat{\varphi} = \sum_{p+q=k} \hat{\varphi}^{(p,q)}$ has the property that each component $\hat{\varphi}^{(p,q)}$ is d-closed and $\bar{\partial}$-closed.

Conversely, if $[\varphi] \in H^{p,q}(M)$ is a $\bar{\partial}$-cohomology class, there exists a representative φ' in this class of $\bar{\partial}$ -cohomology which is d-closed (and $\bar{\partial}$-closed).

Moreover, the d-operator is strict, that is if a form $d\varphi$ has no component of type (p,q) for $p < p_0$, then $d\varphi$ can be written as $d\psi$ where ψ has no component of type (p', q') with $p' < p_0$. These results are global, i.e. they are valid for forms which are C^∞ on M. In general, they would not hold for an open subset of M, for example.

There are two main parts in the proof of these results. The first part is the global theory of harmonic forms which has been already done in chapter 4 for general hermitian compact manifolds. The second part is a purely local calculation which says that the three De Rham-Hodge Laplace operators $\frac{1}{2}\Delta$, $\square$ and $\overline{\square}$ are in fact identical on a kählerian manifold.

5.2 Kählerian manifolds

Definition 5.1. We say that a hermitian metrics h on a complex manifold M is a Kähler metrics or that the manifold is kählerian, if the $(1,1)$-form

$$\omega = i \sum_{i,j=1}^{n} h_{i\bar{j}}\, dz^i \wedge d\bar{z}^j \tag{5.1}$$

is d-closed: $d\omega = 0$. The form ω is called the Kähler form of the metrics.

5.2.1 Local exactness

Lemma 5.1. *For every point $m \in M$, on sufficiently small neighborhoods U of m, any $(1,1)$-form ω which is d-closed, is $\partial\bar{\partial}$-exact so that there exists a function f with $\omega = \partial\bar{\partial} f$.*

Proof. $d\omega = 0$ implies $\partial\omega + \bar{\partial}\omega = 0$, hence $\partial\omega = \bar{\partial}\omega = 0$ because $\partial\omega$ and $\bar{\partial}\omega$ have different types. On sufficiently small neighborhoods U of m, $\omega = \bar{\partial}\varphi$ where φ is a $(1,0)$-form, because of Dolbeault lemma. Then $\partial\omega = 0$ gives $\partial\bar{\partial}\varphi = 0$. So $\partial\varphi$ is a $(2,0)$-form which is holomorphic in U and which is d-closed. By the Poincaré lemma for holomorphic forms $\partial\varphi = \partial\psi$ where ψ is a holomorphic $(1,0)$-form. So $\varphi - \psi$ is a $(1,0)$-form which is ∂-closed; again by Dolbeault lemma, $\varphi - \psi = \partial f$ so that

$$\omega = \bar{\partial}\varphi = \bar{\partial}(\varphi - \psi) = \bar{\partial}\partial f$$

because ψ is holomorphic. □

Examples of Kähler manifolds.

1. Submanifolds of kählerian manifolds.

   Lemma 5.2. *If $W \subset M$ is a complex submanifold of a kählerian manifold, W is also kählerian.*

   Proof. Let h be a kählerian metrics on M, then the restriction $h|_W$ to W is also kählerian. Indeed, W is locally defined by $z_{p+1} = \cdots = z_n = 0$, $h|_W = \sum_{i,j=1}^{n} h_{i\bar{j}}\, dz^i \wedge d\bar{z}^j$ and the Kähler form of $h|_W$ is just $\omega|_W$ which is closed. □

2. $\mathbb{C}^n$ and complex tori.
   $\mathbb{C}^n$ is a kählerian manifold with the standard constant metrics

$$ds^2 = \sum_{j=1}^{n} dz^j d\bar{z}^j$$

and the Kähler form

$$\omega = i \sum_{j=1}^{n} dz^j \wedge d\bar{z}^j.$$

As a consequence, any complex torus $\mathbb{C}^n/L$ (L discrete lattice of rank $2n$ as a real lattice) is kählerian with the constant metrics.

3. Projective space $\mathbb{P}^n$ and projective manifolds.

Lemma 5.3. *$\mathbb{P}^n$ is a kählerian manifold.*

Proof. Let $[Z^0, \ldots, Z^n]$ the homogeous coordinates of $\mathbb{P}^n$, and U_a the affine open set $Z^a \neq 0$, so that in U_a, one has the holomorphic coordinates

$$z_a^j = \frac{Z^j}{Z^a} \qquad j \neq a.$$

Define in U_a:

$$p_a = \log \left(1 + \sum_{j \neq a} \left| z_a^j \right|^2 \right). \tag{5.2}$$

Then in $U_a \cap U_b$

$$p_a - p_b = \log \frac{\left|Z^b\right|^2}{\left|Z^a\right|^2} = \log \left| z_a^b \right|^2.$$

As a consequence on $U_a \cap U_b$

$$\partial\bar{\partial} p_a = \partial\bar{\partial} p_b$$

and this defines a global $(1,1)$-form which is d-closed.

The hermitian metrics in U_0 is given by

$$h_{0,i\bar{j}} = \frac{\partial^2}{\partial z_i \partial \bar{z}_j} \left(\log \left(1 + \left| z_0 \right|^2 \right) \right)$$

where

$$\left| z_0 \right|^2 = \sum_{j=1}^{n} \left| z_0^j \right|^2.$$

More explicitly

$$h_{0,i\bar{j}} = \frac{g_{0,i\bar{j}}}{\left(1 + \left| z_0 \right|^2 \right)^2}$$

with

$$g_{0,i\bar{j}} = \delta_{ij} \left(1 + \left| z_0 \right|^2 \right) - \bar{z}_0^i z_0^j \tag{5.3}$$

hence the metrics is positive definite because

$$\sum_{i=1}^{n} g_{0,i\bar{j}}\zeta^i\bar{\zeta}^j = \left(1+|z_0|^2\right)|\zeta|^2 - |(z_0 \mid \zeta)|^2 \geq |\zeta|^2$$

by Cauchy-Schwarz inequality. □

Lemma 5.4. *Any algebraic projective manifold is kählerian.*

This is a consequence of lemmas 5.1 and 5.2.

4. The Bergman metrics in the ball.
   In the unit ball of $\mathbb{C}^n$

$$B = \left\{ z \in \mathbb{C}^n \mid |z|^2 < 1 \right\}$$

one can define the Bergman metrics

$$h_{i\bar{j}} = \frac{\partial^2}{\partial z_i \partial \bar{z}_j} \log\left(1 - |z|^2\right).$$

It is obviously kählerian because

$$\omega = i\partial\bar{\partial}\log\left(1 - |z|^2\right)$$

is d-closed.

As a consequence, because the Bergman metrics is invariant by biholomorphic transformations of the ball, any manifold which is the quotient B/Γ where Γ is a discrete group of $\mathrm{SU}(n,1)$, is a kählerian manifold.

5.2.2 Cohomological properties of ω

Theorem 5.1. *Let M be a complex compact kählerian manifold, ω its Kähler form.*

1) All the powers $\omega^p (1 \leq p \leq n)$ define non trivial cohomology classes both for d and $\bar{\partial}$. As a consequence

$$H^{2p}(M, \mathbb{C}) \neq 0, \qquad H^{p,p}(M) \neq 0.$$

2) If $W \subset M$ is a closed complex submanifold, its homology class $[W] \in H_{2p}(M, \mathbb{C})$ is not 0.

Proof. 1) We know that, up to a constant factor, ω^n ($n = \dim M$) is the volume form of M, so that

$$\int_M \omega^n > 0.$$

Let us suppose by contradiction that $\omega^p = d\varphi$; then

$$\omega^n = \omega^p \wedge \omega^{n-p} = d\varphi \wedge \omega^{n-p} = d\left(\varphi \wedge \omega^{n-p}\right)$$

because ω^{n-p} is d-closed and then by Stokes theorem $\int_M \omega^n$ would be 0.

If $\omega^p = \bar{\partial}\varphi$, then $\omega^n = \bar{\partial}\left(\varphi \wedge \omega^{n-p}\right)$, but $\varphi \wedge \omega^{n-p}$ would be of type $(n, n-1)$ and thus ∂-closed so $\omega^n = d\left(\varphi \wedge \omega^{n-p}\right)$ again.

As a consequence ω^p cannot be exact for d or $\bar{\partial}$.

2) If W is a complex submanifold of dimension p, then $\omega^p|_W$ is its volume form (up to a constant) so $\int_W \omega^p \neq 0$ and by Stokes theorem, W cannot be the boundary of a chain. □

Theorem 5.2. *One has*

$$\vartheta\omega = 0$$

so that ω *is harmonic for* Δ, $\Box$, $\overline{\Box}$.

Proof. This is a direct calculation using lemma 4.2, formula (4.44) (chapter 4) and the fact that locally $\omega_{i\bar{\jmath}} = \frac{\partial^2 f}{\partial z_i \partial \bar{z}_j}$. □

5.3 Local kählerian geometry

5.3.1 Covariant derivatives

We consider a vector field $\zeta \in C^\infty\left(U, \overline{T}(M)\right)$. In local coordinates (U_a, z_a) or (U_b, z_b) one has

$$\zeta = \sum_{j=1}^{n} \zeta_a^{\bar{\jmath}} \frac{\partial}{\partial \bar{z}_a^j} = \sum_{j=1}^{n} \zeta_b^{\bar{k}} \frac{\partial}{\partial \bar{z}_b^k} \tag{5.4}$$

$$\zeta_b^{\bar{k}} = \sum_{j=1}^{n} \overline{\left(\frac{\partial z_b^k}{\partial z_a^j}\right)} \zeta_a^{\bar{\jmath}}. \tag{5.5}$$

In particular we see that

$$\frac{\partial \zeta_b^{\bar{k}}}{\partial z_a^i} = \sum_{j=1}^{n} \overline{\left(\frac{\partial z_b^k}{\partial z_a^j}\right)} \frac{\partial \zeta_a^{\bar{\jmath}}}{\partial z_a^i} \tag{5.6}$$

In the same way, for a $(0,1)$-form $\varphi \in C^\infty\left(U, \overline{T}^*(M)\right)$

$$\varphi = \sum \varphi_{a\bar{\jmath}} d\bar{z}_a^j = \sum \varphi_{b\bar{k}} d\bar{z}_b^k \tag{5.7}$$

one has

$$\varphi_{b,\bar{k}} = \sum_{j=1}^{n} \overline{\left(\frac{\partial z_a^j}{\partial z_b^k}\right)} \varphi_{a,\bar{\jmath}}$$

so that

$$\frac{\partial \varphi_{b,\bar{k}}}{\partial z_a^i} = \sum_{j=1}^{n} \overline{\left(\frac{\partial z_a^j}{\partial z_{b,k}}\right)} \frac{\partial \varphi_{a,\bar{j}}}{\partial z_a^i}. \tag{5.8}$$

We define the covariant derivative of $\zeta \in C^\infty\left(U, \overline{T}(M)\right)$ or $\varphi \in C^\infty\left(U, \overline{T}^*(M)\right)$ with respect to a vector field $w \in C^\infty\left(U, T(M)\right)$

$$\nabla_w \zeta \in C^\infty\left(U, \overline{T}(M)\right), \qquad \nabla_w \varphi \in C^\infty\left(U, \overline{T}^*(M)\right)$$

by saying that they are vectors or forms with components in a coordinate system (U_a, z_a) given by

$$\begin{aligned} (\nabla_w \zeta)_a^{\bar{j}} &= \sum_{k=1}^{n} w_a^k \frac{\partial \zeta_a^{\bar{j}}}{\partial z_a^k} \\ (\nabla_w \varphi)_{a,\bar{j}} &= \sum_{k=1}^{n} w_a^k \frac{\partial \varphi_{a\bar{j}}}{\partial z_a^k} \end{aligned} \tag{5.9}$$

in particular, for $w = \frac{\partial}{\partial z_a^i}$, we denote

$$(\nabla_w \zeta)_a^{\bar{j}} \equiv \nabla_i \zeta_a^{\bar{j}} = \frac{\partial \zeta_a^{\bar{j}}}{\partial z_a^i}, \quad (\nabla_w \varphi)_{a,\bar{j}} \equiv \nabla_i \varphi_{a,\bar{j}} = \frac{\partial \varphi_{a\bar{j}}}{\partial z_a^i}. \tag{5.10}$$

From (5.5) and (5.8) we see that $\nabla_w \zeta$ or $\nabla_w \varphi$ are well defined vector fields or forms.

Now let us take $\zeta \in C^\infty\left(U, T(M)\right)$. By lowering the indices by means of the hermitian metrics h we obtain a form $H(\zeta) \in C^\infty\left(U, \overline{T}^*(M)\right)$

$$H(\zeta) = \sum_{i,j=1}^{n} h_{a,i\bar{j}} \zeta_a^i d\bar{z}_a^j$$

So if $\zeta \in C^\infty\left(U, T(M)\right)$, we define for $w \in C^\infty\left(U, T(M)\right)$

$$\nabla_w \zeta = H^{-1}\left(\nabla_w H(\zeta)\right). \tag{5.11}$$

In local coordinates (U_a, z_a) for $w = \frac{\partial}{\partial z_a^i}$

$$(\nabla_i \zeta)_a^j = \sum_k h_a^{\bar{k}j} \frac{\partial}{\partial z_a^i} \left(\sum_l h_{a,l\bar{k}} \zeta_a^l\right) \tag{5.12}$$

We denote simply:

$$\nabla_i \zeta_a^j \equiv (\nabla_i \zeta)_a^j .$$

In the same way, if $\varphi \in C^\infty\left(U, T^*(M)\right)$, we define for $w \in C^\infty\left(U, T(M)\right)$

$$\nabla_w \varphi = H\left(\nabla_w H^{-1}(\varphi)\right) \tag{5.13}$$

or in coordinates for $w = \frac{\partial}{\partial z_a^i}$

$$\nabla_i \varphi_{a,j} = \sum_k h_{a,j\bar{k}} \frac{\partial}{\partial z_a^i} \left(\sum_l h_a^{\bar{k}l} \varphi_{a,l} \right). \tag{5.14}$$

For $w \in C^\infty\left(U, \overline{T}(M)\right)$, $\zeta \in C^\infty(U, T(M))$ and $\varphi \in C^\infty(U, T^*(M))$, we can also define

$$\left(\overline{\nabla}_w \varphi\right)_a^j = \sum_k w_a^{\bar{k}} \frac{\partial \zeta_a^j}{\partial \bar{z}_a^k}$$

$$\left(\overline{\nabla}_w \varphi\right)_{a,j} = \sum_k w_a^{\bar{k}} \frac{\partial \varphi_{a,j}}{\partial \bar{z}_a^k}.$$

In particular if $w = \frac{\partial}{\partial \bar{z}_a^k}$, we denote

$$\overline{\nabla}_k \zeta_a^j = \frac{\partial \zeta_a^j}{\partial \bar{z}_a^k} \qquad \overline{\nabla}_k \varphi_{a,j} = \frac{\partial \varphi_{a,j}}{\partial \bar{z}_a^k}$$

Because of the conjugate equations (5.5) or (5.8) one has

$$\overline{\nabla}_w \zeta \in C^\infty(U, T(M)), \qquad \overline{\nabla}_w \zeta \in C^\infty(U, T^*(M)).$$

Finally, we can define $\overline{\nabla}_w \zeta$ and $\overline{\nabla}_w \varphi$ for $\zeta \in C^\infty\left(U, \overline{T}(M)\right)$ or $\varphi \in C^\infty\left(U, \overline{T}^*(M)\right)$ by lowering or raising the indices

$$\overline{\nabla}_w \zeta = H^{-1}\left(\overline{\nabla}_w H(\zeta)\right)$$
$$\overline{\nabla}_w \varphi = H\left(\overline{\nabla}_w H^{-1}(\varphi)\right).$$

The formulas are

$$\begin{aligned} \nabla_i \zeta_a^b &= \frac{\partial \zeta_a^j}{\partial z_a^i} + \sum_{k,l} \left(h_a^{\bar{k}j} \frac{\partial}{\partial z_a^i} \left(h_{a,l\bar{k}}\right) \right) \zeta_a^l \\ \nabla_i \varphi_{a,j} &= \frac{\partial \varphi_{a,j}}{\partial z_a^i} + \sum_{k,l} \left(h_{a,\bar{k}j} \frac{\partial}{\partial z_a^i} \left(h_a^{\bar{k}l}\right) \right) \varphi_{a,l}. \end{aligned} \tag{5.15}$$

We define the Christoffel symbols

$$\Gamma_{il}^j = \sum_k h_a^{\bar{k}j} \frac{\partial}{\partial z_a^i} h_{a,l\bar{k}} \tag{5.16}$$

or:

$$\Gamma_{il}^j = \left(\left(\tfrac{\partial}{\partial z_a^i} h_a \right) h_a^{-1} \right)_l^j \tag{5.17}$$

$$\begin{aligned} \nabla_i \zeta_a^j &= \frac{\partial \zeta_a^j}{\partial z_a^i} + \sum_{l=1}^n \Gamma_{il}^j \zeta_a^l \\ \nabla_i \varphi_{a,j} &= \frac{\partial \varphi_{a,j}}{\partial z_a^i} - \sum_{l=1}^n \Gamma_{ij}^l \varphi_{a,l} \end{aligned} \tag{5.18}$$

where we have used the equation

$$\frac{\partial}{\partial z_a^i}\left(\sum_k h_a^{\bar{k}j} h_{a,l\bar{k}}\right) = \frac{\partial}{\partial z_a^i}\delta_l^j = 0.$$

Also

$$\begin{aligned}\overline{\nabla}_i \zeta_a^{\bar{j}} &= \frac{\partial \zeta_a^{\bar{j}}}{\partial \bar{z}_a^i} + \sum_{l=1}^{n} \overline{\Gamma_{il}^{j}} \zeta_a^{\bar{l}} \\ \overline{\nabla}_i \varphi_{a,\bar{j}} &= \frac{\partial \varphi_{a,\bar{j}}}{\partial \bar{z}_a^i} - \sum_{l=1}^{n} \overline{\Gamma_{ij}^{l}} \varphi_{a,\bar{l}}.\end{aligned} \tag{5.19}$$

We define the covariant derivatives of any tensor field by lowering or raising the indices to reduce to pure derivatives either with respect to $\frac{\partial}{\partial z_a^i}$ or $\frac{\partial}{\partial \bar{z}_a^i}$, and then raising or lowering the indices correspondingly.

For example, one has immediately

Lemma 5.5. *The covariant derivatives of h vanish.*

Proof. For example

$$\begin{aligned}\nabla_i h_{a,k\bar{l}} &= \sum_j h_{a,k\bar{j}} \frac{\partial}{\partial z_a^i}\left(\sum_r h_a^{\bar{j}r} h_{a,r\bar{l}}\right) \\ &= \sum h_{a,k\bar{j}} \frac{\partial}{\partial z_a^i}\left(\delta_{\bar{l}}^{\bar{j}}\right) = 0.\end{aligned}$$

□

Lemma 5.6. *The hermitian metrics h is kählerian if and only if for all indices i, j, k*

$$\Gamma_{jk}^i = \Gamma_{kj}^i. \tag{5.20}$$

Proof. By the definition (5.16)

$$\frac{\partial h_{j\bar{l}}}{\partial z^i} = \sum_k h_{k\bar{l}} \Gamma_{ij}^k.$$

So

$$\begin{aligned}0 = d\omega &= i\sum \frac{\partial h_{j\bar{l}}}{\partial z^i} dz^i \wedge dz^j \wedge d\bar{z}^l + i\sum \frac{\partial h_{j\bar{l}}}{\partial \bar{z}^i} d\bar{z}^i \wedge dz^j \wedge d\bar{z}^l \\ &= \frac{i}{2}\sum h_{k\bar{l}}\left(\Gamma_{ij}^k - \Gamma_{ji}^k\right) dz^i \wedge dz^j \wedge d\bar{z}^l + \\ &+ \frac{i}{2}\sum h_{l\bar{k}}\left(\overline{\Gamma}_{ij}^k - \overline{\Gamma}_{ji}^k\right) d\bar{z}^i \wedge dz^l \wedge d\bar{z}^j.\end{aligned}$$

This implies (5.20). □

5.3.2 Covariant derivatives of differential forms

Let $\varphi \in \mathcal{E}_M^{p,q}(U)$ be written in a coordinate system (U,z) as

$$\varphi = \frac{1}{p!q!}\sum \varphi_{I\bar{J}}\, dz^I \wedge d\bar{z}^J.$$

We have the following lemma.

Lemma 5.7. *Let h be a kählerian metrics. Then*

$$\partial\varphi = \tfrac{1}{p!q!}\textstyle\sum_{i,I,J} \nabla_i \varphi_{I\bar{J}}\, dz^i \wedge dz^I \wedge d\bar{z}^J \tag{5.21}$$

$$\bar{\partial}\varphi = \tfrac{1}{p!q!}\textstyle\sum_{i,I,J} \overline{\nabla}_i \varphi_{I\bar{J}}\, d\bar{z}^i \wedge dz^I \wedge d\bar{z}^J \tag{5.22}$$

$$(\vartheta\varphi)_{IK} = (-1)^{p+1}\textstyle\sum_{j,i} h^{\bar{j}i}\nabla_i \varphi_{I\bar{j}\bar{K}} \quad \textit{(here } |I| = p, |K| = q-1) \tag{5.23}$$

Proof. We prove only (5.23). Using lemma 4.2 of chapter 4

$$(-1)^{p+1}(\vartheta\varphi)^{\bar{I}K} = \sum\left(\frac{\partial}{\partial z_j} + \frac{1}{\det h}\frac{\partial \det h}{\partial z_j}\right)\varphi^{\bar{I}jK} \tag{5.24}$$

On the other hand, by definition, if $|K| = q-1$

$$\sum_{j=1}^n \nabla_j \varphi^{\bar{I}jK} = \sum_{j=1}^n \frac{\partial \varphi^{\bar{I}jK}}{\partial z_j} + \sum_{j,k=1}^n \Gamma^j_{jk}\varphi^{\bar{I}kK} + \sum_{j,k}\Gamma^{k_1}_{jk}\varphi^{\bar{I}jkk_2\ldots k_q} + \cdots$$

But $\Gamma^{k_1}_{jk}$ is symmetric in j,k (lemma 5.6) and $\varphi^{\bar{I}jkk_2\ldots k_q}$ is skew symmetric in j,k, so that

$$\sum_{j=1}^n \nabla_j \varphi^{\bar{I}jK} = \sum_j \frac{\partial \varphi^{\bar{I}jK}}{\partial z_j} + \sum_{j,k}\Gamma^j_{jk}\varphi^{\bar{I}kK}. \tag{5.25}$$

Then, one proves

$$\sum_j \Gamma^j_{jk} = \frac{1}{\det h}\frac{\partial \det h}{\partial z_k} \tag{5.26}$$

to deduce the identity from (5.24) and (5.25). Because the metric tensor $h_{j\bar{k}}$ commutes with the covariant derivative ∇_i, one can lower the indices as one wishes.

To prove (5.26), one knows that $\frac{\partial \det h}{\partial h_{i\bar{j}}}$ is the cofactor of $h_{i\bar{j}}$ so that it is equal to $(\det h)h^{\bar{i}j}$. Then

$$\frac{\partial \det h}{\partial z^i} = \sum_{k,l}\frac{\partial \det h}{\partial h_{k\bar{l}}}\frac{\partial h_{k\bar{l}}}{\partial z^i} = \sum_{k,l}(\det h)h^{\bar{l}k}\frac{\partial h_{k\bar{l}}}{\partial z^i}$$

$$= (\det h)\sum_l \Gamma^l_{il}.$$

Since h is kählerian, $\Gamma^l_{il} = \Gamma^l_{li}$, and we obtain (5.26). □

5.3.3 The operator Λ

The operator Λ is the contraction by the Kähler form.
One has for $\varphi \in \mathcal{E}_M^{p,q}(M)$

$$\varphi = \frac{1}{p!q!} \sum \varphi_{I\bar{J}}\, dz^I \wedge d\bar{z}^J$$

the definition:

$$\Lambda\varphi = \frac{1}{(p-1)!(q-1)!} \sum_{\substack{u,\bar{j} \\ K,\bar{L}}} ih^{\bar{j}i} \varphi_{i\bar{j}K\bar{L}}\, dz^K \wedge d\bar{z}^L \tag{5.27}$$

where $|K| = p-1$, $|L| = q-1$. In particular

$$(\Lambda\varphi)_{K\bar{L}} = (-1)^{p-1} \sum_{i,\bar{j}} ih^{\bar{j}i} \varphi_{iK\bar{j}\bar{L}}. \tag{5.28}$$

It is obvious that Λ is a real operator

$$\Lambda\overline{\varphi} = \overline{\Lambda\varphi}.$$

Theorem 5.3. *One has the relations*

$$[\partial, \Lambda] = i\vartheta \qquad [\bar{\partial}, \Lambda] = -i\bar{\vartheta}. \tag{5.29}$$

Proof. The second relation is a consequence of the first.

1) We calculate $\partial\Lambda\varphi$ using (5.21)

$$(\partial\Lambda\varphi)_{I\bar{L}} = \nabla_{i_1} (\Lambda\varphi)_{i_2\ldots i_p\bar{L}} - \nabla_{i_2} (\Lambda\varphi)_{i_1 i_3\ldots i_p\bar{L}} + \cdots .$$

Here $|L| = q-1$, $|I| = p$. Then we use (5.28)

$$(\partial\Lambda\varphi)_{I\bar{L}} = i(-1)^{p-1} \sum_{i,j} \left[\nabla_{i_1} h^{\bar{j}i} \varphi_{ii_2\ldots i_p\bar{j}\bar{L}} - \nabla_{i_2} h^{\bar{j}i} \varphi_{ii_1 i_3\ldots i_p\bar{j}\bar{L}} + \cdots\right].$$

But ∇ commutes with the metric tensor h, so that:

$$(\partial\Lambda\varphi)_{I\bar{L}} = i(-1)^{p-1} \sum_{i,j} h^{\bar{j}i} \left[\nabla_{i_1} \varphi_{ii_2\ldots i_p\bar{j}\bar{L}} - \nabla_{i_2} \varphi_{ii_1 i_3\ldots i_p\bar{j}\bar{L}} + \cdots\right]. \tag{5.30}$$

2) Then we calculate $\Lambda\partial\varphi$. We use (5.21)

$$(\partial\varphi)_{K\bar{J}} = \nabla_{k_0} \varphi_{k_1\ldots k_p\bar{J}} - \nabla_{k_1} \varphi_{k_0 k_2\ldots k_p\bar{J}} + \cdots$$

with $|K| = p+1$, $|J| = q$. Then we apply (5.28)

$$(\Lambda\partial\varphi)_{I\overline{L}} = (-1)^p i \sum_{i,j} h^{\bar{j}i} (\partial\varphi)_{iI\bar{j}\overline{L}} \qquad |I| = p, |L| = q-1$$

so

$$(\Lambda\partial\varphi)_{I\overline{L}} = (-1)^p i \sum_{i,j} h^{\bar{j}i} \left[\nabla_i \varphi_{I\bar{j}\overline{L}} - \nabla_{i_1}\varphi_{ii_2\ldots i_p\bar{j}\overline{L}} + \cdots\right]. \tag{5.31}$$

So from (5.30), (5.31)

$$(\partial\Lambda\varphi - \Lambda\partial\varphi)_{I\overline{L}} = (-1)^{p+1} i \sum_{i,j} h^{\bar{j}i} \nabla_i \varphi_{I\bar{j}\overline{L}}$$

which by (5.23) of lemma 5.7 is exactly

$$i\,(\vartheta\varphi)_{I\overline{L}}\,.$$

□

5.3.4 The equality of the De Rham-Hodge Laplacians

We come finally to the main theorem.

Theorem 5.4. *On a kählerian manifold one has*

$$\frac{1}{2}\Delta = \Box = \overline{\Box}. \tag{5.32}$$

In particular Δ, its Green operator G and its harmonic projector H respect the type of forms.

Proof. We use the relations of theorem 5.3 to replace ϑ in $\Box$ and $\bar{\vartheta}$ in $\overline{\Box}$ by their expressions in term of ∂, Λ, $\bar{\partial}$

$$\begin{aligned}\Box = \bar{\partial}\vartheta + \vartheta\bar{\partial} &= -i\left[\bar{\partial}\left(\partial\Lambda - \Lambda\partial\right) + \left(\partial\Lambda - \Lambda\partial\right)\bar{\partial}\right] \\ &= -i\left[\bar{\partial}\partial\Lambda - \bar{\partial}\Lambda\partial + \partial\Lambda\bar{\partial} - \Lambda\partial\bar{\partial}\right]\end{aligned}$$

and in the same way

$$\overline{\Box} = \partial\bar{\vartheta} + \vartheta\bar{\partial} = -i\left[\bar{\partial}\partial\Lambda + \partial\Lambda\bar{\partial} - \bar{\partial}\Lambda\partial - \Lambda\partial\bar{\partial}\right]$$

so $\Box = \overline{\Box}$.

Then we use

$$\delta = \vartheta + \bar{\vartheta} \qquad d = \partial + \bar{\partial}$$

to deduce

$$\Delta = \Box + \overline{\Box} + \partial\vartheta + \vartheta\partial + \bar{\partial}\bar{\vartheta} + \bar{\vartheta}\bar{\partial}.$$

But then again by theorem 5.3

$$\partial\vartheta + \vartheta\partial = i\left[\partial\left(\partial\Lambda - \Lambda\partial\right) + \left(\partial\Lambda - \Lambda\partial\right)\partial\right] \equiv 0$$

so

$$\Delta = 2\Box = 2\overline{\Box}.$$

□

5.4 The Hodge decomposition on compact kählerian manifolds

5.4.1 Harmonic forms on compact kählerian manifolds

Theorem 5.5. *Let M be a compact kählerian manifold.*

1) *Let $\omega \in \mathcal{E}^k_M(M)$ be a harmonic form of degree k and let $\omega = \sum_{p+q=k} \omega^{(p,q)}$ be its decomposition in pure types. Then the $\omega^{(p,q)}$ are harmonic forms which are $\bar\partial$-closed and ϑ-closed.*

2) *If $\omega^{(p,q)}$ is a harmonic form of pure type (p,q) for $\Box$ (or $\overline{\Box}$), then it is a harmonic form for Δ.*

Proof. 1) is a consequence of the fact that Δ preserves the types (see theorem 5.4), so if $\Delta\omega = 0$

$$(\Delta\omega)^{(p,q)} = \Delta\omega^{(p,q)} = 0$$

and thus $\Box\omega^{(p,q)} = \overline{\Box}\omega^{(p,q)} = 0$.
2) is a consequence of $2\Box = \Delta$. □

Theorem 5.6. *One can write*

$$\mathcal{H}^k(M) = \bigoplus_{p+q=k} \mathcal{H}^{(p,q)}(M) \tag{5.33}$$

with an orthogonal decomposition. Moreover

$$\mathcal{H}^{(p,q)}(M) = \overline{\mathcal{H}^{(q,p)}(M)} \tag{5.34}$$

Proof. The orthogonal decomposition is a trivial consequence of theorem 5.5, while (5.34) is obtained by the isomorphism

$$\omega^{(p,q)} \in \mathcal{H}^{(p,q)}(M) \to \overline{\omega^{(p,q)}} \in \mathcal{H}^{(q,p)}(M)$$

because $\Box = \overline{\Box}$. □

We can rephrase theorem 5.6 as follows.

Theorem 5.7. *Let M be a compact kählerian manifold*

1) *If $[\varphi] \in H^k(M, \mathbb{C})$, and $\varphi \in [\varphi]$ is a d-closed form, then there exists a form ψ such that in the decomposition:*

$$\varphi + d\psi = \sum_{p+q=k} \omega^{(p,q)}$$

the $\omega^{(p,q)}$ are both d-closed and $\bar{\partial}$-closed and thus define cohomology classes for the $\bar{\partial}$-operator.

2) *If $[\varphi^{(p,q)}] \in H^{p,q}(M)$ is a class of $\bar{\partial}$-cohomology of pure type (p, q), there exists ψ of type $(p, q-1)$ such that $\varphi + \bar{\partial}\psi$ is $\bar{\partial}$-closed and d-closed and thus defines a cohomology class for d.*

Proof. 1) Start with $[\varphi]$. There exists a form $\omega \in [\varphi]$ which is harmonic. So $\omega = \varphi + d\psi$ and its decomposition in pure types has the required property.
2) Start with a class of $\bar{\partial}$-cohomology $[\varphi^{(p,q)}]$. There exists $\omega^{(p,q)} \in [\varphi^{(p,q)}]$ which is harmonic for $\Box$ but also for $\Delta = \frac{1}{2}\Box$. So $\omega^{(p,q)}$ is d-closed. $\Box$

Corollary 5.1. *Let M be a compact kählerian manifold. Any holomorphic form $\varphi \in H^0(M, \Omega^p_M)$ is d-closed and harmonic.*

Proof. The form φ is of type $(p, 0)$, so $\vartheta\varphi = 0$ trivially. But φ is holomorphic so $\bar{\partial}\varphi = 0$. Thus $\Box\varphi = 0$, so $\Delta\varphi = 0$ and φ is harmonic and d-closed. $\Box$

5.5 The pure Hodge structure on cohomology

Let us recall from chapter 2 the definition of the Hodge filtration F and its conjugate $\overline{F}$. If M is a complex manifold, one can write a k-differential form φ on M in local complex coordinates as

$$\varphi = \sum_{|I|+|J|=k} \varphi_{I\bar{J}}\, dz^I \wedge d\bar{z}^J \tag{5.35}$$

We say that φ has type $\geq p$, if one has $|I| \geq p$ in in the sum (5.35) (for every system of complex coordinates on M). We define $F^p\mathcal{E}^k_M$ as the subsheaf of $\mathcal{E}^k_M$ of k-forms of type $\geq p$. This defines a decreasing filtration and d respects this filtration:

$$d(F^p\mathcal{E}^k_M) \subset F^p\mathcal{E}^{k+1}_M$$

The conjugate filtration is defined by saying that $\varphi \in \overline{F}^q\mathcal{E}^k_M$ if $|J| \geq q$ in the sum (5.35) or in other words

$$\overline{F}^q\mathcal{E}^k_M = \overline{F^q\mathcal{E}^k_M}.$$

The above filtrations induce filtrations on the cohomology $H^k(M,\mathbb{C})$, which we still denote F^p and $\overline{F}^q$.

Theorem 5.6 and theorem 5.7 can be restated as

Theorem 5.8. *Let M be a compact Kähler manifold. The cohomology groups $H^k(M,\mathbb{C})$, equipped with the Hodge filtrations F^p and $\overline{F}^q$, carry a pure Hodge structure of weight k:*

$$H^k(M,\mathbb{C}) = \bigoplus_{p+q=k} A^{p,q}(M)$$

where $A^{p,q}(M) = F^pH^k(M,\mathbb{C}) \cap \overline{F}^q H^k(M,\mathbb{C})$ are isomorphic to the Dolbeault groups $H^{p,q}(M)$.

Proof. The natural linear map

$$\mathcal{H}^{(p,q)}(M) \to F^pH^k(M,\mathbb{C}) \cap \overline{F}^q H^k(M,\mathbb{C})$$

is injective because of theorem 5.6, and is surjective because the harmonic projector H respects the types of the forms (theorem 5.4). The theorem follows from theorem 5.6. □

5.5.1 Strictness for d

Theorem 5.9. *Let M be a compact Kähler manifold. The differential*

$$d\colon \Gamma\left(M,\mathcal{E}^k_M\right) \to \Gamma\left(M,\mathcal{E}^{k+1}_M\right)$$

is strict for the Hodge filtration $F^p\Gamma\left(M,\mathcal{E}^{\cdot}_M\right)$.

Proof. The statement means that if $\omega \in F^p\Gamma\left(M,\mathcal{E}^{p+q}_M\right)$ and $\omega = d\psi$, $\psi \in \Gamma\left(M,\mathcal{E}^{p+q-1}_M\right)$, one can find $\phi \in F^p\Gamma\left(M,\mathcal{E}^{p+q-1}_M\right)$ with $\omega = d\phi$. By hypothesis, we can write

$$\omega = \sum_{k\geq 0} \omega^{p+k,q-k}$$

where $\omega^{p+k,q-k}$ is of type $(p+k,q-k)$; also ψ decomposes as

$$\psi = \sum_{r+s=p+q-1} \psi^{r,s}$$

into forms of type (r,s). Then the equation $\omega = d\psi$ reads as

$$\omega^{p,q} = \partial\psi^{p-1,q} + \bar\partial\psi^{p,q-1} \tag{5.36}$$

$$0 = \partial\psi^{p-2,q+1} + \bar\partial\psi^{p-1,q} \tag{5.37}$$

$$0 = \partial\psi^{p-3,q+2} + \bar\partial\psi^{p-2,q+1} \tag{5.38}$$

$$\cdots$$

$$0 = \partial\psi^{0,p+q-1} + \bar\partial\psi^{1,p+q-2} \tag{5.39}$$

$$0 = \bar\partial\psi^{0,p+q-1} \tag{5.40}$$

By (5.40), $\psi^{0,p+q-1}$ defines a $\bar\partial$-cohomology class of $H^{0,p+q-1}(M)$, hence by theorem 5.8, an element of $H^{p+q-1}(M,\mathbb{C})$; which means that $\psi^{0,p+q-1}$ can be written as the sum of the $\bar\partial$ of a form of type $(0,p+q-2)$ and a d-closed form of type $(0,p+q-1)$:

$$\psi^{0,p+q-1} = \alpha^{0,p+q-1} + \bar\partial\theta^{0,p+q-2}, \qquad d\alpha^{0,p+q-1} = 0 \tag{5.41}$$

It follows that

$$\omega = d\left(\psi - \alpha^{0,p+q-1}\right) = d\left(\psi - \psi^{0,p+q-1} + \bar\partial\theta^{0,p+q-2}\right) \tag{5.42}$$

Now, $\alpha^{0,p+q-1}$ is d-closed, so for degree reasons $\partial\alpha^{0,p+q-1} = d\alpha^{0,p+q-1} = 0$, and from (5.39) and (5.41) we deduce

$$0 = \partial\bar\partial\theta^{0,p+q-2} + \bar\partial\psi^{1,p+q-2} = \bar\partial\left(-\partial\theta^{0,p+q-2} + \psi^{1,p+q-2}\right)$$

so that again there exists a d-closed form $\alpha^{1,p+q-2}$ and a form $\theta^{1,p+q-3}$ such that

$$-\partial\theta^{0,p+q-2} + \psi^{1,p+q-2} = \alpha^{1,p+q-2} + \bar\partial\theta^{1,p+q-3} \tag{5.43}$$

so that using (5.42) and (5.43)

$$\omega = d\left(\psi - \alpha^{0,p+q-1} - \alpha^{1,p+q-2}\right) = d\left(\psi - \psi^{0,p+q-1} - \psi^{1,p+q-2} + \bar\partial\theta^{1,p+q-3}\right)$$

etc. $\cdots$.

Finally, we arrive at

$$\begin{aligned}\omega &= d\left(\psi - \psi^{0,p+q-1} - \psi^{1,p+q-2} + \cdots - \psi^{p-1,q} + \bar\partial\theta^{p-1,q-1}\right)\\ &= d\left(\sum_{k\geq 1}\psi^{p+k-1,q-k} + \bar\partial\theta^{p-1,q-1}\right)\end{aligned}$$

But

$$d\bar\partial\theta^{p-1,q-1} = -d\partial\theta^{p-1,q-1}$$

so that

$$\omega = d\phi$$

with

$$\phi = \sum_{k\geq 1}\psi^{p+k-1,q-k} - \partial\theta^{p-1,q-1} \in F^p\Gamma\left(M,\mathcal{E}_M^{p+q-1}\right)$$

□

By theorem 1.1 of chapter 1 the above theorem can be restated as

Theorem 5.10. *Let M be a compact Kähler manifold. The spectral sequence associated to the Hodge filtration F on the complex of global sections $\Gamma\left(M,\mathcal{E}_M^{\cdot}\right)$ degenerates at E_1.*

5.5.2 The case of closed forms of pure type

Theorem 5.11. *Let M be a compact Kähler manifold. If ω is a form on M of type (p,q) which is d-exact, there exists a form α of type $(p-1,q-1)$ such that*

$$\omega = \partial\bar{\partial}\alpha \tag{5.44}$$

In particular, we can write

$$\omega = d\beta, \quad \beta \text{ of type } (p, q-1) \tag{5.45}$$

$$\omega = d\gamma, \quad \gamma \text{ of type } (p-1, q) \tag{5.46}$$

Proof. Equations (5.45) and (5.46) follow from (5.44), taking $\beta = -\partial\alpha$ and respectively $\gamma = \bar{\partial}\alpha$. Let us start with ω of type (p,q) which is d-exact, so that it is ∂- and $\bar{\partial}$-closed. Using harmonic theory (theorem 4.5 of chapter 4) one decomposes ω as

$$\omega = h^{p,q} + \bar{\partial}\vartheta G\omega \tag{5.47}$$

where $h^{p,q}$ is harmonic, ϑ is the adjoint of $\bar{\partial}$ and G is the Green operator. Since the harmonic forms are orthogonal to the d-exact and the $\bar{\partial}$-exact forms, one has $h^{p,q} = 0$, so

$$\omega = \bar{\partial}\psi^{p,q-1}, \quad \psi^{p,q-1} = \vartheta G\omega \tag{5.48}$$

(we recall that G and ϑ respect types). Using the fact that ω is ∂-closed, and (5.48) , we find that $\partial\psi^{p,q-1}$ is $\bar{\partial}$-closed, so we write the Hodge decomposition of $\partial\psi^{p,q-1}$ as

$$\partial\psi^{p,q-1} = h^{p+1,q-1} + \bar{\partial}\psi^{p+1,q-2}$$

with $h^{p+1,q-1}$ harmonic, hence orthogonal to the images of ∂ and of $\bar{\partial}$, so that $h^{p+1,q-1} = 0$ in the previous equation. Then

$$\partial\psi^{p,q-1} = \bar{\partial}\psi^{p+1,q-2} \tag{5.49}$$

It follows that $\partial\psi^{p+1,q-2}$ is $\bar{\partial}$-closed, so by the same reasoning

$$\partial\psi^{p+1,q-2} = \bar{\partial}\psi^{p+2,q-3}$$

until we obtain

$$\partial\psi^{p+q-2,1} = \bar{\partial}\psi^{p+q-1,0} \tag{5.50}$$

Then $\partial\psi^{p+q-1,0}$ is a form of type $(p+q,0)$ which is $\bar{\partial}$-closed by the above equation; such a form is holomorphic, and thus harmonic for $\Box$ and $\overline{\Box}$, by corollary 5.1; and in particular it is orthogonal to the image of ∂. But $\partial\psi^{p+q-1,0}$ is also in the image of ∂, so it is zero:

$$\partial\psi^{p+q-1,0} = 0$$

Using Hodge decomposition for $\psi^{p+q-1,0}$ we deduce

$$\psi^{p+q-1,0} = h^{p+q-1,0} + \partial\phi^{p+q-2,0} \tag{5.51}$$

Then, using (5.50) and (5.51)

$$\partial\psi^{p+q-2,1} = \bar{\partial}\psi^{p+q-1,0} = \bar{\partial}\partial\phi^{p+q-2,0}$$

(harmonic forms are ∂-closed) so

$$\partial\left(\psi^{p+q-2,1} + \bar{\partial}\phi^{p+q-2,0}\right) = 0$$

so again

$$\psi^{p+q-2,1} + \bar{\partial}\phi^{p+q-2,0} = h^{p+q-1,1} + \partial\phi^{p+q-3,1} \tag{5.52}$$

etc. $\cdots$ so finally, using (5.49) we arrive at

$$\partial\psi^{p,q-1} = \bar{\partial}\partial\phi^{p,q-2}$$

so that

$$\partial\left(\psi^{p,q-1} + \bar{\partial}\phi^{p,q-2}\right) = 0$$

and using Hodge decomposition

$$\psi^{p,q-1} + \bar{\partial}\phi^{p,q-2} = h^{p,q-1} + \partial\phi^{p-1,q-1} \tag{5.53}$$

so by (5.53) and (5.48)

$$\omega = \bar{\partial}\psi^{p,q-1} = \bar{\partial}\partial\phi^{p-1,q-1}$$

which completes the proof. □

Chapter 6

The theory of residues on a smooth divisor

6.1 Introduction

In this chapter, we shall explain the theory of residues for a smooth hypersurface. The theory has a long history. It starts from the Cauchy formula in classical holomorphic function theory. In the 19th century, it is used to construct the theory of abelian integrals of the 1st, 2nd and 3rd kinds on a compact Riemann surface. The idea is to realize the compact Riemann surface as an algebraic curve in $\mathbb{P}^2$, possibly with "standard" singularities and to obtain the various differentials of degree 1 as residues. From that construction, one could deduce all the classical theory of algebraic curves. Picard and Poincaré defined the notion of residue of rational forms on algebraic surfaces, in particular studied the periods of rational differentials and proved that they are periods of abelian integrals on the polar curve. The modern theory was defined by Leray to study the Cauchy problem of holomorphic partial differential equations. Actually, Leray constructed a residue theory for a hypersurface which is the union of divisors with normal crossing. But the main result, namely that the cohomology of the complementary of a hypersurface can be realized with forms with poles of order at most 1 was proved by Leray only for a smooth hypersurface.

6.2 Forms with logarithmic singularities

Definition 6.1. Let M be a complex manifold of dimension n, and D be a smooth hypersurface. Let φ be a C^∞ k-form on $M \setminus D$. We say that φ has logarithmic singularities on D if for any point $m_0 \in D$, there exists a neighborhood U of m_0 in M, where D is defined by a local equation $s = 0$:

$$D \cap U = \{\, m \in U \mid s(m) = 0 \,\} \tag{6.1}$$

and the form φ can be written in $U \setminus D$ as

$$\varphi = \frac{ds}{s} \wedge \omega + \theta \tag{6.2}$$

where ω and θ are C^∞ forms on U of degree $k-1$ and k respectively.

We denote $\mathcal{E}^k_M\langle\log D\rangle$ the sheaf on M of C^∞ k-forms, which have logarithmic singularities on D (and are C^∞ outside D).

One has obviously

Lemma 6.1. *Let m be a point of M.*

(i) If $m \notin D$, $\mathcal{E}^k_{M,m}\langle\log D\rangle = \mathcal{E}^k_{M,m}$.

(ii) If $m \in D$, $\mathcal{E}^k_{M,m}\langle\log D\rangle$ is the $\mathcal{E}^0_{M,m}$-module generated by the kth exterior products of the differentials, $\frac{dz_1}{z_1}$, $dz_2,\dots, dz_n$, $d\bar{z}_1,\dots, d\bar{z}_n$ where $z_1 = 0$ is a local equation of D near m.

Definition 6.2 (of the residue form)**.** Let $\varphi \in \Gamma\left(M, \mathcal{E}^k_M\langle\log D\rangle\right)$ be a k-form with logarithmic singularities on D. One writes locally φ as in (6.2). Then one defines

$$\operatorname{Res}\varphi = \omega|_D.$$

Lemma 6.2.

(i) $\operatorname{Res}\varphi$ *is well defined as an element* $\operatorname{Res}\varphi \in \Gamma\left(D, \mathcal{E}^{k-1}_D\right)$.

(ii) Res *commutes with d:* $\operatorname{Res} d\varphi = d\operatorname{Res}\varphi$.

Proof.

(i) If $s' = 0$ is another equation of D, then $s = s'f$ where f is holomorphic and invertible. But

$$\frac{ds}{s} = \frac{ds'}{s'} + \frac{df}{f}$$

so

$$\varphi = \frac{ds}{s} \wedge \omega + \theta = \frac{ds'}{s'} \wedge \omega + \frac{df}{f} \wedge \omega + \theta.$$

Hence $\varphi = \frac{ds'}{s'} \wedge \omega + \theta'$ with the same ω. Suppose now that one has two decompositions of φ, namely

$$\varphi = \frac{ds}{s} \wedge \omega + \theta = \frac{ds}{s} \wedge \omega' + \theta'.$$

Then one deduces

$$\frac{ds}{s} \wedge (\omega - \omega') = \theta' - \theta$$

and thus $\frac{ds}{s} \wedge (\omega - \omega')$ is a smooth form. Choose a system of coordinates $(z_1, \dots, z_n)$ where $s = z_1$; an easy calculation shows that $\omega - \omega'$ is a linear combination of $dz_1 \wedge \alpha$, $z_1\beta$, so that $\omega|_D = \omega'|_D$.

(ii) We write φ as in (6.2). Then

$$d\varphi = \frac{ds}{s} \wedge d\omega + d\theta$$

hence $\operatorname{Res} d\varphi = d\omega|_D = d \operatorname{Res} \varphi$. □

As a consequence of (ii) we deduce

Lemma 6.3. *Let* $\varphi \in \Gamma\left(M, \mathcal{E}_M^k\langle \log D\rangle\right)$.

(i) If φ is a d-closed form (on $M \setminus D$), then $\operatorname{Res}\varphi$ *is a d-closed form on D.*

(ii) If φ is in $d\Gamma\left(M, \mathcal{E}_M^{k-1}\langle \log D\rangle\right)$, *then* $\operatorname{Res}\varphi$ *is exact on D.*

(iii) Res induces a morphism in cohomology.

$$\operatorname{Res}: \frac{\ker\left\{d\colon \Gamma\left(M, \mathcal{E}_M^k\langle \log D\rangle\right) \to \Gamma\left(M, \mathcal{E}_M^{k+1}\langle \log D\rangle\right)\right\}}{d\Gamma\left(M, \mathcal{E}_M^{k-1}\langle \log D\rangle\right)} \to H^{k-1}(D, \mathbb{C})$$

6.3 The long exact homology residue sequence

We choose a riemannian metric on M. Let $m_0 \in M$ and u an unit tangent vector at m_0; we denote by $\gamma(s \mid u, m_0)$ the geodesic through m_0 tangent to u (where s is the arc length).

Definition 6.3 (of tubular neighborhoods)**.** Let D be a hypersurface of M.

We define a tubular neighborhood T_ε as the set of points of M which are at a distance $< \varepsilon$ of D. This neighborhood can be described as follows: from any point $m_0 \in D$, we draw the geodesic $\gamma(s \mid u, m_0)$ where u is an unit tangent vector orthogonal to D. We assume that ε is less than the injectivity radius of the exponential maps at all points $m_0 \in D$. Then, $\overline{T}_\varepsilon$ is the set of all points $\gamma(t \mid u, m_0)$ for $t \leq \varepsilon$, u orthogonal to D at m_0, and $m_0 \in D$.

There is an obvious retraction

$$r\colon \overline{T}_\varepsilon \to D \tag{6.3}$$

such that

$$r(\gamma(t \mid u, m_0)) = m_0$$

and a retraction

$$\rho\colon M \setminus D \to M \setminus T_\varepsilon \tag{6.4}$$

which associates to any point $\gamma(t \mid u, m_0) \in \overline{T}_\varepsilon$ with $t \neq 0$ the point $\gamma(\varepsilon \mid u, m_0)$ and to any point of $M \setminus \overline{T}_\varepsilon$ the same point.

6.3.1 The long exact homology sequence

Theorem 6.1. *One has a long exact homology sequence*

$$\begin{aligned} \cdots \longrightarrow H_p(M \setminus D, \mathbb{C}) \xrightarrow{i_*} H_p(M, \mathbb{C}) \xrightarrow{q} H_{p-2}(D, \mathbb{C}) \xrightarrow{\delta} \\ \longrightarrow H_{p-1}(M \setminus D, \mathbb{C}) \xrightarrow{i_*} \cdots \end{aligned} \tag{6.5}$$

We give only the description of the morphisms

(i) The morphism i_* is induced by the canonical embedding $i\colon M \setminus D \to M$.

(ii) To define q, consider a p-singular cycle C on M. Then one takes the intersection class $[C.D]$, which has dimension $p-2$.

(iii) Finally to define the morphism δ, one starts with a $(p-2)$-cycle C on D. It induces a p-chain in $\overline{T}_\varepsilon$ by taking all the arcs $(\gamma(t \mid u, m_0))$ for $t \leq \varepsilon$, $m_0 \in C$ and u orthogonal to m_0 (and unit length). This defines a chain $\tilde{C}$ in $\overline{T}_\varepsilon$ and one takes the boundary $\partial\tilde{C}$ in $\partial\overline{T}_\varepsilon$. Then by definition:

$$\delta[C] = [\partial\tilde{C}].$$

6.3.2 The residue formula

Theorem 6.2. *Let $[\gamma] \in H_p(S)$ be a homology class and φ be a d-closed $(p+1)$-form on $M \setminus D$, which has polar singularities of order at most 1 on D. Then one has*

$$\int_{\delta[\gamma]} \varphi = 2i\pi \int_\gamma \operatorname{Res} \varphi. \tag{6.6}$$

Proof. We can assume that φ has a logarithmic singularity. It is sufficient to prove (6.6) for a simplex σ which is contained in a sufficiently small open set so that D has the equation $z_1 = 0$, $(z_1, \ldots, z_n)$ being holomorphic coordinates. Hence we can write

$$\varphi = \frac{dz_1}{z_1} \wedge \omega + \theta.$$

Locally we can use the standard hermitian metric, so that $\delta_\varepsilon\sigma$ becomes a union of circles with center in σ and radius ε in a certain z_1-plane. Moreover ε is as small as we wish. Now, one has :

$$\delta_\varepsilon\sigma - \delta_{\varepsilon'}\sigma = \partial C$$

where C is a union of annuli centered on points of σ in the corresponding z_1-plane between the radius ε and ε' and thus

$$\int_{\delta_\varepsilon\sigma} \varphi - \int_{\delta_{\varepsilon'}\sigma} \varphi = \int_{\partial C} \varphi = \int_C d\varphi = 0.$$

But

$$\lim_{\varepsilon\to 0}\int_{\delta_\varepsilon\sigma}\theta = 0$$

and for $\varepsilon \to 0$

$$\lim_{\varepsilon\to 0}\int_{\delta_\varepsilon\sigma}\frac{dz_1}{z_1}\wedge\omega = 2i\pi\int_\sigma\omega = 2i\pi\int_\sigma \operatorname{Res}\varphi.$$

□

6.4 The residue sequence in cohomology and the Gysin morphism

6.4.1 Definition of the sequence

Taking the transpose of the long exact residue sequence in homology (theorem 6.2), we deduce

Theorem 6.3. *Let D be a smooth hypersurface in M; one has a long exact sequence*

$$\cdots \longrightarrow H^{p-1}(M\setminus D,\mathbb{C}) \xrightarrow{res} H^{p-2}(D,\mathbb{C}) \xrightarrow{{}^t q} H^p(M,\mathbb{C}) \xrightarrow{i^*} \\ \longrightarrow H^p(M\setminus D,\mathbb{C}) \xrightarrow{res} \cdots \tag{6.7}$$

Proof. We have just to identify the transpose of the morphisms of the sequence (6.5); i^* is the induced morphism in cohomology by the embedding $i\colon M\setminus D \to M$.

The morphism res is the transpose of the morphism δ of the sequence (6.5). In particular if $[\varphi]$ is a class of cohomology given by a d-closed form on $M\setminus D$ which has logarithmic singularities on D, we have

$$res[\varphi] = 2i\pi\,[\operatorname{Res}\varphi] \tag{6.8}$$

because of the residue formula (6.6) and the realization of the duality between homology and De Rham cohomology using the integration on chains. Finally ${}^t q$ is the transpose of the intersection morphism $q\colon H_p(M,\mathbb{C}) \to H_{p-2}(D,\mathbb{C})$. □

Definition 6.4. ${}^t q\colon H^p(D,\mathbb{C}) \to H^{p+2}(M,\mathbb{C})$ is called the Gysin morphism.

6.4.2 Construction of the Gysin morphism

We consider the line bundle $L(D) \to M$ associated to D. Then, by definition, there exists a holomorphic section $\sigma\colon M \to L(D)$ such that

$$D = \left\{ m \in M \mid \sigma(m) = 0 \right\}. \tag{6.9}$$

We can also introduce a hermitian metrics denoted $\| \ \|$ on $L(D)$ and we define

$$\eta_D = -\frac{1}{2i\pi} \partial \log \|\sigma\|^2. \tag{6.10}$$

In a small neighborhood U of $m_0 \in D$ where one can find coordinates $(z_1, \dots, z_n)$ on U such that $z_1 = 0$ is the equation of D in U, we see that

$$\eta_{D|_U} = -\frac{1}{2i\pi}\frac{dz_1}{z_1} + \eta'_D \tag{6.11}$$

where η'_D is C^∞ in U. We also define

$$\Omega_D = -\frac{1}{2i\pi} \bar{\partial}\partial \log \|\sigma\|^2 = \bar{\partial}\eta'_D. \tag{6.12}$$

This form is a $(1,1)$-form on M (it is C^∞ because of (6.11)) and it represents the Chern class $c_1\left(L(D)\right)$ of the bundle $L(D)$.

Let us now define

$$\gamma_{D,M} : H^p(D,\mathbb{C}) \to H^{p+2}(M,\mathbb{C})$$

We start from a cohomology class $[\alpha] \in H^p(D,\mathbb{C})$ and a d-closed p-form $\alpha \in [\alpha]$. Because the tubular neighborhood T_ε retracts on D, so that $H^p\left(T_\varepsilon, \mathbb{C}\right) \simeq H^p(D,\mathbb{C})$, there exists a d-closed p-form α' in T_ε such that $\alpha'|_D = \alpha$. Using a function χ which is 1 on $T_{\varepsilon/2}$ and which has its support in T_ε, one can find an extension $\tilde{\alpha} = \chi\alpha'$ of α to M, which is d-closed in $T_{\varepsilon/2}$. Now we define

$$\begin{aligned} \beta &= d\left(\tilde{\alpha} \wedge \eta_D\right) \\ &= d\tilde{\alpha} \wedge \eta_D + (-1)^p \tilde{\alpha} \wedge \Omega_D. \end{aligned} \tag{6.13}$$

This is a $(p+2)$-form on M. Because $d\tilde{\alpha} = 0$ on $T_{\varepsilon/2}$, β has no singularities on D and is d-closed, so it induces an element

$$[\beta] \in H^{p+2}\left(M,\mathbb{C}\right)$$

and we define

$$[\beta] = \gamma_{D,M}\left([\alpha]\right). \tag{6.14}$$

Lemma 6.4. *$\gamma_{D,M}$ is exactly ${}^tq$, namely if $\Gamma \in H_p(M,\mathbb{C})$ then*

$$\int_{\Gamma.D} \alpha = \int_\Gamma \gamma_{D,M}(\alpha). \tag{6.15}$$

Proof. Because $\tilde{\alpha}$ has its support in T_ε, β has also its support in T_ε, so that the integral

$$I = \int_\Gamma \gamma_{D,M}(\alpha) = \int_\Gamma d\left(\tilde{\alpha} \wedge \eta_D\right)$$

depends only on the relative homology class of $\Gamma \mod M \setminus \overline{T}_\varepsilon$, i.e. it depends only on the image Γ' of Γ in $H_p\left(M, M \setminus \overline{T}_\varepsilon, \mathbb{C}\right)$. Moreover the retraction $r \colon \overline{T}_\varepsilon \to D$ induces an isomorphism in homology. So finally Γ' is homologous to $\widetilde{\Gamma.D}$ where $\widetilde{\Gamma.D}$ is obtained as the union of small disks of radius ε centered on $\Gamma.D$. So

$$I = \int_{\widetilde{\Gamma.D}} d\left(\tilde{\alpha} \wedge \eta_D\right) = \lim_{\varepsilon' \to 0} \int_{\widetilde{\Gamma.D} \setminus T_{\varepsilon'}} d\left(\tilde{\alpha} \wedge \eta_D\right).$$

But $\partial\left(\widetilde{\Gamma.D} \setminus \overline{T}_{\varepsilon'}\right)$ is contained in $\partial T_{\varepsilon'} \cup \partial T_\varepsilon$. The part in ∂T_ε gives 0 because $\tilde{\alpha}|_{M \setminus T_\varepsilon} = 0$. So

$$I = \lim_{\varepsilon' \to 0} \left(- \int_{\widetilde{\Gamma.D} \cap \partial T_{\varepsilon'}} \tilde{\alpha} \wedge \eta_D \right).$$

Then using (6.11), we obtain

$$I = \int_{\Gamma.D} \alpha$$

which proves (6.15). □

Theorem 6.4. *Let M be a Khälerian manifold, D a smooth hypersurface of M, $j : D \to M$ the natural embedding. The Gysin morphism $\gamma_{D,M}$: $H^p(D, \mathbb{C}) \to H^{p+2}(M, \mathbb{C})$ can be obtained via Poincaré duality from the pull-back*

$$j^* : H^{2n-p-2}(M, \mathbb{C}) \to H^{2n-p-2}(D, \mathbb{C})$$

It follows that $\gamma_{D,M}$ is real, commutes to conjugation, and is a morphism of pure Hodge structures of degree 1.

Proof. The first assertion is a consequence of lemma 6.4 and the definition of the mapping q in (6.5). Using Serre duality we find that the dual of j^* sends $H^{r,s}(D)$ to $H^{r+1,s+1}(M)$, which implies that $\gamma_{D,M}$ is a morphism of pure Hodge structures of degree 1. □

Chapter 7

Complex spaces

7.1 Complex analytic varieties and complex spaces

We refer to [GF], [GR] for generalities and details on complex spaces.
Let $U \subset \mathbb{C}^n$ be an open set, and $\mathcal{O}_U$ be the sheaf of holomorphic functions on U: for V open in U, by definition

$$\mathcal{O}_U(V) = \{ f\colon V \to \mathbb{C} \text{ holomorphic} \}$$

Then $\mathcal{O}_U$ is a sheaf of local $\mathbb{C}$-algebras.

Theorem 7.1 (Oka's theorem). *The sheaf $\mathcal{O}_U$ is coherent.*

Definition 7.1. A (complex) analytic subset of $U \subset \mathbb{C}^n$ is a closed subset $S \subset U$ with the following property; for every $x \in S$ there exist a neighborhood V of x in U and a finite number of holomorphic functions $f_1, \dots, f_p$ on V such that $S \cap V = \{ z \in V \mid f_1(z) = \cdots = f_p(z) = 0 \}$.

An intersection, or a finite union, of analytic subsets of U is analytic. Note also that U itself is, trivially, an analytic subset. An analytic subset $S \subset U$ is *globally defined in U* if it is the set of the zeroes of holomorphic functions defined on all of U: $S = \{ z \in U \mid g_1(z) = \cdots = g_s(z) = 0 \}$ where $g_1, \dots, g_s$ are holomorphic on U.

To an analytic subset $S \subset U$ we associate a subsheaf of local $\mathbb{C}$-algebras $\mathcal{I}_S \subset \mathcal{O}_U$ defined as

$$\mathcal{I}_S(V) = \{ f \in \mathcal{O}_U(V) \mid f|_{S \cap V} \equiv 0 \}$$

in other words, the subsheaf of holomorphic functions vanishing on S.

Theorem 7.2 (Cartan coherence theorem). *The sheaf $\mathcal{I}_S$ is coherent, or equivalently, finitely generated.*

We define

$$\mathcal{O}_S = (\mathcal{O}_U / \mathcal{I}_S)\,|_S$$

which is a coherent sheaf of local rings on S. It is easy to see that $\mathcal{O}_S$ injects into the sheaf $\mathcal{C}_S$ of continuous functions on S as a subsheaf. The coherent $\mathbb{C}$-ringed space $(S, \mathcal{O}_S)$ is called a *complex analytic variety*, or simply a *complex variety.* A section of $\mathcal{O}_S$ on an open subset $Z \subset S$ is called a *holomorphic function on* Z. If Z is an open subset of S, the $\mathbb{C}$-ringed space $(Z, \mathcal{O}_Z)$ where $\mathcal{O}_Z = \mathcal{O}_S|_Z$, is a complex variety.

Let $S \subset U$ be a complex variety. An *analytic subset* of S is a closed subset T such that for every $x \in T$ there exist a neighborhood V of x in S and a finite number of holomorphic functions $f_1, \dots, f_p$ on V such that $T \cap V = \{ z \in V \colon f_1(z) = \cdots = f_p(z) = 0 \}$.

Then T is an analytic subset of U.

To an analytic subset $T \subset S$ we associate the subsheaf of rings $\mathcal{I}_T \subset \mathcal{O}_S$ defined as

$$\mathcal{I}_T(V) = \{ f \in \mathcal{O}_S(V) \colon f|_{T \cap V} \equiv 0 \}$$

that is, the subsheaf of holomorphic functions on S vanishing on T. Then T, equipped with the sheaf $\mathcal{O}_T = (\mathcal{O}_S/\mathcal{I}_T)|_T$ is a complex variety, called a *complex closed subvariety of* S. Let us remark that T is also a complex subvariety of U.

Definition 7.2. A complex (analytic) space X is a $\mathbb{C}$-ringed space $(X, \mathcal{O}_X)$ with the following properties.

(1) X is a metrizable, countable at infinity, topological space.

(2) For every $x \in X$ there exist an open neighborhood V of x in X and an isomorphism of ringed spaces $(V, \mathcal{O}_X|_V) \simeq (S, \mathcal{O}_S)$, where S is a complex variety.

In other words, a complex space behaves locally as a complex variety. The sheaf $\mathcal{O}_X$ is coherent; it injects into the sheaf $\mathcal{C}_X$ of continuous functions on X as a subsheaf.

A section of $\mathcal{O}_X$ on an open subset $V \subset S$ is called a *holomorphic function on* V.

A morphism $f \colon X \to Y$ of complex spaces is by definition a morphism of ringed spaces $f \colon (X, \mathcal{O}_X) \to (Y, \mathcal{O}_Y)$, so that if $g \colon V \to \mathbb{C}$ is a holomorphic function on an open set $V \subset Y$, $g \circ f \colon f^{-1}(V) \to \mathbb{C}$ is holomorphic.

Theorem 7.3 (Examples).

(1) A complex manifold M, equipped with the sheaf $\mathcal{O}_M$ of holomorphic functions on M, is a complex space. We refer to a complex manifold also as a smooth complex space.

(2) A complex variety is a complex space.

(3) (Complex open subspaces). An open subset V of a complex space X is a complex space (the sheaf being the restriction $\mathcal{O}_X|_V$ to V).

(4) (Complex closed subspaces). Let X be a complex space. An analytic subset of X is a closed subset E such that for every $x \in E$ there exist a

neighborhood V of x in X and a finite number of holomorphic functions $f_1, \ldots, f_p$ on V such that $E \cap V = \{ z \in V \mid f_1(z) = \cdots = f_p(z) = 0 \}$. The subsheaf of ideals $\mathcal{I}_E \subset \mathcal{O}_X$ is defined as

$$\mathcal{I}_E(V) = \{ f \in \mathcal{O}_X(V) \mid f|_{E \cap V} \equiv 0 \} .$$

Then E, equipped with the sheaf $\mathcal{O}_E = (\mathcal{O}_X / \mathcal{I}_E)|_E$ is a complex space, called a complex closed subspace *of X.*

(5) *An intersection, or a finite union, of complex subspaces of X is a complex subspace.*

(6) *(The support of a coherent $\mathcal{O}_X$-module is a complex subspace.) Let X be a complex space and $\mathcal{F}$ be a coherent $\mathcal{O}_X$-module. Then* $\operatorname{supp}(\mathcal{F})$ *is a complex closed subspace of X.*

(7) *(The subspace defined by an ideal of finite type.) Let X be a complex space and $\mathcal{J} \subset \mathcal{O}_X$ be an ideal of finite type. The subspace defined by $\mathcal{J}$ is by definition the support $Y =$* $\operatorname{supp}(\mathcal{O}_X/\mathcal{J})$. *Let V be an open subset of X such that $\mathcal{J}$ is generated on V by the holomorphic functions $g_1, \ldots, g_s$; then $Y \cap V = \{ x \in V \colon g_1(x) = \cdots = g_s(x) = 0 \}$. The ideal $\mathcal{J}$ is contained in the ideal $\mathcal{I}_Y$, but in general is different. The ideal $\mathcal{I}_Y$ appears as the maximal coherent ideal which defines Y.*

(8) *Let $f \colon X \to Y$ be a morphism of complex spaces. If T is a subspace of Y, $f^{-1}(T)$ is a subspace of X.*

Remark. By definition, our complex spaces are *reduced*, that is, for every open set $V \subset X$ the ring $\mathcal{O}_X(V)$ has no nilpotents elements (because it is a subring of $\mathcal{C}_X(V)$), so that holomorphic functions on V are continuous functions. We do not need the general notion of complex space, which allows nilpotent elements in the structure sheaf. When referring to a complex space, we always mean a reduced complex space.

One of the main properties of (reduced) complex spaces is that they are generically smooth. More precisely, we say that a point x of a complex space X is *smooth (or simple, or regular)* if there is an open neighborhood of x which is a manifold, and *singular* otherwise. Then

Theorem 7.4. *Let X be a complex space. The subset U of the smooth points of X is open and dense in X. Moreover its complement* $\operatorname{sing}(X)$ *(the set of singular points) is a nowhere dense complex subspace of X.*

A complex space X is *reducible* if there are two proper, nonempty complex subspaces X_1, X_2, with $X = X_1 \cup X_2$, and *irreducible* otherwise. Any space X has a decomposition into irreducible components:

Theorem 7.5. *Let X be a complex space, $U = X \setminus \operatorname{sing} X$. Let $U = \bigsqcup_j U_j$ be the decomposition into distinct connected components. Then*

(1) the closure X_j of U_j in X is an irreducible complex subspace of X;

(2) $X = \bigcup_j X_j$;

(3) every irreducible complex subspace of X is contained in one of the X_j;

(4) X is irreducible if and only if there is only one X_j, or if and only if $X \setminus \operatorname{sing} X$ is connected.

The spaces X_j are called the irreducible components of X.

If X is compact, the number of irreducible components is finite.

Let $f: X \to Y$ be a proper surjective morphism of complex spaces. If X is irreducible, Y is irreducible.

The *dimension* $\dim X$ of an irreducible complex space X is the dimension of the connected manifold $X \setminus \operatorname{sing} X$; in general $\dim X$ is the supremum of the dimensions of its irreducible components.

If $Y \subset X$ is a nowhere dense complex subspace, $\dim Y < \dim X$.

Throughout all the present book we will make the assumption that all the complex spaces we are considering have finite dimension. Nevertheless, many results hold also for complex spaces whose dimension is not finite.

7.2 Coherent sheaves on complex spaces

If $(X, \mathcal{O}_X)$ is a complex space, by a coherent sheaf on X we mean a coherent sheaf of $\mathcal{O}_X$-modules.

Theorem 7.6 (Grauert's direct image theorem). *Let $f: X \to Y$ be a proper morphism of complex spaces, $\mathcal{F}$ a coherent sheaf on X. The higher direct image sheaves $\mathcal{R}^k f_* \mathcal{F}$ are coherent $\mathcal{O}_Y$-modules for any $k \geq 0$.*

If $f: X \to Y$ is a proper morphism of complex spaces and Z is a subspace of X, $f(Z)$ is a subspace of Y (it is in fact the support of the sheaf $f_* \mathcal{O}_Z$, which is coherent by the above theorem).

Corollary 7.1. *Let X be a compact complex space, $\mathcal{F}$ a coherent sheaf on X. The cohomology groups $H^p(X, \mathcal{F})$ are finite dimensional $\mathbb{C}$-vector spaces.*

In fact, we consider the trivial morphism $f: X \to \{*\}$, where $\{*\}$ is a point. Then the direct images of the sheaf $\mathcal{F}$ on X are the cohomology groups $H^p(X, \mathcal{F})$. For them, the coherence means that they are finite dimensional $\mathbb{C}$-vector spaces.

7.3 Modifications and blowing-up

Definition 7.3. A modification is a morphism $f\colon \tilde{X} \to X$ of complex spaces with the following properties:

(1) f is proper and surjective;

(2) there exists a nowhere dense subspace $E \subset X$ such that the restriction $f|_{\tilde{X}\backslash f^{-1}(E)}$ is an isomorphism $\tilde{X}\backslash f^{-1}(E) \simeq X\backslash E$. The space $\tilde{E} = f^{-1}(E)$ is called the exceptional space of the modification.

We will often denote a modification as $f\colon (\tilde{X}, \tilde{E}) \to (X, E)$.

Definition 7.4. Let X be a complex space, $\mathcal{J} \subset \mathcal{O}_X$ an ideal of finite type, Y the subspace of X defined by $\mathcal{J}$. The blowing-up of X centered at $\mathcal{J}$ is the complex space $\tilde{X}$, provided with a morphism $f\colon \tilde{X} \to X$, characterized by the following properties:

(a) the ideal $f^*(\mathcal{J})\,\mathcal{O}_{\tilde{X}} \subset \mathcal{O}_{\tilde{X}}$ generated by the functions $\{s \circ f, s \in \mathcal{J}\}$ is an invertible sheaf; in particular the subspace of $\tilde{X}$ defined by $f^*(\mathcal{J})\,\mathcal{O}_{\tilde{X}}$, that is, $f^{-1}(Y)$, is a divisor;

(b) if $h\colon Z \to X$ is a morphism of complex spaces such that the ideal $h^*(\mathcal{J})\,\mathcal{O}_Z$ of $\mathcal{O}_Z$ is invertible, there exists a unique morphism $g\colon Z \to \tilde{X}$ with $f \circ g = h$.

The property (a) implies that f induces an isomorphism $\tilde{X}\backslash f^{-1}(Y) \simeq X\backslash Y$; hence f is a modification whose exceptional space is $\tilde{Y} = f^{-1}(Y)$.

When the ideal $\mathcal{J}$ is the ideal $\mathcal{I}_Y$ of the holomorphic functions on X vanishing on Y, we say also that *f (or $\tilde{X}$) is the blowing-up of X along Y.*

The blowing-up of a manifold M along a submanifold N (see chapter 2) is a particular case af the above definition; in this case the ideal $\mathcal{J}$ is the ideal $\mathcal{I}_N$ of the holomorphic functions vanishing on N.

The property (b) implies that the blowing-up, if it exists, is unique. Hence the existence and the construction of the blowing-up are a local problem on X: if we construct the blowing-up $\tilde{U}_j$ centered at $\mathcal{J}|_{U_j}$ of every open set U_j of a covering of X, then the $\tilde{U}_j$ will glue together to form the blowing-up $\tilde{X}$. Hence, it is sufficient to construct the blowing-up on an open neighborhood U of a point of X where the ideal $\mathcal{J}|_U$ is generated by a finite number of functions $f_0, \ldots, f_r \in \Gamma(U, \mathcal{J})$. We consider the sheaf of algebras $\mathcal{A} = \bigoplus_{m \geq 0} \mathcal{J}^m$ (by definition $\mathcal{J}^0 = \mathcal{O}_X$). $\mathcal{A}$ is an algebra over $\mathcal{O}_X$, generated on U, as an algebra, by $f_0, \ldots, f_r$. Precisely there is a surjective morphism of $\mathcal{O}_U$-algebras $\phi\colon \mathcal{O}_U[T_0, \ldots, T_r] \to \mathcal{A}|_U$ sending T_j to f_j. Let $\mathcal{M}$ be the kernel of ϕ, so that we have an exact sequence

$$0 \longrightarrow \mathcal{M} \longrightarrow \mathcal{O}_U[T_0, \ldots, T_r] \longrightarrow \mathcal{A}|_U \longrightarrow 0$$

It can be proved that, after replacing U by a smaller neighborhood, $\mathcal{M}$ is generated as an $\mathcal{O}_U[T_0,\ldots,T_r]$-algebra by a finite set $g_1 = g_1(x,T_0,\ldots,T_r),\ldots,$ $g_s = g_s(x,T_0,\ldots,T_r) \in \Gamma(U,\mathcal{O}_U[T_0,\ldots,T_r])$, homogeneous with respect to the variables $T_0,\ldots,T_r$.

Let $\mathbb{P}^r$ be the complex projective space with coordinates $t_0,\ldots,t_r$. An element $g = g(x,T_0,\ldots,T_r) \in \Gamma(U,\mathcal{O}_U[T_0,\ldots,T_r])$ detects a holomorphic polynomial $g(x,t_0,\ldots,t_r)$ on $\mathbb{P}^r$, homogeneous with respect to the variables $t_0,\ldots,t_r$ and depending holomorphically on $x \in U$. Let $\tilde{U} \subset U \times \mathbb{P}^r$ be the subspace defined by

$$\tilde{U} = \left\{ (x,t) \in U \times \mathbb{P}^r \;\middle|\; g_1(x,t_0,\ldots,t_r) = \cdots = g_s(x,t_0,\ldots,t_r) = 0 \right\}$$

Then $\tilde{U}$, equipped with the first projection $f\colon \tilde{U} \to U$, is the blowing-up of U centered at $\mathcal{J}|_U$. In order to understand better the construction, let us note that the polynomials $f_iT_j - f_jT_i \in \mathcal{O}_U[T_0,\ldots,T_r]$ are mapped to zero through ϕ, so that they belong to $\mathcal{M}$; hence $\tilde{U}$ is contained in the space

$$\tilde{U}' = \left\{ (x,t) \in U \times \mathbb{P}^r \;\middle|\; f_i(x)t_j - f_j(x)t_i = 0, \quad 0 \le i,j \le r \right\}$$

In general, $\tilde{U}$ is smaller than $\tilde{U}'$. But if U is smooth and we are blowing up U along a smooth submanifold, the polynomials $f_iT_j - f_jT_i$ generate $\mathcal{M}$ and we find $\tilde{U} = \tilde{U}'$; that is we recover the blowing-up introduced in chapter 2.

Another important notion in the theory of the blowing-up is the *strict transform*, which relates the blowing-up of a space X and the blowing-up of a subspace Y along the same subspace Z ($Z \subset Y \subset X$).

Theorem 7.7. *Let $f\colon \tilde{X} \to X$ be the blowing-up of X along a subspace Y (i.e. the blowing-up of X centered at the ideal $\mathcal{I}_Y$ of holomorphic functions vanishing on Y). Let $Z \subset X$ be a subspace such that $Y \subset Z$. Let $\tilde{Z}$ be the closure of $f^{-1}(Z \setminus Y)$ in $\tilde{X}$. Then $\tilde{Z}$ is a complex subspace of $\tilde{X}$, called the strict transform of Z through f. Then $f(\tilde{Z}) = Z$, and the restriction $f|_{\tilde{Z}}\colon \tilde{Z} \to Z$ is the blowing-up of Z along Y.*

There are examples of modifications which are not blowing-up. The next theorem relates the modifications to the blowing-up.

We say that a sequence of morphisms of complex spaces

$$\cdots \longrightarrow X_k \xrightarrow{f_k} X_{k-1} \xrightarrow{f_{k-1}} \cdots \xrightarrow{f_1} X_0 = X$$

is *locally finite* if over each relatively compact open subset of X, all but finitely many f_k are isomorphisms. It is easy to see that a composition of a locally finite sequence of morphisms is a well defined morphism $f\colon X' \to X$, where X' is a complex space. If each f_k is a modification, f is also a modification.

Theorem 7.8 (Hironaka Chow lemma). *Let $f\colon X' \to X$ be a modification. There is a commutative diagram*

$$\begin{array}{ccc} X'' & \xrightarrow{\quad q \quad} & X' \\ & {\scriptstyle p}\searrow \quad \swarrow{\scriptstyle f} & \\ & X & \end{array}$$

where p and q are the composition of locally finite sequences of blowing-up along smooth centers. In particular every modification f is dominated by a locally finite sequence of blowing-up. If X is compact, or a relatively compact open set of a complex space, p and q are obtained by finite sequences of blowing-up along smooth centers.

The following is one of the most celebrated and difficult results in the theory of complex spaces.

Theorem 7.9 (Hironaka desingularization theorem). *Let X be a complex space. There exists a modification $f\colon (\tilde{X}, \tilde{E}) \to (X, \operatorname{sing}(X))$ such that $\tilde{X}$ is smooth (i.e. a complex manifold), $\tilde{E} \subset \tilde{X}$ is a divisor with normal crossing, and f is the composition of a locally finite sequence of blowing-up along smooth subspaces. In particular if X is compact, or a relatively compact open set of a complex space, f is obtained by a finite sequence of blowing-up along smooth subspaces.*

Another useful result is that every coherent ideal $\mathcal{J} \subset \mathcal{O}_X$ on a complex space X can be made invertible on a suitable modification of X, which implies also that any subspace $Y \subset X$ can be transformed into a divisor with normal crossing.

Theorem 7.10. *Let X be a complex space, $\mathcal{J} \subset \mathcal{O}_X$ a coherent ideal. There exists a modification $f\colon \tilde{X} \to X$, which is the composition of a locally finite sequence of blowing-up along smooth subspaces, such that the ideal $f^*(\mathcal{J})\,\mathcal{O}_{\tilde{X}} \subset \mathcal{O}_{\tilde{X}}$ is invertible. In particular, if $Y \subset X$ is a subspace, there is a modification f as above, such that $\tilde{X}$ is smooth, and $f^{-1}(Y)$ is a divisor with normal crossing. If X is compact, or a relatively compact open set of a complex space, f can be obtained by a finite sequence of blowing-up along smooth subspaces.*

Let us finally state a result, which can be useful in the sequel, which is a simple consequence of the construction of the blowing-up.

Lemma 7.1. *Let $f\colon X' \to X$ be a surjective morphism of complex spaces, $\mathcal{I} \subset \mathcal{O}_X$ a coherent ideal, $\mathcal{J} = f^*(\mathcal{I})\,\mathcal{O}_{X'} \subset \mathcal{O}_{X'}$ the lifted ideal, $g\colon \tilde{X} \to X$ (resp. $g'\colon \tilde{X}' \to X'$) the blowing-up of X (resp. X') centered at $\mathcal{I}$ (resp. $\mathcal{J}$).*

Then there is a commutative diagram

$$\begin{array}{ccc} \tilde{X}' & \xrightarrow{\tilde{f}} & \tilde{X} \\ {\scriptstyle g'}\big\downarrow & & \big\downarrow{\scriptstyle g} \\ X' & \xrightarrow{f} & X \end{array}$$

We remark that if $\mathcal{I} = \mathcal{I}_Y$ is the ideal of a subspace $Y \subset X$, so that g is the blowing-up of X along Y, we cannot conclude that g' is the blowing-up of X' along $f^{-1}(Y)$, because the ideal $\mathcal{J}$ defines $f^{-1}(Y)$, but does not necessarily coincide with the ideal of functions vanishing on $f^{-1}(Y)$.

7.4 Algebraic and projective varieties, Moishezon spaces

Algebraic varieties (over the complex numbers) can be considered as particular complex spaces, whose local equations are polynomials. Actually, the very definition requires the use of the Zariski topology instead of the usual one (the strong topology). On $\mathbb{C}^n$ it is the topology whose closed sets are the algebraic subsets, defined as the set of common zeroes of a finite number of polynomials:

$$\{ z \in \mathbb{C}^n \mid p_1(z) = \cdots = p_k(z) = 0 \} .$$

The Zariski topology is quasi-compact, non Hausdorff. Closed and open subsets, neighborhoods... for the Zariski topology will be called Z-closed, Z-open, Z-neighborhoods...; they are obviously close, open, neighborhoods ... in the strong topology. Let U be a Z-open subset of $\mathbb{C}^n$; a rational function on U is a holomorphic function f on U which is locally a quotient of polynomials: for every point $x \in U$ there exist a Z-neighborhood V of x and polynomials $p(z)$, $q(z)$ (with $q(z) \neq 0$ on V) such that $f|_V = p/q$. The restriction of a rational function to Z-open subsets is rational, so that we can define the sheaf of the rational functions on U, which we denote $\mathcal{A}_U$.

The sheaf $\mathcal{A}_U$ is coherent.

An algebraic subset of $U \subset \mathbb{C}^n$ is a Z-closed subset $S \subset U$. Then it is automatically locally defined by rational functions: for every $x \in S$ there exist a Z-neighborhood V of x in U and a finite number of rational functions $f_1, \ldots, f_p$ on V such that $S \cap V = \{ z \in V \mid f_1(z) = \cdots = f_p(z) = 0 \}$. An intersection, or a finite union, of algebraic subsets of U is algebraic. Note that U itself is, trivially, an algebraic subset. An algebraic subset $S \subset U$ is *globally defined in* U if it is the set of the zeroes of rational functions defined on all of U: $S = \{ z \in U \mid g_1(z) = \cdots = g_s(z) = 0 \}$ where $g_1, \ldots, g_s$ are rational on U.

To an algebraic subset $S \subset U$ we associate a subsheaf of rings $\mathcal{I}_S \subset \mathcal{A}_U$ defined as

$$\mathcal{I}_S(V) = \{ f \in \mathcal{A}_U(V) \mid f|_{S \cap V} \equiv 0 \}$$

The sheaf $\mathcal{I}_S$ is coherent, or equivalently, finitely generated.

Hence the quotient $\mathcal{A}_U/\mathcal{I}_S$ is a coherent sheaf of local rings on U. On the other hand, if $x \in U \setminus S$, we have by definition $\mathcal{I}_{S,x} = \mathcal{A}_{U,x}$, so that the restriction of $\mathcal{A}_U/\mathcal{I}_S$ to $U \setminus S$ is identically zero. We define

$$\mathcal{A}_S = (\mathcal{A}_U/\mathcal{I}_S)\,|_S$$

which is a coherent sheaf of local $\mathbb{C}$-algebras on S. The coherent $\mathbb{C}$-ringed space $(S, \mathcal{A}_S)$ is called an *algebraic affine variety*, or simply an *affine variety*. A section of $\mathcal{A}_S$ on an open subset $T \subset S$ is called *a rational function on* T. If T is an open subset of S, the $\mathbb{C}$-ringed space $(T, \mathcal{A}_T)$ where $\mathcal{A}_T = \mathcal{A}_S|_T$, is an affine variety.

Let $S \subset U$ be an affine variety. An *algebraic subset* of S is a Z-closed subset T; this is equivalent to saying that for every $x \in T$ there exist a Z-neighborhood V of x in S and a finite number of rational functions $f_1, \ldots, f_p$ on V such that $T \cap V = \{ z \in V \mid f_1(z) = \cdots = f_p(z) = 0 \}$.

Then T is an algebraic subset of U.

An *algebraic variety* X is a ringed space $(X, \mathcal{A}_X)$ with the following properties. X is a quasi compact topological space, and for every $x \in X$ there exist an open neighborhood V of x in X and an isomorphism of $\mathbb{C}$-ringed spaces $(V, \mathcal{A}_X|_V) \simeq (S, \mathcal{A}_S)$, where S is an affine algebraic variety.

A section of $\mathcal{A}_X$ on a Z-open subset $V \subset S$ is called *a rational function on* V.

One has also obvious notions of algebraic subvariety of an algebraic variety and of morphism between two algebraic varieties.

The projective spaces $\mathbb{P}^n_{\mathbb{C}}$ are algebraic varieties. A *projective (algebraic) variety* is an algebraic variety isomorphic to an algebraic subvariety of some $\mathbb{P}^n_{\mathbb{C}}$.

As we have seen, the general definitions on algebraic varieties as $\mathbb{C}$-ringed spaces and those on complex spaces are similar. The holomorphic functions must be replaced by the rational functions, and the strong topology by the Zariski topology. Everybody knows that there are huge differences between the two theories; however, for the purposes of the present book, an algebraic variety will be considered as an important special case of complex space, which enjoys particularly good properties. Strictly speaking, to any algebraic variety $(X, \mathcal{A}_X)$ we can associate a (unique) complex space $(X^{\mathrm{an}}, \mathcal{A}_X^{\mathrm{an}})$. Going back to the beginning, if U is a Zariski open subset of $\mathbb{C}^n$, U^{an} is the set U equipped with the strong topology, and $\mathcal{A}_U^{\mathrm{an}}$ is the sheaf $\mathcal{O}_U$ of holomorphic functions. If X is an algebraic subvariety S of U, then S is also a complex subspace of U^{an}, which we denote S^{an} when provided with the strong topology; in this case $\mathcal{A}_S^{\mathrm{an}}$ is $\mathcal{O}_{S^{\mathrm{an}}}$, the sheaf of holomorphic functions on S^{an}. Finally, in the general case we cover X with open sets V isomorphic to affine algebraic varieties, so that $(V^{an}, \mathcal{O}_{V^{an}})$ have been defined, and we construct $(X^{an}, \mathcal{A}_X^{an})$ by gluing together the V^{an} and the $\mathcal{O}_{V^{an}}$. Analogously, to every morphism $f\colon X \to Y$ of algebraic varieties we associate a (unique) morphism of complex spaces $f^{an} : X^{an} \to Y^{an}$.

Hence when dealing with an algebraic variety, we will leave in the background the sheaf $\mathcal{A}_X$. By *an algebraic variety we will always mean a complex space of type* $(X^{an}, \mathcal{A}_X^{an})$, *and by an algebraic morphism between algebraic varieties a morphism of the type* $f^{an} : X^{an} \to Y^{an}$. Analogously, a *projective variety* is a complex space $(X^{an}, \mathcal{A}_X^{an})$ such that X is projective.

In our language, we can state the following properties and definitions.

- A Zariski open set of an algebraic variety X^{an} is an open set of type U^{an}, where U is a Z-open set of X.
- An algebraic variety is compact if it is compact as a complex space (strong topology). A projective variety is compact.
- A quasi-projective variety is an algebraic variety isomorphic to a Z-open set of a projective variety.
- Every affine, or quasi-projective variety X can be compactified: there exists a compact algebraic variety $\bar{X}$ such that X is isomorphic to a Z-open subset of $\bar{X}$.
- The product of two algebraic (resp. projective) varieties is algebraic (resp. projective).
- Any complex subspace of a projective variety is a projective variety (theorem of Chow).
- The irreducible components of an algebraic (resp. projective) variety are algebraic (resp. projective).
- If $f : X \to Y$ is an algebraic morphism of algebraic varieties, for $T \subset Y$ algebraic subvariety, $f^{-1}(T)$ is algebraic subvariety of X; if f is proper and Z is an algebraic subvariety of X, $f(Z)$ is an algebraic subvariety of Y.
- The subspace sing(X) of the singular points of an algebraic (resp. projective) variety X is an algebraic (resp. projective) subvariety, in particular is Z-closed. An *algebraic manifold* is an algebraic variety with no singularities.
- The blowing-up of an algebraic (resp. projective) variety along an algebraic (resp. projective) subvariety is algebraic (resp. projective). An analogous result holds for the strict transforms.

There are algebraic versions of the Hironaka Chow lemma, the Hironaka desingularization theorem, and of theorem 7.10. They are even stronger, because the algebraic varieties are quasi compact for the Zariski topology.

Theorem 7.11 (Algebraic Chow lemma). *Let $f\colon X' \to X$ be an algebraic modification of algebraic varieties. There is a commutative diagram*

$$\begin{array}{ccc} X'' & \xrightarrow{\quad q \quad} & X' \\ & {\scriptstyle p}\searrow \quad \swarrow{\scriptstyle f} & \\ & X & \end{array}$$

of algebraic varieties and morphisms, where p and q are the composition of finite sequences of blowing-up along smooth algebraic subvarieties. Moreover X'' can be taken quasi projective (projective if X is compact).

Theorem 7.12 (Hironaka algebraic desingularization theorem). *Let X be an algebraic variety. There exists a modification $f\colon (\tilde{X}, \tilde{E}) \to (X, E)$ of algebraic varieties and morphisms such that $\tilde{X}$ is smooth (i.e. an algebraic manifold), $\tilde{E} \subset \tilde{X}$ is a divisor with normal crossing, $E = \operatorname{sing}(X)$ and f is the composition of a finite sequence of blowing-up along smooth algebraic subvarieties. Moreover $\tilde{X}$ can be taken quasi projective (projective if X is compact).*

Theorem 7.13. *Let $(X, \mathcal{A}_X)$ be an algebraic variety, $\mathcal{J} \subset \mathcal{A}_X$ a coherent ideal. There exists a modification $f\colon \tilde{X} \to X$, which is the composition of a finite sequence of algebraic blowing-up, such that the ideal $f^*(\mathcal{J})\,\mathcal{A}_{\tilde{X}} \subset \mathcal{O}_{\tilde{X}}$ is invertible. In particular, if $Y \subset X$ is an algebraic subvariety, there is a modification f as above, such that $\tilde{X}$ is a quasi projective (projective if X is compact) manifold and $f^{-1}(Y)$ is a divisor with normal crossing.*

One has to be very careful in dealing with algebraic varieties as complex spaces. In particular let us point out the following fact.

Let $f\colon X \to Y$ be a modification of complex spaces. If X is algebraic, Y is not necessarily algebraic, and conversely. Neither projectivity or compactness can help in this context. That means in particular that there are compact complex spaces X, not necessarily algebraic, such there exists a modification $f\colon X' \to X$ where X' is algebraic (or even projective, after the algebraic Chow lemma). Such spaces are called *Moishezon spaces* (they where defined by Moishezon in [M] as compact complex spaces carrying many global meromorphic functions; the original definition is equivalent to ours).

Compact algebraic varieties are obviously Moishezon spaces.

One of the main theorems proved by Moishezon is the following.

Theorem 7.14. *Let X' be a compact Moishezon space, X a complex space, $f\colon X' \to X$ a morphism of complex spaces. Then $f(X')$ is a Moishezon space.*

- A complex space is Moishezon if and only if each irreducible component of X is Moishezon.

- If $g\colon \tilde{X} \to X$ is a modification, $\tilde{X}$ is Moishezon if and only if X is Moishezon.

- Every subspace of a Moishezon space is Moishezon.
- Every Moishezon space has a projective desingularization.

7.5 (B)-Kähler spaces

In the present paragraph we deal with a class of compact complex spaces (the Kähler spaces) which carry some properties of compact Kähler manifolds especially useful in the theory of mixed Hodge structures. The class has been introduced by Fujiki [F] with the name class $\mathscr{C}$, as the class of compact complex spaces bimeromorphic to a compact Kähler manifold. On the other hand, a definition of Kähler space can be given by means of plurisubharmonic functions on X, mimicking one of the definitions of a Kähler manifold. Fortunately, Varouchas in [V] has proved that all the possible reasonable definitions of a Kähler space coincide. The fundamental theorem of Varouchas is the following.

Theorem 7.15. *Let M be a compact Kähler manifold, X a complex space, $f\colon M \to X$ a surjective morphism. There exists a compact Kähler manifold X' and a modification $g\colon X' \to X$.*

Let us recall the following

Theorem 7.16. *The blowing-up of a compact Kähler manifold along a submanifold is a Kähler manifold.*

(In the case of a blowing-up at a point, an easy proof can be found in [BL]).

Definition 7.5. A (B)-Kähler space is a compact complex space X such that there exists a modification $X' \to X$ where X' is a compact Kähler manifold, or, equivalently, there exists a surjective morphism $M \to X$, M a compact Kähler manifold.

Remark. We have decided (following [A1]) to use the expression "(B)-Kähler space" (whose meaning is: bimeromorphic to a Kähler manifold) instead of "belonging to the class $\mathscr{C}$" because it is clearly simpler. Note that the simplest expression "Kähler space" would be misleading. In fact the compact Kähler manifolds are obviously (B)-Kähler spaces, but a (B)-Kähler space which is a manifold is not necessarily a Kähler manifold.

Examples.

1. A complex space is (B)-Kähler if and only if each irreducible component of X is (B)-Kähler.

2. A compact algebraic variety X, or a Moishezon space, is a (B)-Kähler space: there is a modification $X' \to X$, X' being a projective, hence Kähler, manifold.

3. If $f\colon X \to Y$ is a surjective morphism of compact complex spaces and X is (B)-Kähler, Y is (B)-Kähler.

Theorem 7.17. *Let X be a (B)-Kähler space.*

(i) If $g\colon \tilde{X} \to X$ is a modification, $\tilde{X}$ is (B)-Kähler.

(ii) Every subspace Y of X is (B)-Kähler.

Proof. Let $f\colon M \to X$ be a modification, M being a compact Kähler manifold. By Hironaka-Chow lemma there is a commutative diagram

$$\begin{array}{ccc} M_1 & \xrightarrow{\;q\;} & M' \\ & {\scriptstyle p}\searrow \quad \swarrow{\scriptstyle f} & \\ & X & \end{array}$$

where p and q are the composition of finite sequences of blowing-up along smooth centers. At each step, the blowing-up in the sequence defining q is a Kähler manifold, so by theorem 7.16 M_1 is a Kähler manifold. Hence replacing M by M_1 and f by p we can suppose that f itself is the composition of finite sequence of blowing-up along smooth centers.

We prove (i). It is easy (for example using lemma 7.1) to construct a commutative diagram

$$\begin{array}{ccc} T & \xrightarrow{\;\tilde{f}\;} & \tilde{X} \\ {\scriptstyle g'}\downarrow & & \downarrow{\scriptstyle g} \\ M & \xrightarrow{\;f\;} & X \end{array}$$

where g' and $\tilde{f}$ are modifications. By Hironaka-Chow lemma there is a commutative diagram

$$\begin{array}{ccc} M' & \xrightarrow{\;u\;} & T \\ & {\scriptstyle v}\searrow \quad \swarrow{\scriptstyle g'} & \\ & X & \end{array}$$

where u and v are compositions of finite sequences of blowing-up along smooth centers. It follows by theorem 7.16 that M' is a compact Kähler manifold. The composition $\tilde{f} \circ u\colon M' \to \tilde{X}$ is a modification, hence $\tilde{X}$ is (B)-Kähler.

Next we prove (ii). We can suppose that Y is irreducible. By theorem 7.10 there is a modification $h\colon \tilde{M} \to M$ obtained by a finite sequence of blowing-up along smooth subspaces, such that $D = h^{-1}\left(f^{-1}(Y)\right)$ is a divisor with normal crossing. Then, $\tilde{M}$ is a compact Kähler manifold, and the irreducible components D_j of D, as submanifolds of $\tilde{M}$, are Kähler. There is a j such that $(f \circ h)(D_j) = Y$, hence Y is (B)-Kähler. $\square$

The strictness of the differential d on global differential forms with respect to the Hodge filtration and the equivalent degeneration at E_1 of the Hodge spectral sequence holds for manifolds which are (B)-Kähler.

Theorem 7.18. *Let X be a compact (B)-Kähler manifold. The differential*

$$d\colon \Gamma\left(X, \mathcal{E}_X^k\right) \to \Gamma\left(X, \mathcal{E}_X^{k+1}\right)$$

is strict for the Hodge filtration $F^p\Gamma\left(X, \mathcal{E}_X^k\right)$. Equivalently, the spectral sequence associated to filtration $F^p\Gamma\left(X, \mathcal{E}_X^{\cdot}\right)$ on the complex $\Gamma\left(X, \mathcal{E}_X^{\cdot}\right)$ degenerates at E_1.

Proof. We prove the degeneration of the spectral sequence. There exists a modification $f\colon \tilde{X} \to X$ such that $\tilde{X}$ is a compact Kähler manifold. The first term of the spectral sequence $\left(E_r^{m,k}(X), d_r\right)$ is:

$$E_1^{m,k}(X) = H^k\left(X, \Omega_X^{-m}\right)$$

Let us consider also the spectral sequence $(E_r^{m,k}(\tilde{X}), \tilde{d}_r)$ for $\tilde{X}$, with $E_1^{m,k}(\tilde{X}) = H^k(\tilde{X}, \Omega_{\tilde{X}}^{-m})$. There are linear mappings $f_r^{m,k}\colon E_r^{m,k}(X) \to E_r^{m,k}(\tilde{X})$, induced by the pullback f^*, which commute with d_r and $\tilde{d}_r$. By theorem 5.10 of chapter 5, $E_r^{m,k}(\tilde{X})$ degenerates at E_1, that is, $\tilde{d}_r = 0$, for $r \geq 1$ and $E_r^{m,k}(\tilde{X}) = E_1^{m,k}(\tilde{X})$, for $r \geq 1$.

The mappings $f_1^{m,k}\colon E_1^{m,k}(X) \to E_1^{m,k}(\tilde{X})$ coincide with the pullback, through f, of cohomology classes: $H^k\left(X, \Omega_X^{-m}\right) \to H^k(\tilde{X}, \Omega_{\tilde{X}}^{-m})$, hence they are injective by theorem 4.8 of chapter 4.

It follows that d_1 is the restriction of $\tilde{d}_1$ to a subspace, so that $d_1 = 0$. We deduce that $E_2^{m,k}(X) = E_1^{m,k}(X)$ and $f_2^{m,k} = f_1^{m,k}$, so that $f_2^{m,k}$ is injective too. Again we find $d_2 = 0$. Continuing in the same way we prove by induction that $f_r^{m,k}$ is injective and $d_r = 0$ for every $r \geq 1$. $\square$

7.6 Semianalytic and subanalytic sets

Let $U \subset \mathbb{R}^n$ be an open set, and $\mathcal{O}_U^{\mathbb{R}}$ be the sheaf of real analytic functions on U: for V open in U, by definition

$$\mathcal{O}_X^{\mathbb{R}}(V) = \{\, f\colon V \to \mathbb{R} \text{ analytic} \,\}$$

Then $\mathcal{O}_X^{\mathbb{R}}$ is a sheaf of local $\mathbb{R}$-algebras.

A (real) analytic subset of $U \subset \mathbb{R}^n$ is a closed subset $S \subset U$ with the following property; for every $x \in S$ there exist a neighborhood V of x in U and a finite number of real analytic functions $f_1, \dots, f_p$ on V such that $S \cap V = \{ z \in V \mid f_1(z) = \cdots = f_p(z) = 0 \}$.

An intersection, or a finite union, of analytic subsets of U is analytic. Note also that U itself is, trivially, an analytic subset.

To an analytic subset $S \subset U$ we associate a subsheaf of local $\mathbb{R}$-algebras $\mathcal{I}_S \subset \mathcal{O}_U^{\mathbb{R}}$ defined as

$$\mathcal{I}_S(V) = \{ f \in \mathcal{O}_U^{\mathbb{R}}(V) \mid f|_{S \cap V} \equiv 0 \}$$

in other words, the subsheaf of analytic functions vanishing on S.

The quotient $\mathcal{O}_U^{\mathbb{R}}/\mathcal{I}_S$ is a sheaf of local $\mathbb{R}$-algebras on U. On the other hand, if $x \in U \setminus S$, we have by definition $\mathcal{I}_{S,x} = \mathcal{O}_{U,x}^{\mathbb{R}}$, so that the restriction of $\mathcal{O}_U^{\mathbb{R}}/\mathcal{I}_S$ to $U \setminus S$ is identically zero. We define

$$\mathcal{O}_U^{\mathbb{R}} = \left(\mathcal{O}_U^{\mathbb{R}}/\mathcal{I}_S \right)|_S$$

which is a sheaf of local rings on S. It is easy to see that $\mathcal{O}_U^{\mathbb{R}}$ injects into $\mathcal{C}_S$ as a subsheaf. The $\mathbb{R}$-ringed space $(S, \mathcal{O}_S^{\mathbb{R}})$ is called a *real analytic variety*, or simply an *analytic variety*. A section of $\mathcal{O}_S^{\mathbb{R}}$ on an open subset $Z \subset S$ is called *an analytic function on* Z. *A (real) analytic space* X is a $\mathbb{R}$-ringed space $(X, \mathcal{O}_X^{\mathbb{R}})$ with the following property. For every $x \in X$ there exist an open neighborhood V of x in X and an isomorphism of $\mathbb{R}$-ringed spaces $(V, \mathcal{O}_X^{\mathbb{R}}|_V) \simeq (S, \mathcal{O}_S^{\mathbb{R}})$, where S is an analytic variety.

In other words, an analytic space behaves locally as an analytic variety.

A section of $\mathcal{O}_X^{\mathbb{R}}$ on an open subset $V \subset S$ is called *an analytic function on* V.

A morphism $f \colon X \to Y$ of analytic spaces is by definition a morphism of $\mathbb{R}$-ringed spaces $f \colon (X, \mathcal{O}_X^{\mathbb{R}}) \to (Y, \mathcal{O}_Y^{\mathbb{R}})$.

A real analytic manifold is a real analytic space.

Let X be a complex space. A *real analytic function* on an open set $U \subset X$ can be defined as the real (or the imaginary) part of a holomorphic function on U. The analytic functions on open sets of X form a sheaf, which we denote $\mathcal{O}_X^{\mathbb{R}}$. It is a sheaf of local $\mathbb{R}$-algebras. It is easy to see that $(X, \mathcal{O}_X^{\mathbb{R}})$ is a real analytic space, *the analytic space underlying the complex space* X.

Remark. The structure sheaf of an analytic space is not necessarily coherent. It is coherent if the space is a manifold.

Let X be an analytic space. For each $x \in X$, let $S(x)$ be the smallest family of germs at x of subsets of X such that:

(1) A, $B \in S(x)$ implies $A \cup B$ and $A \setminus B \in S(x)$;

(2) if f is an analytic function on a neighborhood of x, the germ at x of the subset $\{f > 0\}$ belongs to $S(x)$.

A subset M of X is called *semianalytic* if its germ at every $x \in X$ belongs to $S(x)$. Roughly speaking, a semianalytic set is, in a neighborhood of every point $x \in X$, a finite union of subsets defined by analytic inequalities and equalities.

Locally finite unions and intersections, complements, closures, interiors and boundaries of semianalytic sets are semianalytic. Analytic subspaces are semianalytic.

Let M be a semianalytic subset of X. A point of M is *q-simple, or q-regular* if there is a neighborhood U of x in M such that the ringed space $\left(M, \mathcal{O}^{\mathbb{R}}_{X|M}\right)$ is isomorphic to $\left(V, \mathcal{O}^{\mathbb{R}}_{\mathbb{R}^q|V}\right)$, V an open subset of $\mathbb{R}^q$. In other words, U is in a natural way a real analytic manifold.

The set M^* of simple points of M is a real analytic manifold, open and dense in M. The dimension of M is, by definition, the dimension of the manifold M^*.

Suppose $\dim M = p$. Then $\dim \overline{M} = p$, $\dim \overline{M} \setminus M < p$. The semianalytic set $bM = \overline{M} \setminus M$ is called *the border* of M; bM is closed, if M is locally closed. If M^* is the set of p-simple points of M, then $\left(M^*, \mathcal{O}^{\mathbb{R}}_{X|M^*}\right)$ is a real analytic manifold of dimension p, and $\operatorname{sing}(M) = M \setminus M^*$ is semianalytic in X, $\dim \operatorname{sing}(M) < p$.

We point out that if M is an analytic space, by the above assertion $\operatorname{sing}(M)$ is a semianalytic subset; but it is not analytic in general. Nevertheless, there exists a nowhere dense analytic subspace Y of M, such that $\operatorname{sing}(M) \subset Y$.

One of the deepest results about the topology of semianalytic subsets is the theorem of Łojasiewicz [Ło] [Gi]. We need it in the following form.

Theorem 7.19. *Let X be an analytic space, $M \subset X$ a semianalytic subset. There exists a triangulation of X compatible with M. That is, there exists a locally finite simplicial complex $\mathcal{V}$ in an affine space $\mathbb{R}^m$ and a homeomorphism $\tau\colon \mathcal{V} \to X$ such that*

(a) for any open simplex $\sigma \in \mathcal{V}$, $\tau(\sigma)$ is is semianalytic in X and has only simple points;

(b) M is the image of a subcomplex $\mathcal{M} \subset \mathcal{V}$.

As an important consequence we obtain

Theorem 7.20. *Let X be an analytic space (in particular a complex space).*

(i) Let $M \subset X$ be a semianalytic subset. There exists a fundamental system of neighborhoods V of M in X such that M is a deformation retract of V.

(ii) Every point of X has a fundamental system of contractible neighborhoods.

In fact, the properties (i) and (ii) go back, through the above theorem, to analogous, well known properties of simplicial complexes.

Let $f\colon X \to Y$ be a morphism of analytic spaces. The inverse image of a semianalytic subset of Y is easily seen to be semianalytic in X. On the contrary, there are examples of proper mappings f as above, and analytic subspaces Z of X such that $f(Z)$ is not semianalytic in Y. For this reason Hironaka [H3] has introduced the notion of *subanalytic subset* of an analytic space. We do not give here the exact definition, instead we collect the main properties of the subanalytic subsets. Roughly speaking, the family of subanalytic subsets of an analytic space X is the smallest family $\mathcal{S}$ of subsets of X, such that

1. $\mathcal{S}$ contains all the images $f(T)$, where $f\colon Z \to X$ is a proper morphism of analytic spaces and $T \subset Z$ is a semianalytic subset.

2. $\mathcal{S}$ is stable with respect to the following operations: locally finite unions and intersections, complement, closure, interior and boundary, inverse images and images by proper morphisms of analytic spaces.

A semianalytic subset is subanalytic. A subanalytic subset of X is locally connected, its connected components are subanalytic and form a locally finite family in X. Let M be a subanalytic subset of X. A point of M is *q-simple, or q-regular* if there is a neighborhood U of x in M such that the ringed space $\left(M, \mathcal{O}_X^{\mathbb{R}}|_M\right)$ is isomorphic to $\left(V, \mathcal{O}_{\mathbb{R}^q}^{\mathbb{R}}|_V\right)$, V an open subset of $\mathbb{R}^q$.

The set M^* of simple points of M is a real analytic manifold, open and dense in M. The dimension of M is, by definition, the dimension of the manifold M^*.

Suppose $\dim M = p$. Then $\dim \overline{M} = p$, $\dim \overline{M} \setminus M < p$. The subanalytic set $bM = \overline{M} \setminus M$ is called *the border* of M; bM is closed, if M is locally closed. If M^* is the set of p-simple points of M, then $\left(M^*, \mathcal{O}_X^{\mathbb{R}}|_{M^*}\right)$ is a real analytic manifold of dimension p, and $\operatorname{sing}(M) = M \setminus M^*$ is subanalytic in X, $\dim \operatorname{sing}(M) < p$.

Hence the family of the subanalytic subsets has all the good properties of the semianalytic subsets (including the triangulation theorem 7.19: see [Ve]) and furthermore the images of subanalytic sets by proper analytic morphisms are subanalytic.

7.7 The Borel-Moore homology of a complex space

We refer to [BHae], [BMo] for more details on the contents of the present section.

We denote by K the ring $\mathbb{Z}$ or one of the fields $\mathbb{Q}$, $\mathbb{R}$, $\mathbb{C}$. Let X be a paracompact, locally compact topological space; K_X (or simply K if no confusion arises) will denote the corresponding constant sheaf. Then for a family of supports Φ in X the cohomology groups $H^k_\Phi(X, K)$ are defined. On the other hand, there are many interesting cohomology groups defined on X, which share the same notations H^k, like the singular cohomology groups and the Borel-Moore cohomology groups. For a general X, such cohomology spaces do not necessarily coincide with each other. Fortunately, it is the case if X is a locally closed subanalytic subset of some analytic space, in particular if X is a real or complex analytic space, and this is essentially a consequence of the triangulation theorem 7.19 and its generalization to subanalytic subsets [Ve].

Given a paracompact, locally compact topological space X and a family of supports Φ in X *the Borel-Moore homology modules with coefficients in K and supports in Φ* consist of K-modules $H^\Phi_k(X, K)$ (simply $H_k(X) = H_k(X, K)$ if Φ is the family of all closed subsets of X) whose main properties are:

(1) If $U \subset V$ are open subsets of X, there are restrictions $j^{V,U}: H_k(V, K) \to H_k(U, K)$, so that the assignment $U \mapsto H_k(U, K)$ is a presheaf. If A is closed in X, there is a natural morphism $i_{A,X}: H_k(A, K) \to H_k(X, K)$.

(2) If A is a closed subset of X and $U = X \setminus A$ there is a natural exact sequence

$$\cdots H_k(A, K) \xrightarrow{i_{A,X}} H_k(X, K) \xrightarrow{j^{X,U}} H_k(U, K) \xrightarrow{\partial_{U,A}} H_{k-1}(A, K) \cdots \qquad (7.1)$$

where $\partial_{U,A}$ is called the *boundary, or connecting, homomorphism.*

(3) Every continuous map $f: X \to Y$ induces a natural push-down morphism $f_*: H^c_k(X, K) \to H^c_k(Y, K)$, (where ^{c} means compact supports); if f is proper, it induces also $f_*: H_k(X, K) \to H_k(Y, K)$.

(4) If X is a differentiable manifold, the Borel-Moore homology modules with compact supports $H^c_k(X, K)$ coincide with the k-th modules of singular homology; in particular they are homotopy invariants.

(5) If X is a connected differentiable manifold of dimension p, the presheaf $U \mapsto H_p(U, K)$ is a sheaf $\mathcal{K}$, which turns out to be locally isomorphic to the constant sheaf K, and it is globally isomorphic to the constant sheaf if and only if X is orientable. The module $H_p(X, K)$ is 0 if X is non orientable, and it is isomorphic to K if X is orientable. In the orientable case, a nonzero element of $H_p(X, K)$ defines an orientation and a multiplicity (i.e. a constant) on X.

Let X be as in (5), ω a (real or complex) differential form of degree p on X, and c a class in $H_p(X, K)$. Then either $c = 0$ or c defines an orientation on X and a multiplicity $k \in K$. We define the integral of ω on c by the formula

$$\int_c \omega = k \int_X \omega$$

where the integral in the second member is performed according to the defined orientation. If X is not connected, we define $\int_c \omega$ as a sum over the connected components of X. The above integral converges under some conditions on ω, for example if ω has compact support.

Let X be an analytic manifold, M a locally closed subanalytic set of X of dimension p, c a class in $H_p(M, K)$. Let ω be a p-form defined on an open neighborhood U of M. Then we can define $\int_c \omega$ as follows. Let M^* be the sap of p-regular points of M, which is open and dense in M. M^* is an analytic manifold of dimension p, and it embeds into U: $i\colon M^* \to U$. Then we consider the class $c^* = j^{M,M^*}(c) \in H_p(M^*, K)$ and the p-form $i^*\omega$ on M^* and we define

$$\int_c \omega = \int_{c^*} i^*\omega$$

We point out that if ω has compact support, the support of $i^*\omega$ is not necessarily compact. The following theorem, which goes back to Lelong [Le], Herrera [He] and Dolbeault and Poly [DP], ensures the convergence of the integral.

Theorem 7.21. *Let $M' \subset M$ be closed subanalytic subsets of an analytic variety X, such that $M \setminus M'$ is a smooth, closed submanifold of $X \setminus M'$. Let c be a class in $H_p(M \setminus M', K)$ and ω a differentiable p-form on X. We suppose that either ω has compact support, or M is compact. Then the integral $\int_c i^*\omega$, where i is the embedding of $M \setminus M'$ into $X \setminus M'$, converges.*

The above theorem clearly applies to locally closed subanalytic sets M', M, replacing X by any open set U such that M is contained and closed in U, provided that the restriction of ω to U has compact support.

7.8 Subanalytic chains

Let X be an analytic space, M a closed subanalytic subset of X (possibly $M = X$), p a nonnegative integer. A *p-subanalytic prechain on M with coefficients in K* is a pair (N, c), where N is a locally closed subanalytic subset of X contained in M, $\dim N \leq p$, and $c \in H_p(N) = H_p(N, K)$. The set of the p-subanalytic prechains on M with coefficients in K will be denoted $\mathcal{PS}_p(M)$ (or $\mathcal{PS}_p(M, K)$ if it is necessary to mention K). Let us remark that if $\dim N < p$, then necessarily $c = 0$.

We define commutative and associative operations on $\mathcal{PS}_p(X)$ as follows.

$$(N_1, c_1) + (N_2, c_2) = \left(N, c^1 + c^2\right) \tag{7.2}$$

where $N = N_1 \cup N_2 \setminus L$, $L = bN_1 \cup bN_2$ and c^s is the image of c_s by the compositions

$$H_p(N_s) \xrightarrow{j^{N_s, N_s \setminus L}} H_p(N_s \setminus L) \xrightarrow{i_{N_s \setminus L, N}} H_p(N) \tag{7.3}$$

The above maps are defined because $N_s \setminus L$ is open in N_s, and is closed in N.

The product by elements $k \in K$ is defined by $k(N,c) = (N, kc)$.

We define on $\mathcal{PS}_p(M)$ the following equivalence relation:

$$(N_1, c_1) \equiv (N_2, c_2) \Longleftrightarrow (N_1, c_1) + (N_2, -c_2) = (N, 0)$$

The quotient $\mathcal{S}_p(M)$ (also denoted by $\mathcal{S}_p(M,K)$) of $\mathcal{PS}_p(M)$ by $\equiv$ carries an induced K-module structure. The elements of $\mathcal{S}_p(M,K)$ are called *p-subanalytic chains of M with coefficients in K.*

Using the exact sequence (7.1) corresponding to the pair $(\overline{N}, bN)$ we obtain maps

$$\mathcal{PS}_p(M,K) \to \mathcal{PS}_{p-1}(M,K), \quad (N,c) \mapsto (bN, \partial_{N,bN}(c))$$

which are linear and compatible with the relation $\equiv$, so that they induce boundary homomorphisms

$$\partial\colon \mathcal{S}_p(M,K) \to \mathcal{S}_{p-1}(M,K)$$

satisfying $\partial \circ \partial = 0$. We obtain the complex $(\mathcal{S}_\cdot(M,K), \partial)$ of the subanalytic chains on M.

Let now $V \subset M$ be an open subset of M and U any open set in X such that $U \cap M = V$. Then V is a closed subanalytic subset of the analytic space U, so that the complex $\mathcal{S}_\cdot(V,K)$ is defined and does not depend on the choice of U. If $W \subset V$ is also open in M, a restriction morphism $r_{V,W}\colon \mathcal{S}_p(V,K) \to \mathcal{S}_p(W,K)$, compatible with boundaries, is induced by the maps

$$\mathcal{PS}_p(V,K) \to \mathcal{PS}_p(W,K), \quad (N,c) \mapsto \left(N \cap W, j^{N, N\cap W}(c)\right)$$

The assignment $V \mapsto \mathcal{S}_p(U,K)$ defines a presheaf on M, which can be proved to be a sheaf. We obtain a complex of sheaves $(\mathcal{S}_\cdot, \partial)$ on M. The sheaf $\mathcal{S}_p$ will be denoted by $\mathcal{S}_{M,p}$ when we need to mention the space M, and will be called the sheaf of germs of subanalytic p-chains on M.

For an open set U of M let $S_p(U) = S_p(U,K) = \Gamma_c(U, \mathcal{S}_{M,p})$ be the K-module of the chains with compact support in U. Let (N,c) be a p-subanalytic chain in U; then c is a section on N of the sheaf $\mathcal{K}$ defined by $\mathcal{K}(W) = H_p(W)$, hence its support is defined. The class of (N,c) in $\mathcal{S}_{M,p}(U)$ has compact support in U if the support of c is relatively compact in U; this happens in particular if N is relatively compact in U. If c has relatively compact support, there exists a chain (N', c'), equivalent to (N,c), such that N' is relatively compact. Hence we are allowed to deal only with chains (N,c) with $\overline{N}$ compact in U.

If $V \subset U$ is also open, there is an obvious injection (the extension by zero) $S_p(V) \to S_p(U)$. If $\alpha \in S_p(U)$, then $\partial\alpha \in S_{p-1}(U)$.

The presheaf on M defined by $U \mapsto \hom_K(S_p(U, K), K)$ is a sheaf, which we denote by $\mathcal{S}^p$. An element of $\mathcal{S}^p(V)$ is called *a p-subanalytic cochain on U with coefficients in K*. The boundaries ∂ give by transposition differentials $\delta\colon \mathcal{S}^p \to \mathcal{S}^{p+1}$ such that $\delta \circ \delta = 0$, hence $(\mathcal{S}^{\cdot}, \delta)$ is a complex. The duality between the $S_p(U)$ and $\mathcal{S}^p(U)$ is given by a bilinear K-valued pairing: $\langle\ ,\ \rangle$, such that (by definition of δ)

$$\langle \alpha, \delta\omega \rangle = \langle \partial\alpha, \omega \rangle \tag{7.4}$$

for $\alpha \in S_{p+1}(U)$ and $\omega \in \mathcal{S}^p(U)$. The sheaf $\mathcal{S}^p$ will be denoted by $\mathcal{S}^p_M$ when we need to mention the space M, and will be called the sheaf of germs of subanalytic p-cochains on M. The main results of [BH], [DP] can be summarized in the following

Theorem 7.22. *Let M be a closed subanalytic subset of an analytic space X, and Φ a paracompactifying family of supports on M. Then*

(1) The sheaves $\mathcal{S}_{M,p}$ are soft, and the sheaves $\mathcal{S}^p_M$ are flabby.

(2) There are natural isomorphisms

$$H_p^{\Phi}(M, K) \simeq H_p\left(\Gamma_\Phi\left(M, \mathcal{S}_{M,\cdot}\right)\right)$$

between the Borel-Moore homology of M with supports in Φ and the homology of the complex of global sections with supports in Φ of $(\mathcal{S}_M, \cdot, \partial)$.

(3) If K is a field, the complex $(\mathcal{S}^{\cdot}_M, \delta)$ is a resolution of the constant sheaf K_M, so that there are isomorphisms

$$H^p_\Phi(M, K) \simeq H^p\left(\Gamma_\Phi\left(M, \mathcal{S}^{\cdot}_M\right)\right)$$

where the right hand side is the cohomology of the complex $\left(\Gamma_\Phi\left(M, \mathcal{S}^{\cdot}_M\right), \delta\right)$ of global sections with supports in Φ. In this case, the pairing $\langle\ ,\ \rangle$ on $S_p \times \mathcal{S}^p$ induces a nondegenerate pairing

$$H_p^c(M, K) \times H^p(M, K) \to K \tag{7.5}$$

The original proof of [BH] deal with semianalytic chains. The advantage of considering subanalytic chains is that the sheaves $\mathcal{S}_p$ of subanalytic chains behave well with respect to proper mappings. In fact, let $f\colon X \to Y$ be a proper morphism of analytic spaces. If (N, c) is a subanalytic p-prechain on X, $f(N)$ is a subanalytic closed subset of Y and the push-down f_*c is a class in $H_p(f(N))$, hence $(f(N), f_*c)$ is subanalytic p-prechain on Y; the construction passes to subanalytic p-chains, so that f induces morphisms

$$f_*\colon f_*\mathcal{S}_{X,p} \to \mathcal{S}_{Y,p}$$

of sheaves of K-modules, commuting with boundaries.

If $f\colon X \to Y$ is a not necessarily a proper morphism, it is possible to define the push-down fo' chains with compact support. Such chains have representative prechains (N, c) with $\bar{N}$ compact. Then $(f(N), f_* c)$ is subanalytic p-prechain on Y with compact support, since $f|_{\overline{N}}$ is proper. We obtain morphisms

$$f_*\colon S_{X,p}\left(f^{-1}(V)\right) \to S_{Y,p}(V)$$

for any an open set V in Y. By duality we also obtain pullback

$$f^*\colon \mathcal{S}_Y^p \to f_*\mathcal{S}_X^p$$

commuting with δ. In particular

$$\langle c, f^*\omega\rangle = \langle f_* c, \omega\rangle$$

for any $c \in S_{X,p}\left(f^{-1}(V)\right)$ and $\omega \in \mathcal{S}_Y^p(V)$, V an open set in Y.

7.9 Integration of forms on complex subanalytic chains

Let (N, c) be a p-subanalytic prechain of an analytic manifold X with coefficients in $\mathbb{C}$, and ω a p-differential form on X. Let N^* be the set of p-regular points of N, $i\colon N^* \to X$ the embedding. Let $c^* = j^{N,N^*}(c) \in H_p(N^*, \mathbb{C})$. Then we define

$$\int_{(N,c)} \omega = \int_c \omega = \int_{c^*} i^*\omega \in \mathbb{C}$$

which converges, by theorem 7.16, if N is relatively compact in X, or c, or ω, has compact support. If (N_1, c_1) is a p-subanalytic prechain equivalent to (N, c), the integrals $\int_{(N,c)} \omega$ and $\int_{(N_1,c_1)} \omega$ are equal. This means that for every chain with compact support $\eta \in S_p(X, \mathbb{C})$ with coefficients in $\mathbb{C}$, the integral $\int_\eta \omega$ is well defined. Moreover:

Theorem 7.23 (Stokes formula). *Let η be p-subanalytic chain with coefficients in $\mathbb{C}$ and compact support of an analytic manifold X, ω a p-differential form on X. Then*

$$\int_\eta d\omega = \int_{\partial\eta} \omega.$$

The Stokes theorem can be given the following interpretation. The integration of differential forms on chains with compact support is a morphism of sheaves $j^p\colon \mathcal{E}_X^p \to \mathcal{S}_X^p$ which turns out to be injective. The Stokes formula reads as $j^{p+1}(d\omega) = \delta j^p(\omega)$, that is $j^\cdot : \mathcal{E}_X^\cdot \to \mathcal{S}_X^\cdot$ is a morphism of complexes of sheaves. Since both the complexes are resolutions of $\mathbb{C}_X$, we conclude that the pairing (7.5) between homology and cohomology is also given by integration of forms on chains with compact support.

7.10 The Mayer-Vietoris sequence for modifications

Let us consider a modification of complex (or real analytic) spaces given by the commutative diagram :

$$\begin{array}{ccc} \tilde{E} & \xrightarrow{i} & \tilde{X} \\ {\scriptstyle q}\downarrow & & \downarrow{\scriptstyle \pi} \\ E & \xrightarrow{j} & X \end{array}$$

where $E \subset X$ is a nowhere dense closed subspace, $j\colon E \to X$ is the natural inclusion, and π is a proper modification inducing an isomorphism $\tilde{X} \setminus \tilde{E} \simeq X \setminus E$. Let K be the ring $\mathbb{Z}$ or one of the fields $\mathbb{Q}$, $\mathbb{R}$, $\mathbb{C}$. We want to introduce the Mayer-Vietoris sequence for cohomology and prove that it is exact. We apply formula (3.20) of chapter 3 to the constant sheaf K and to the pairs (E, X) and $(\tilde{E}, \tilde{X})$, and we relate them with natural pullback maps, obtaining a commutative diagram with exact rows:

$$\begin{array}{ccccccc} \cdots H^k(\tilde{X}, K) & \xrightarrow{i^*} & H^k(\tilde{E}, K) & \xrightarrow{\tilde{\delta}} & H^{k+1}_{\phi}(\tilde{X} \setminus \tilde{E}, K) & \xrightarrow{\tilde{\rho}_*} & H^{k+1}(\tilde{X}, K) \cdots \\ {\scriptstyle \pi^*}\uparrow & & {\scriptstyle q^*}\uparrow & & {\scriptstyle g}\uparrow & & {\scriptstyle \pi^*}\uparrow \\ \cdots H^k(X, K) & \xrightarrow{j^*} & H^k(E, K) & \xrightarrow{\delta} & H^{k+1}_{\phi}(X \setminus E, K) & \xrightarrow{\rho_*} & H^{k+1}(X, K) \cdots \end{array}$$

where $g = (\pi|_{\tilde{X} \setminus \tilde{E}})^*$ is an isomorphism, and Φ is the family of closed sets in $\tilde{X}$ contained in $\tilde{X} \setminus \tilde{E}$, which is also the family of closed sets in X contained in $X \setminus E$. We define the *Mayer-Vietoris sequence*:

$$\begin{aligned} \cdots \longrightarrow H^k(X, K) &\xrightarrow{(\pi^*, j^*)} H^k(\tilde{X}, K) \oplus H^k(E, K) \xrightarrow{\Theta} \\ &\longrightarrow H^k(\tilde{E}, K) \xrightarrow{\Psi} H^{k+1}(X, K) \longrightarrow \cdots \end{aligned} \tag{7.6}$$

where:

$$\Psi = \rho_* \circ g^{-1} \circ \tilde{\delta}, \qquad \Theta = (-1)^k (i^* - q^*)$$

It is easy to see that the above sequence is exact. For example we prove that $\ker (\pi^*, j^*) = \operatorname{im} \Psi$. Let $\xi \in H^{k+1}(X, K)$ such that $\pi^* (\xi) = 0$ and $j^* (\xi) = 0$; from the second equality we find $\xi = \rho_* (\mu)$ for some $\mu \in H^{k+1}_{\Phi} (X \setminus E, K)$; let $\tilde{\mu} = g (\mu)$; we have $\tilde{\rho}_* (\tilde{\mu}) = \pi^* (\rho_* (\mu)) = \pi^* (\xi) = 0$; hence there exists $\theta \in H^k(\tilde{E}, K)$ with $\tilde{\delta} (\theta) = \tilde{\mu}$; it follows $\xi = \rho_* \left(g^{-1} (\tilde{\mu})\right) = \Psi (\theta)$. Conversely, if $\xi = \Psi (\theta)$ for some $\theta \in H^k(\tilde{E}, K)$, $\pi^* (\xi) = (\pi^* \circ \rho_* \circ g^{-1} \circ \tilde{\delta}) (\theta) = (\tilde{\rho}_* \circ g \circ g^{-1} \circ \tilde{\delta}) (\theta) = (\tilde{\rho}_* \circ \tilde{\delta}) (\theta) = 0$, and $j^* (\xi) = (j^* \circ \rho_* \circ g^{-1} \circ \tilde{\delta}) (\theta) = 0$.

Part II

Differential forms on complex spaces

Chapter 1

The basic example

1.1 Introduction

Let X be a complex space. We denote by $\mathbb{C}_X$ the constant sheaf on X. When X is smooth we denote by $\mathcal{E}^{\cdot}_X$ the De Rham complex of differential forms on X. There exists a resolution of singularities of X, i.e. a commutative diagram:

$$\begin{array}{ccc} \tilde{E} & \xrightarrow{\ i\ } & \tilde{X} \\ {\scriptstyle q}\downarrow & & \downarrow{\scriptstyle \pi} \\ E & \xrightarrow{\ j\ } & X \end{array} \tag{1.1}$$

where $E \subset X$ is a nowhere dense closed subspace, containing the singularities of X, $j\colon E \to X$ is the natural inclusion, $\tilde{X}$ is a smooth manifold and π is a proper modification inducing an isomorphism $\tilde{X} \setminus \tilde{E} \simeq X \setminus E$.

Throughout all the present chapter we will consider a complex space X with the following property. There exists a resolution of singularities of X as in the diagram (1.1) with the additional assumption that *E and $\tilde{E}$ are smooth manifolds*. For simplicity we will call such an X a *quasi-smooth* complex space. In the case of a quasi-smooth space X it is quite easy to understand how to construct a complex $\Lambda^{\cdot}_X$ of "differential forms" on X which is a resolution of the constant sheaf $\mathbb{C}_X$. The main idea is very simple. On the manifolds $\tilde{X}$, E and $\tilde{E}$ everything is well known: the definition of differential forms, of the differential of a form, the theorem of De Rham, the notion of pullback of forms (together with its important property of commuting with differentials), and classical Hodge theory. Then the diagram (1.1) allows us to give definitions for X and prove the expected theorems. Roughly speaking, an object on X is triple of objects on $\tilde{X}$, E and $\tilde{E}$. The manifold $\tilde{E}$ plays a role different from $\tilde{X}$ and E, in the sense that the object which is the $\tilde{E}$-component must be "shifted" by one. We will say that $\tilde{X}$ and E have rank 0, while $\tilde{E}$ has rank 1. For example, a k-differential form on X is a triple of forms: a k-form on $\tilde{X}$, a k-form on E and a $(k-1)$-form on $\tilde{E}$. Of course X is not a "direct sum" (whatever this could mean) of $\tilde{X}$, E and $\tilde{E}$, so, in order to describe properties of X we have to perturb the triples by using the maps i, j, q, π in the diagram

(1.1). In the case of differential forms, it is the definition of the differential d of a form which encodes the relations among $\tilde{X}$, E and $\tilde{E}$ in the above mentioned diagram, by means of the two pullback i^* and q^*.

As we will see in the next chapter, the differential forms we introduce here for a quasi-smooth X are a particular case. On one hand, they suffice to describe De Rham theory and Hodge-Deligne theory on X; on the other hand, if we consider a map $f\colon X \to Y$ of complex spaces, X and Y quasi-smooths, we are not able in general to pullback to X the forms on Y. In fact if

$$\begin{array}{ccc} \tilde{F} & \longrightarrow & \tilde{Y} \\ \big\downarrow & & \big\downarrow \\ F & \longrightarrow & Y \end{array} \tag{1.2}$$

is a diagram for Y corresponding to (1.1), in general the morphism f does not extend to a morphism $\tilde{f}\colon \tilde{X} \to \tilde{Y}$, and moreover there is no relationship between $f^{-1}(F)$ and E.

Neverthless, by means of the (particular) forms we consider in this chapter we can define two filtrations on the cohomology of X, W and F, such that $H^k(X,\mathbb{C})$ carries a mixed Hodge structure (under the assumption that $\tilde{X}$, E and $\tilde{E}$ are compact Kähler manifolds). This is a consequence of the following important property: the spectral sequence of the cohomology spaces, corresponding to the filtration W, degenerates at level 2. The proof of this property in our basic situation is simply due to degree reasons, while the analogous result in the general case (see chapter 3) will require much more work.

1.2 A resolution of $\mathbb{C}_X$

Let X be a complex space. We denote by $\mathbb{C}_X$ the constant sheaf on X. When X is smooth we denote by $\mathcal{E}^{\cdot}_X$ the De Rham complex of differential forms on X.

Let us consider a commutative diagram (1.1).

In the above situation one has

Proposition 1.1. *Let $U \subset X$ be an open neighborhood of a point $x \in E$.*

1. *Let $\eta_1, \eta_2 \in H^k\left(\pi^{-1}(U),\mathbb{C}\right)$ be two cohomology classes whose restrictions to $H^k\left(\pi^{-1}(U)\cap \tilde{E},\mathbb{C}\right)$ coincide. There exists an open neighborhood $V \subset U$ of x such that the restrictions of η_1, η_2 to $H^k\left(\pi^{-1}(V),\mathbb{C}\right)$ coincide.*

2. *Let $\theta \in H^k\left(\pi^{-1}(U)\cap \tilde{E},\mathbb{C}\right)$. There exists an open neighborhood $V \subset U$ of x and $\eta \in H^k\left(\pi^{-1}(V),\mathbb{C}\right)$ inducing $\theta|_{\pi^{-1}(V)\cap\tilde{E}}$.*

Proof. We treat the case $k \geq 1$ (the case $k = 0$ being trivial). Let us consider the diagram

$$\begin{array}{ccc} \pi^{-1}(U) \cap \tilde{E} & \xrightarrow{i} & \pi^{-1}(U) \\ {\scriptstyle q}\downarrow & & \downarrow{\scriptstyle \pi} \\ U \cap E & \xrightarrow{j} & U \end{array} \tag{1.3}$$

induced by the diagram (1.1). The Mayer-Vietoris sequence associated to (1.3) is

$$\begin{aligned} \cdots \longrightarrow H^k(U, \mathbb{C}) \xrightarrow{\Phi} H^k(U \cap E, \mathbb{C}) \oplus H^k\left(\pi^{-1}(U), \mathbb{C}\right) \xrightarrow{\Theta} \\ \longrightarrow H^k\left(\pi^{-1}(U) \cap \tilde{E}, \mathbb{C}\right) \xrightarrow{\Psi} H^{k+1}(U, \mathbb{C}) \longrightarrow \cdots \end{aligned} \tag{1.4}$$

Let us remember that if $\sigma \in H^k(U, \mathbb{C})$ (resp. $\alpha \in H^k(U \cap E, \mathbb{C})$) with $k \geq 1$, there is a smaller neighborhood V of x, such that $\sigma|_V = 0$ (resp. $\alpha|_{V \cap E} = 0$).

First we prove 1. By assumption we have $\Theta(0, \eta_1) = \Theta(0, \eta_2)$ hence by the exactness of the sequence (1.4) there is $\sigma \in H^k(U, \mathbb{C})$ such that $\eta_1 - \eta_2 = \Phi(\sigma)$. Since $\sigma|_V = 0$ on a smaller neighborhood V of x, it follows that the restrictions of η_1, η_2 to $H^k\left(\pi^{-1}(V), \mathbb{C}\right)$ coincide.

Next we prove 2. Let $\sigma = \Psi(\theta)$. Again $\sigma|_W = 0$ on a smaller neighborhood W of x so that by (1.4) there exists $(\alpha, \beta) \in H^k(W \cap E, \mathbb{C}) \oplus H^k\left(\pi^{-1}(W), \mathbb{C}\right)$ with $\Theta(\alpha, \beta) = \theta$. Let $V \subset W$ be an open neighborhood of x such that $\alpha|_{V \cap E} = 0$. It follows that $\theta|_{\pi^{-1}(V) \cap \tilde{E}}$ is the restriction of $\beta|_V$. □

From now on let us consider a quasi-smooth complex space X, that is, in the diagram (1.1) E and $\tilde{E}$ are smooth manifolds. In the case of a quasi-smooth space X it is easy to construct a complex $\Lambda^{\cdot}_X$ of "differential forms" on X which is a resolution of the constant sheaf $\mathbb{C}_X$. The complex $\Lambda^{\cdot}_X$ is defined by

$$\Lambda^{\cdot}_X = \pi_* \mathcal{E}^{\cdot}_{\tilde{X}} \oplus j_* \mathcal{E}^{\cdot}_E \oplus (j \circ q)_* \mathcal{E}^{\cdot}_{\tilde{E}}(-1)$$

More precisely, let U be an open set of X, $\tilde{U} = \pi^{-1}(U) \subset \tilde{X}$; then

$$\Lambda^k_X(U) = \mathcal{E}^k_{\tilde{X}}(\tilde{U}) \oplus \mathcal{E}^k_E(U \cap E) \oplus \mathcal{E}^{k-1}_{\tilde{E}}(\tilde{U} \cap \tilde{E})$$

that is, a k-differential form on U is nothing else that a triple of forms (ω, σ, θ) on $\tilde{U} \subset \tilde{X}$, $U \cap E \subset E$, $\tilde{U} \cap \tilde{E} \subset \tilde{E}$ respectively, such that ω and σ have the same degree k, while θ has degree $k - 1$. The definition makes sense because $\tilde{X}$, E and $\tilde{E}$ are smooth, so the notion of differential form on them is well known.

The differential is defined by

$$\begin{aligned} d\colon \Lambda^k_X &\to \Lambda^{k+1}_X \quad \text{with} \\ \Lambda^k_X &= \pi_*\mathcal{E}^k_{\tilde{X}} \oplus j_*\mathcal{E}^k_E \oplus (j\circ q)_*\,\mathcal{E}^{k-1}_{\tilde{E}}, \\ \Lambda^{k+1}_X &= \pi_*\mathcal{E}^{k+1}_{\tilde{X}} \oplus j_*\mathcal{E}^{k+1}_E \oplus (j\circ q)_*\,\mathcal{E}^{k}_{\tilde{E}}, \\ d(\omega,\sigma,\theta) &= \big(d\omega, d\sigma, d\theta + (-1)^k(i^*\omega - q^*\sigma)\big) \end{aligned} \tag{1.5}$$

where $i^*\colon \mathcal{E}^k_{\tilde{X}} \to \mathcal{E}^k_{\tilde{E}}$ and $q^*\colon \mathcal{E}^k_E \to \mathcal{E}^k_{\tilde{E}}$ are the usual pullback of forms.

A trivial computation shows that $d^2 = 0$ so that $\Lambda^{\cdot}_X$ is a complex.

We define the natural augmentation

$$\mathbb{C}_X \to \Lambda^0_X$$

$$c \mapsto (c, c, 0)$$

which will make $\Lambda^{\cdot}_X$ a resolution of $\mathbb{C}_X$.

Notation. From now on we will write for simplicity $\Lambda^{\cdot}_X = \mathcal{E}^{\cdot}_{\tilde{X}} \oplus \mathcal{E}^{\cdot}_E \oplus \mathcal{E}^{\cdot}_{\tilde{E}}(-1)$ instead of $\pi_*\mathcal{E}^{\cdot}_{\tilde{X}} \oplus j_*\mathcal{E}^{\cdot}_E \oplus (j\circ q)_*\,\mathcal{E}^{\cdot}_{\tilde{E}}(-1)$.

Theorem 1.1. *The complex $\Lambda^{\cdot}_X = \mathcal{E}^{\cdot}_{\tilde{X}} \oplus \mathcal{E}^{\cdot}_E \oplus \mathcal{E}^{\cdot}_{\tilde{E}}(-1)$ is a resolution of $\mathbb{C}_X$. Moreover each $\Lambda^k_X = \mathcal{E}^k_{\tilde{X}} \oplus \mathcal{E}^k_E \oplus \mathcal{E}^{k-1}_{\tilde{E}}$ is a fine sheaf.*

Proof. Let $x \in X$; if $x \notin E$, X is smooth in a neighborhood of x and $\Lambda^{\cdot}_X$ coincides near x with the De Rham complex $\mathcal{E}^{\cdot}_X$, and the conclusion follows from Poincaré lemma.

Let $x \in E$; let $(\omega,\sigma,\theta) \in \Lambda^k_X(U)$, i.e. $\omega \in \mathcal{E}^k_X\left(\pi^{-1}(U)\right)$, $\sigma \in \mathcal{E}^k_E\,(U \cap E)$, $\theta \in \mathcal{E}^{k-1}_{\tilde{E}}\left(\pi^{-1}(U) \cap \tilde{E}\right)$, with $\big(d\omega, d\sigma, d\theta + (-1)^k(i^*\omega - q^*\sigma)\big) = (0,0,0)$. Then $d\sigma = 0$ implies, by Poincaré lemma on E, that $\sigma = d\sigma'$ (after possibly shrinking U). Then we have

$$d\left(\theta - (-1)^k q^*\sigma'\right) = -(-1)^k i^*(\omega)$$

which implies, because $d\omega = 0$, that ω gives a class in $H^k\left(\pi^{-1}(U), \mathbb{C}\right)$ whose restriction to $H^k\left(\pi^{-1}(U) \cap \tilde{E}, \mathbb{C}\right)$ is zero; by proposition 1.1 (after shrinking U) we can write $\omega = d\omega'$ with $\omega' \in \mathcal{E}^{k-1}_{\tilde{X}}\left(\pi^{-1}(U)\right)$; it follows

$$d\left[\theta + (-1)^k i^*\omega' - (-1)^k q^*\sigma'\right] = 0$$

thus $\theta + (-1)^k i^*\omega' - (-1)^k q^*\sigma'$ gives a class in $H^{k-1}\left(\pi^{-1}(U) \cap \tilde{E}, \mathbb{C}\right)$. Again by proposition 1.1 (after possibly shrinking U once more) we can write

$$\theta + (-1)^k i^*\omega' - (-1)^k q^*\sigma' = (-1)^{k-1} i^*\omega'' + d\theta'$$

where $\omega'' \in \mathcal{E}^{k-1}_{\tilde{X}}\left(\pi^{-1}(U)\right)$, $d\omega'' = 0$, and $\theta' \in \mathcal{E}^{k-2}_{\tilde{E}}\left(\pi^{-1}(U) \cap \tilde{E}\right)$ (here we suppose of course $k \geq 2$: the case $k \leq 1$ needs minor modifications). As a consequence

$$(\omega, \sigma, \theta) = d(\omega' + \omega'', \sigma', \theta')$$

Finally, for every k , $\Lambda^k_X = \mathcal{E}^k_{\tilde{X}} \oplus \mathcal{E}^k_E \oplus \mathcal{E}^{k-1}_{\tilde{E}}$ is a fine sheaf; in fact it is a direct sum of direct images of fine sheaves. □

It follows from the above theorem that the cohomology of X can be described by means of sections of the complex $\Lambda^{\cdot}_X$:

$$H^k(X, \mathbb{C}) = \frac{\ker\left\{d\colon \Gamma\left(X, \Lambda^k_X\right) \to \Gamma\left(X, \Lambda^{k+1}_X\right)\right\}}{d\Gamma\left(X, \Lambda^{k-1}_X\right)} \tag{1.6}$$

Definition 1.1. The triple $(\tilde{X}, E, \tilde{E})$ is called the hypercovering of the quasi smooth space X corresponding to the diagram (1.1). We say that $\tilde{X}$ and E have rank 0, while $\tilde{E}$ has rank 1.

The above description of the cohomology by means of sections of the complex $\Lambda^{\cdot}_X$ is strictly related to the Mayer-Vietoris sequence associated to the diagram (1.1). In fact we define morphisms of complexes

$$\Psi\colon \theta \in \mathcal{E}^{k-1}_{\tilde{E}} \mapsto (0, 0, \theta) \in \Lambda^k_X$$
$$\Phi\colon\ (\omega, \sigma, \theta) \in \Lambda^k_X \mapsto (\omega, \sigma) \in \mathcal{E}^k_{\tilde{X}} \oplus \mathcal{E}^k_E$$
$$\Theta\colon\ (\omega, \sigma) \in \mathcal{E}^k_{\tilde{X}} \oplus \mathcal{E}^k_E \mapsto (-1)^k\left(i^*\omega - q^*\sigma\right) \in \mathcal{E}^k_{\tilde{E}}$$

The above morphisms commute with the respective differentials, so that they induce morphisms in cohomologies. Then

Theorem 1.2. *The morphisms Ψ, Φ, Θ induce in cohomology the Mayer-Vietoris sequence:*

$$\begin{aligned} \cdots \longrightarrow H^k(X, \mathbb{C}) \xrightarrow{\ \Phi\ } H^k(\tilde{X}, \mathbb{C}) \oplus H^k(E, \mathbb{C}) \xrightarrow{\ \Theta\ } \\ \longrightarrow H^k(\tilde{E}, \mathbb{C}) \xrightarrow{\ \Psi\ } H^{k+1}(X, \mathbb{C}) \longrightarrow \cdots \end{aligned} \tag{1.7}$$

Proof. The proof is trivial. □

Remark. The projection

$$(\omega, \sigma, \theta) \in \Lambda^k_X \mapsto \theta \in \mathcal{E}^{k-1}_{\tilde{E}}$$

does not commute to differentials.

1.3 The weight filtration W

Let M be a complex manifold. The *weight filtration* on the De Rham complex $\mathcal{E}^{\cdot}_M$ is the trivial increasing filtration

$$W_m\mathcal{E}^k_M = \mathcal{E}^k_M \quad \text{for } m \geq 0, \qquad W_m\mathcal{E}^k_M = 0 \quad \text{for } m < 0$$

Let X be a quasi-smooth space. We define an increasing *weight filtration* W_m on $\Lambda^{\cdot}_X = \mathcal{E}^{\cdot}_{\tilde{X}} \oplus \mathcal{E}^{\cdot}_E \oplus \mathcal{E}^{\cdot}_{\tilde{E}}(-1)$ by

$$W_m\Lambda^k_X = W_m\mathcal{E}^k_{\tilde{X}} \oplus W_m\mathcal{E}^k_E \oplus W_{m+1}\mathcal{E}^{k-1}_{\tilde{E}} \tag{1.8}$$

The filtration W_m has the following properties :

i. $W_m\Lambda^k_X = \Lambda^k_X \quad \text{for } m \geq 0.$

ii. $W_m\Lambda^k_X = 0 \quad \text{for } m < -2.$

iii. $(\Lambda^{\cdot}_X, d)$ is a filtered complex, namely

$$d\left(W_m\Lambda^k_X\right) \subset W_m\Lambda^{k+1}_X$$

The fact that d respects the filtration is a consequence of the formula (1.5) for d, and the fact that the pullback i^* and q^* preserve the filtration.

From the definition (1.8) we deduce immediately:

Lemma 1.1. *We have*

$$W_m\Lambda^k_X = \begin{cases} \Lambda^k_X & \text{if } m \geq 0 \\ (0) \oplus (0) \oplus \mathcal{E}^{k-1}_{\tilde{E}} & \text{if } m = -1 \\ 0 & \text{if } m \leq -2. \end{cases} \tag{1.9}$$

Hence the graded complex with respect the filtration W_m is given by

$$\frac{W_m\Lambda^k_X}{W_{m-1}\Lambda^k_X} = \begin{cases} \mathcal{E}^k_{\tilde{X}} \oplus \mathcal{E}^k_E & \text{if } m = 0 \\ (0) \oplus (0) \oplus \mathcal{E}^{k-1}_{\tilde{E}} & \text{if } m = -1 \\ 0 & \text{if } m \neq 0, m \neq -1. \end{cases} \tag{1.10}$$

The filtration W_m induces a filtration, still denoted W, on the complex of global sections:

$$W_m\Gamma\left(X, \Lambda^k_X\right) = \Gamma\left(X, W_m\Lambda^k_X\right) \tag{1.11}$$

Since the sheaves $\frac{W_m\Lambda^{\cdot}_X}{W_{m-1}\Lambda^{\cdot}_X}$ are fine, we have the natural isomorphism of complexes

$$\frac{\Gamma\left(X, W_m\Lambda^{\cdot}_X\right)}{\Gamma\left(X, W_{m-1}\Lambda^{\cdot}_X\right)} = \Gamma\left(X, \frac{W_m\Lambda^{\cdot}_X}{W_{m-1}\Lambda^{\cdot}_X}\right) \tag{1.12}$$

The filtration W passes to the the cohomology:

$$W_m H^k(X,\mathbb{C}) = \frac{\ker\left\{d\colon \Gamma\left(X, W_m\Lambda_X^k\right) \to \Gamma\left(X, W_m\Lambda_X^{k+1}\right)\right\}}{d\Gamma\left(X,\Lambda_X^{k-1}\right)\cap\Gamma\left(X, W_m\Lambda_X^k\right)} \tag{1.13}$$

Then

Theorem 1.3. *The morphisms Φ and Θ in the Mayer-Vietoris sequence* (1.7) *are morphisms of filtered spaces for the filtration W; moreover Φ is a strict morphism for the shifted filtrations $W_m H^k(\tilde{E},\mathbb{C}) \to W_{m-1}H^{k+1}(X,\mathbb{C})$.*

Proof. The proof is trivial. The morphisms Φ, Θ respect the filtration W because that they do so at level of complexes (as to Θ, we recall that the pull-back respect the degrees of the forms). The fact that Φ is a strict morphism for the shifted filtrations is obvious by the definition. □

As a consequence:

Theorem 1.4. *For X quasi-smooth:*

$$W_m H^k(X,\mathbb{C}) = H^k(X,\mathbb{C}) \quad \text{for } m \geq 0$$

$$W_{-1}H^k(X,\mathbb{C}) = \operatorname{im}\left\{\Psi\colon H^{k-1}(\tilde{E},\mathbb{C}) \to H^k(X,\mathbb{C})\right\} = \frac{H^{k-1}(\tilde{E},\mathbb{C})}{\operatorname{im}\Theta}$$

$$W_m H^k(X,\mathbb{C}) = 0 \quad \text{for } m \leq -2$$

while the graded cohomology is:

$$\frac{W_0 H^k(X,\mathbb{C})}{W_{-1}H^k(X,\mathbb{C})} = \frac{H^k(X,\mathbb{C})}{\operatorname{im}\Psi}$$

$$\frac{W_{-1} H^k(X,\mathbb{C})}{W_{-2}H^k(X,\mathbb{C})} = \frac{H^{k-1}(\tilde{E},\mathbb{C})}{\operatorname{im}\Theta}$$

and zero otherwise.

Proof. The classes of $H^k(X,\mathbb{C})$ which are in $W_{-1}H^k(X,\mathbb{C})$ have, by definition, a representative of the type $(0,0,\theta)$, where θ is a $(k-1)$-form on $\tilde{E}$. Hence it is clear that $W_{-1}H^k(X,\mathbb{C}) = \operatorname{im}\Psi = \frac{H^{k-1}(\tilde{E},\mathbb{C})}{\operatorname{im}\Theta}$. The rest of the statement follows from theorem 1.3. □

1.4 The spectral sequence of the filtration W

For the complex $\Gamma\left(X,\Lambda_X^{\cdot}\right)$ and its filtration W_m as in (1.11), we can construct the corresponding spectral sequence denoted by $E_r^{m,k}$ with first term

$$E_1^{m,k} = H^k\left(X, \frac{W_m\Lambda_X^{\cdot}}{W_{m-1}\Lambda_X^{\cdot}}\right) = H^k\left(\frac{\Gamma\left(X, W_m\Lambda_X^{\cdot}\right)}{\Gamma\left(X, W_{m-1}\Lambda_X^{\cdot}\right)}\right) \tag{1.14}$$

where the second equality follows from (1.12).

Then $E_1^{m,k}$ appears as the cohomology of degree k of the complex

$$\cdots \longrightarrow \frac{\Gamma\left(X, W_m \Lambda_X^k\right)}{\Gamma\left(X, W_{m-1} \Lambda_X^k\right)} \longrightarrow \frac{\Gamma\left(X, W_m \Lambda_X^{k+1}\right)}{\Gamma\left(X, W_{m-1} \Lambda_X^{k+1}\right)} \longrightarrow \cdots$$

whose differential is induced by d. By lemma 1.1, formula (1.10), the above complex is identically 0 for $m \neq 0, -1$, and $E_1^{0,k}$ is the cohomology of degree k of the complex

$$\cdots \longrightarrow \Gamma(\tilde{X}, \mathcal{E}_{\tilde{X}}^k) \oplus \Gamma(E, \mathcal{E}_E^k) \longrightarrow \Gamma(\tilde{X}, \mathcal{E}_{\tilde{X}}^{k+1}) \oplus \Gamma(E, \mathcal{E}_E^{k+1}) \longrightarrow \cdots \qquad (m=0)$$

where the differentials are the direct sums of the differentials on each term; while $E_1^{-1,k}$ is the cohomology of degree $k-1$ of the complex

$$\cdots \longrightarrow \Gamma(\tilde{E}, \mathcal{E}_{\tilde{E}}^{k-1}) \longrightarrow \Gamma(\tilde{E}, \mathcal{E}_{\tilde{E}}^k) \longrightarrow \cdots \qquad (m=-1)$$

where the differentials are the De Rham differentials on $\tilde{E}$. As a consequence:

$$E_1^{m,k} = \begin{cases} 0 \quad \text{if } m \neq 0, -1 \\ H^k(\tilde{X}, \mathbb{C}) \oplus H^k(E, \mathbb{C}) \quad \text{if } m = 0 \\ H^{k-1}(\tilde{E}, \mathbb{C}) \quad \text{if } m = -1 \end{cases} \tag{1.15}$$

Hence the only nonzero differential d_1 is

$$d_1 \colon E_1^{0,k} = H^k(\tilde{X}, \mathbb{C}) \oplus H^k(E, \mathbb{C}) \to E_1^{-1,k+1} = H^k(\tilde{E}, \mathbb{C}) \tag{1.16}$$

It is clear that

$$d_1([\omega], [\sigma]) = \left[(-1)^k (i^*\omega - q^*\sigma)\right] = \Theta([\omega], [\sigma]) \tag{1.17}$$

where Θ appears in the Mayer-Vietoris sequence (1.7).

The second term of the spectral sequence is

$$E_2^{m,k} = \frac{\ker\left\{ d_1 \colon E_1^{m,k} \to E_1^{m-1,k+1} \right\}}{\operatorname{im}\left\{ d_1 \colon E_1^{m+1,k-1} \to E_1^{m,k} \right\}}$$

hence from above we easily get

$$E_2^{m,k} = \begin{cases} 0 \quad \text{if } m \neq 0, -1 \\ \dfrac{W_0 H^k(X, \mathbb{C})}{W_{-1} H^k(X, \mathbb{C})} = \dfrac{H^k(X, \mathbb{C})}{\operatorname{im} \Psi} = \ker \Theta \quad \text{if } m = 0 \\ \dfrac{W_{-1} H^k(X, \mathbb{C})}{W_{-2} H^k(X, \mathbb{C})} = \dfrac{H^{k-1}(\tilde{E}, \mathbb{C})}{\operatorname{im} \Theta} \quad \text{if } m = -1 \end{cases} \tag{1.18}$$

where Ψ, Θ are morphisms in the Mayer-Vietoris sequence (1.7), so that

$$E_2^{m,k} = \frac{W_m H^k(X,\mathbb{C})}{W_{m-1} H^k(X,\mathbb{C})} \tag{1.19}$$

For degree reasons, $d_2 \colon E_2^{m,k} \to E_2^{m-2,k+1}$ is always zero. Still for degree reasons we have:

Theorem 1.5. *The differential*

$$d_r \colon E_r^{m,k} \to E_r^{m-r,k+1}$$

is identically zero for $r \geq 2$, that is, the spectral sequence degenerates at the level 2, in particular

$$E_r^{m,k} = E_2^{m,k} \quad \textit{for } r \geq 2.$$

1.5 The filtrations F^p and $\bar{F}^q$

In this section, we define the filtration by the types on the complexes $\Lambda_X^{\cdot}$.

We define a decreasing filtration (*the Hodge filtration, or the filtration by the types*) F^p on $\Lambda_X^{\cdot}$, by the formula

$$F^p\left(\Lambda_X^k\right) = F^p\left(\mathcal{E}_{\tilde{X}}^k\right) \oplus F^p\left(\mathcal{E}_E^k\right) \oplus F^p\left(\mathcal{E}_{\tilde{E}}^{k-1}\right) \tag{1.20}$$

where $F^p(\mathcal{E}_{\tilde{X}}^k)$, $F^p\left(\mathcal{E}_E^k\right)$, $F^p(\mathcal{E}_{\tilde{E}}^{k-1})$ are the standard Hodge filtrations for the usual De Rham complex on a manifold.

F^p defines a decreasing filtration and a filtered complex for d

$$\begin{gathered} \cdots \subset F^{p+1}\left(\Lambda_X^k\right) \subset F^p\left(\Lambda_X^k\right) \subset \cdots \\ d\left(F^p\left(\Lambda_X^k\right)\right) \subset F^p\left(\Lambda_X^{k+1}\right) \end{gathered} \tag{1.21}$$

The conjugate filtration $\bar{F}^q$ is defined as

$$\bar{F}^q \Lambda_X^k = \overline{F^q \Lambda_X^k} \tag{1.22}$$

The formula (1.5) for d proves that d preserves F^p, using the fact that i^* and q^* preserve F^p and that d preserves F^p in the standard De Rham case.

F^p, $\bar{F}^q$ induce filtrations on the complex $\Gamma\left(X, \Lambda_X^{\cdot}\right)$ and also filtrations on the cohomology $H^k(X, \mathbb{C})$.

1.6 Mixed Hodge structures on the cohomology and on the spectral sequence

In this section, we consider a quasi-smooth complex space X such that $\tilde{X}$, E and $\tilde{E}$ in the diagram (1.1) are *compact Kähler manifolds.*

Lemma 1.2. *The filtrations F^p, $\bar{F}^q$ induce on $E_1^{0,k}$, $E_1^{-1,k}$, a pure Hodge structure of weights k and $k-1$:*

$$E_1^{m,k} = \bigoplus_{p+q=k+m} \left(E_1^{m,k}(X)\right)^{p,q}$$

where

$$\left(E_1^{m,k}(X)\right)^{p,q} = \begin{cases} 0 & \text{if } m \neq 0, -1 \\ H^{p,q}(\tilde{X}) \bigoplus H^{p,q}(E) \quad (p+q=k) & \text{if } m = 0 \\ H^{p,q}(\tilde{E}) \quad (p+q=k-1) & \text{if } m=-1 \end{cases} \tag{1.23}$$

Moreover the (only non zero) differential $d_1 : E_1^{0,k} \to E_1^{-1,k+1}$ is a morphism of pure Hodge structures of weight k, and so is a strict morphism for the filtrations F^p (or $\bar{F}^q$).

Proof. The proof follows immediately from the formula (1.15) and the standard Hodge theory for the compact Kähler manifolds $\tilde{X}$, E and $\tilde{E}$. □

Theorem 1.6. *Under the assumptions that $\tilde{X}$, E and $\tilde{E}$ are compact Kähler manifolds, the cohomology spaces $H^k(X, \mathbb{C})$, provided with the weight filtration W shifted by $-k$, carry a mixed Hodge structure whose graded quotients are given by the formula* (1.18).

(The shift of W by $-k$ is needed to normalize (1.18); in the shifted filtration $W'_m = W_{m-k}$ the quotient $\frac{W'_m H^k(X,\mathbb{C})}{W'_{m-1} H^k(X,\mathbb{C})}$ has weight m, as expected).

The term $E_2^{m,k}$ of the spectral sequence associated to the filtration W_m on $\Gamma\left(X, \Lambda_X^{\cdot}\right)$ has a pure Hodge structure of weight $m+k$ for the filtrations induced by $E_1^{m,k}$:

$$E_2^{m,k} = \bigoplus_{p+q=k+m} \left(E_2^{m,k}\right)^{p,q}$$

where

$$\left(E_2^{m,k}\right)^{p,q} = \frac{\ker\left\{ d_1 : \left(E_1^{m,k}\right)^{p,q} \to \left(E_1^{m-1,k+1}\right)^{p,q} \right\}}{\operatorname{im}\left\{ d_1 : \left(E_1^{m+1,k-1}\right)^{p,q} \to \left(E_1^{m,k}\right)^{p,q} \right\}}$$

Since the spectral sequence degenerates at the level 2 (by theorem 1.5 or simply by (1.18)), we have

$$E_2^{m,k} = \frac{W_m H^k(X,\mathbb{C})}{W_{m-1}H^k(X,\mathbb{C})}$$

which implies that $H^k(X,\mathbb{C})$ has a mixed Hodge structure.
To be more precise, the morphism

$$\Theta = (-1)^k(i^* - q^*) : H^k(\tilde{X},\mathbb{C}) \oplus H^k(E,\mathbb{C}) \to H^k(\tilde{E},\mathbb{C})$$

respects the type of forms so that it induces morphisms

$$\Theta^{p,q} : H^{p,q}(\tilde{X}) \oplus H^{p,q}(E) \to H^{p,q}(\tilde{E})$$

and gives the decompositions

$$\ker\Theta = \bigoplus_{p+q=k} \ker\Theta^{p,q}$$

$$\operatorname{im}\Theta = \bigoplus_{p+q=k} \operatorname{im}\Theta^{p,q}$$

so that by (1.18) we obtain:

$$\frac{W_0 H^k(X,\mathbb{C})}{W_{-1}H^k(X,\mathbb{C})} = \bigoplus_{p+q=k} \ker\Theta^{p,q}$$

and

$$\frac{W_{-1} H^k(X,\mathbb{C})}{W_{-2}H^k(X,\mathbb{C})} = \bigoplus_{p+q=k-1} \frac{H^{p,q}(\tilde{E})}{\operatorname{im}\Theta^{p,q}}$$

Remark. We point out that the statement of theorem 1.6 is not yet complete. More precisely, the graded spaces $\frac{W_m H^k(X,\mathbb{C})}{W_{m-1}H^k(X,\mathbb{C})}$ are isomorphic to $E_2^{m,k}(X)$, which carry a pure Hodge structure of weight $m+k$ for the filtration F induced by $E_1^{m,k}(X)$ ($E_2^{m,k}(X)$ is a quotient of a subspace of $E_1^{m,k}(X)$) . On the other hand, the filtration F on $H^k(X,\mathbb{C})$ induces a filtration on the quotient $\frac{W_m H^k(X,\mathbb{C})}{W_{m-1}H^k(X,\mathbb{C})}$. We have not shown that the the two filtrations, under the isomorphism (1.19), coincide (and the same for $\bar{F}$). This will be made clear in the general case (chapter 3).

As it will be seen in chapter 3, the mixed Hodge structure in the above theorem 1.6 is unique.

1.7 Chains and homology

For any real analytic space M we denote by $\mathcal{S}_{M,k}$ the sheaf of germs of subanalytic k-chains on M (with coefficients in $\mathbb{C}$), and $\partial\colon \mathcal{S}_{M,k} \to \mathcal{S}_{M,k-1}$ the boundary operator.

Let X be a quasi-smooth complex space, with the corresponding diagram (1.1). We define *the sheaf of k-chains dual to* $\Lambda_X^k = \mathcal{E}_{\tilde{X}}^k \oplus \mathcal{E}_E^k \oplus \mathcal{E}_{\tilde{E}}^{k-1}$ as

$$\mathcal{C}_{X,k} = \pi_* \mathcal{S}_{\tilde{X},k} \oplus j_* \mathcal{S}_{E,k} \oplus (j \circ q)_* \mathcal{S}_{\tilde{E},k-1}$$

which we simply write

$$\mathcal{C}_{X,k} = \mathcal{S}_{\tilde{X},k} \oplus \mathcal{S}_{E,k} \oplus \mathcal{S}_{\tilde{E},k-1}$$

and the boundary operator as

$$\begin{gathered}\partial\colon \mathcal{C}_{X,k} \to \mathcal{C}_{X,k-1} \quad \text{with}\\ \mathcal{C}_{X,k} = \mathcal{S}_{\tilde{X},k} \oplus \mathcal{S}_{E,k} \oplus \mathcal{S}_{\tilde{E},k-1},\\ \mathcal{C}_{X,k-1} = \mathcal{S}_{\tilde{X},k-1} \oplus \mathcal{S}_{E,k-1} \oplus \mathcal{S}_{\tilde{E},k-2},\\ \partial(\alpha,\beta,\gamma) = \left(\partial\alpha + (-1)^k i_*\gamma, \partial\beta - (-1)^k q_*\gamma, \partial\gamma\right) \end{gathered} \tag{1.24}$$

where i_* and q_* are the push-down of chains. It is clear that $\partial\partial = 0$, so that we obtain a co-complex $(\mathcal{C}_{X,\cdot}, \partial)$.

For an open set U of X let $C_k(U) = \Gamma_c\left(U, \mathcal{C}_{X,k}\right)$ be the $\mathbb{C}$-module of the chains with compact support in U. Then

$$C_k(U) = S_{\tilde{X},k}\big(\tilde{U}\big) \oplus S_{E,k}\left(U \cap E\right) \oplus S_{\tilde{E},k-1}\big(\tilde{U} \cap \tilde{E}\big)$$

where $S_{\tilde{X},k}\big(\tilde{U}\big)$, $S_{E,k}\left(U \cap E\right)$, $S_{\tilde{E},k-1}\big(\tilde{U} \cap \tilde{E}\big)$ are the respective chains with compact support.

Theorem 1.7. *The homology of the complex of global sections $(C.(X), \partial)$ is the Borel-Moore homology with compact supports of X:*

$$H_k^c\left(X, \mathbb{C}\right) = \frac{\ker\left\{\partial\colon C_k(X) \to C_{k-1}(X)\right\}}{\partial C_{k+1}(X)} \tag{1.25}$$

Proof. The presheaf on X defined by $U \mapsto \hom_{\mathbb{C}}\left(C_k\left(U, \mathbb{C}\right)\right)$ is a sheaf, which we denote by $\mathcal{C}_X^k$. The boundaries ∂ give by transposition differentials $\delta\colon \mathcal{C}_X^k \to \mathcal{C}_X^{k+1}$ such that $\delta \circ \delta = 0$, hence $\left(\mathcal{C}_X^{\cdot}, \delta\right)$ is a complex. It is enough to prove that $\left(\mathcal{C}_X^{\cdot}, \delta\right)$ is a resolution of $\mathbb{C}_X$.

The sheaf $\mathcal{C}_X^k$ decomposes as

$$\mathcal{C}_X^k = \mathcal{S}_{\tilde{X}}^k \oplus \mathcal{S}_E^k \oplus \mathcal{S}_{\tilde{E}}^{k-1} \tag{1.26}$$

and by part I, chapter 7, $\mathcal{S}^{\cdot}_{\tilde{X}}$, $\mathcal{S}^{\cdot}_{E}$ and $\mathcal{S}^{\cdot}_{\tilde{E}}$ are resolutions of the constant sheaf on $\tilde{X}$, E and $\tilde{E}$ respectively.

The augmentation $\mathbb{C}_X \to \mathcal{C}^0_X$ is given by $c \mapsto (c, c, 0)$.

The duality between $(\omega, \sigma, \theta) \in \mathcal{C}^k_X$ and $(\alpha, \beta, \gamma) \in \mathcal{C}_{X,k}$ is given by

$$\langle(\omega, \sigma, \theta), (\alpha, \beta, \gamma)\rangle = \langle\omega, \alpha\rangle + \langle\sigma, \beta\rangle + \langle\theta, \gamma\rangle \tag{1.27}$$

and δ is defined by

$$\langle\delta(\omega, \sigma, \theta), (\alpha, \beta, \gamma)\rangle = \langle(\omega, \sigma, \theta), \partial(\alpha, \beta, \gamma)\rangle \tag{1.28}$$

By (1.24), (1.27), (1.28) and using the pullback-pushdown formula

$$\langle i^*\omega, \gamma\rangle = \langle\omega, i_*\gamma\rangle, \quad \langle q^*\sigma, \gamma\rangle = \langle\sigma, q_*\gamma\rangle$$

we easily obtain

$$\delta(\omega, \sigma, \theta) = (\delta\omega, \delta\sigma, \delta\theta + (-1)^k(i^*\omega - q^*\sigma) \tag{1.29}$$

The above formula looks exactly like the formula for d in (1.5). Hence, *mutatis mutandis*, the proof that $(\mathcal{C}^{\cdot}_X, \delta)$ is a resolution of $\mathbb{C}_X$ follows almost word by word the proof of theorem 1.1. □

1.8 Integration of forms on chains

There is a duality between forms in Λ^k_X and chains in $\mathcal{C}_{X,k}$, given by integration: for $(\alpha, \beta, \gamma) \in \mathcal{C}_{X,k}$ and $(\omega, \sigma, \theta) \in \Lambda^k_X$ we put

$$\int_{(\alpha,\beta,\gamma)} (\omega, \sigma, \theta) = \int_\alpha \omega + \int_\beta \sigma + \int_\gamma \theta \tag{1.30}$$

where the terms on the right hand side are the usual integrals of forms on subanalytic chains on the manifolds $\tilde{X}$, E, $\tilde{E}$.

Lemma 1.3 (Stokes theorem). *Let $\lambda \in \Gamma\left(X, \Lambda^k_X\right)$ and $\eta \in \Gamma\left(X, \mathcal{C}_{X,k}\right)$ such that λ, or η, has compact support. Then*

$$\int_\eta d\lambda = \int_{\partial\eta} \lambda \tag{1.31}$$

Proof. Let $\lambda = (\omega, \sigma, \theta)$, $\eta = (\alpha, \beta, \gamma)$; we must prove:

$$\int_{(\alpha,\beta,\gamma)} d(\omega, \sigma, \theta) = \int_{\partial(\alpha,\beta,\gamma)} (\omega, \sigma, \theta) \tag{1.32}$$

In fact

$$\int_{(\alpha,\beta,\gamma)} d(\omega,\sigma,\theta) = \int_{\alpha} d\omega + \int_{\beta} d\sigma + \int_{\gamma} \left(d\theta + (-1)^k \left(i^*\omega - q^*\sigma\right)\right)$$

By Stokes theorem for the manifolds $\tilde{X}$, E and $\tilde{E}$

$$\int_{\alpha} d\omega = \int_{\partial\alpha} \omega, \quad \int_{\beta} d\sigma = \int_{\partial\beta} \sigma, \quad \int_{\gamma} d\theta = \int_{\partial\gamma} \theta$$

on the other hand

$$\begin{aligned}\int_{\partial(\alpha,\beta,\gamma)} (\omega,\sigma,\theta) &= \int_{\partial\alpha} \omega + (-1)^k \int_{i_*\gamma} \omega + \int_{\partial\beta} \sigma - (-1)^k \int_{q_*\gamma} \sigma + \int_{\partial\gamma} \theta = \\ &= \int_{\partial\alpha} \omega + (-1)^k \int_{\gamma} i^*\omega + \int_{\partial\beta} \sigma - (-1)^k \int_{\gamma} q^*\sigma + \int_{\partial\gamma} \theta\end{aligned}$$

which implies (1.32). □

As a consequence we obtain the following analogous of one of the classical theorems of De Rham.

Theorem 1.8. *Let X be a quasi-smooth complex space. Then the integration formula* (1.30) *induces the natural duality between the cohomology $H^k(X,\mathbb{C})$ and the homology $H_k^c(X,\mathbb{C})$.*

Chapter 2

Differential forms on complex spaces

2.1 Introduction

By a complex space we mean a reduced, not necessarily irreducible, complex analytic space.

Let X be a complex space. We denote by $\mathbb{C}_X$ the constant sheaf on X. When X is smooth we denote by $\mathcal{E}^{\cdot}_X$ the De Rham complex of differential forms on X. Our goal is to define a complex $\Lambda^{\cdot}_X$ on X (in fact, as we shall see, a family of complexes) which replaces, in the singular case, the De Rham complex. Let us consider a resolution of singularities of X, i.e. a commutative diagram:

$$\begin{array}{ccc} \tilde{E} & \xrightarrow{\ i\ } & \tilde{X} \\ {\scriptstyle q}\downarrow & & \downarrow{\scriptstyle \pi} \\ E & \xrightarrow{\ j\ } & X \end{array} \tag{2.1}$$

where $E \subset X$ is a nowhere dense closed subspace, containing the singularities of X, $j\colon E \to X$ is the natural inclusion, $\tilde{X}$ is a smooth manifold and π is a proper modification inducing an isomorphism $\tilde{X} \setminus \tilde{E} \simeq X \setminus E$.

In chapter 1 we have treated the case where E and $\tilde{E}$ are smooth, so that on them the De Rham complexes $\mathcal{E}^{\cdot}_E$ and $\mathcal{E}^{\cdot}_{\tilde{E}}$ are available, and we have defined a section of $\Lambda^{\cdot}_X$ as a triple of forms: a p-form on on $\tilde{X}$, a p-form on E and a $(p-1)$-form on $\tilde{E}$. In the general case it is then natural to proceed as follows. First we define the complexes $\Lambda^{\cdot}_E$ on E, and $\Lambda^{\cdot}_{\tilde{E}}$ on $\tilde{E}$ by induction on the dimension of the spaces (the dimensions of E and $\tilde{E}$ are strictly less than $\dim X$). Then we define a section of Λ^p_X as a triple (ω, σ, θ) where ω is a section of $\mathcal{E}^p_{\tilde{X}}$, σ a section of Λ^p_E, and θ a section of $\Lambda^{p-1}_{\tilde{E}}$.

The next step would be to define the differential $d(\omega, \sigma, \theta)$, mimicking the definition in formula (1.5) of chapter 1. We run immediately into the first difficulty, since we need to be able *to pull back a section $\omega \in \mathcal{E}^p_{\tilde{X}}$ (resp. $\sigma \in \Lambda^p_E$) to $i^*\omega \in \Lambda^p_{\tilde{E}}$ (resp. $q^*\sigma \in \Lambda^p_{\tilde{E}}$), such that d commutes to the pullback.*

In general the above requirement cannot be fulfilled. The complexes $\Lambda^{\cdot}_E$ and $\Lambda^{\cdot}_{\tilde{E}}$ have been constructed from a desingularization E_1 of E, and, respectively,

$\tilde{E}_1$ of $\tilde{E}$, hence there are no *a priori* relationships among $\tilde{X}$, E_1 and $\tilde{E}_1$.

On the other hand, the diagram (2.1) is not unique, as well as the analogous diagrams for E and $\tilde{E}$. Therefore we are led naturally to introduce, for every space X, a family $\mathcal{R}(X)$ of complexes $\Lambda^{\cdot}_X$ instead of a single one, so that, once a diagram (2.1) has been fixed, we can choose suitable complexes $\Lambda^{\cdot}_E \in \mathcal{R}(E)$ and $\Lambda^{\cdot}_{\tilde{E}} \in \mathcal{R}(\tilde{E})$ such that the pullback $\mathcal{E}^{\cdot}_{\tilde{X}} \to \Lambda_{\tilde{E}}$ and $\Lambda^{\cdot}_E \to \Lambda^{\cdot}_{\tilde{E}}$ exist and commute to differentials. In the end we can define $\Lambda^{\cdot}_X$ together with $d\colon \Lambda^p_X \to \Lambda^{p+1}_X$ and prove that it is a resolution of $\mathbb{C}_X$.

Because we are working by induction on $\dim X$, we must build a consistent set of properties, definitions, constructions, existence and uniqueness theorems for spaces and mappings, concerning the families $\mathcal{R}(X)$ of admissible complexes $\Lambda^{\cdot}_X$, and the families $\mathcal{R}(f)$ of admissible pullback $\Lambda^{\cdot}_Y \to \Lambda^{\cdot}_X$ for maps $f\colon X \to Y$.

It is important to remark that the definition of the complexes $\Lambda^{\cdot}_X$ and the definition of the pullback $\Lambda^{\cdot}_Y \to \Lambda^{\cdot}_X$ must be given step by step, during the induction procedure.

Let us be more precise. For every X we define a family of complexes $\mathcal{R}(X) = \{\Lambda^{\cdot}_X\}$ and for every morphism $f\colon X \to Y$ a family $\mathcal{R}(f)$ of morphisms of complexes between the $\Lambda^{\cdot}_Y \in \mathcal{R}(Y)$ and some of the $\Lambda^{\cdot}_X \in \mathcal{R}(X)$, more precisely morphisms $\Lambda^{\cdot}_Y \to f_*\Lambda^{\cdot}_X$ which we simply denote $\Lambda^{\cdot}_Y \to \Lambda^{\cdot}_X$ and call *(admissible) pullback* with the following properties.

(I) $\Lambda^{\cdot}_X$ is a fine resolution of $\mathbb{C}_X$.

(II) For $p > 2\dim X$, $\Lambda^p_X = 0$.

(III) If X is smooth, the ordinary De Rham complex $\mathcal{E}^{\cdot}_X$ belongs to $\mathcal{R}(X)$, and for every morphism $f\colon X \to Y$ between smooth complex manifolds the ordinary De Rham pullback $f^*\colon \mathcal{E}^{\cdot}_Y \to f_*\mathcal{E}^{\cdot}_X$ is an admissible pullback.

(IV) There exists a smooth, open, dense analytic subset $U \subset X$ such that the restriction $\Lambda^{\cdot}_X|_U$ is the ordinary De Rham complex $\mathcal{E}^{\cdot}_U$. Here analytic means that the complement of U in X is an analytic subspace of X.

The family of pullback will satisfy the following properties.

(C) (Composition). Let $g\colon Z \to X$, $f\colon X \to Y$ be two morphisms, $\alpha\colon \Lambda^{\cdot}_Y \to \Lambda^{\cdot}_X$, $\beta\colon \Lambda^{\cdot}_X \to \Lambda^{\cdot}_Z$ two pullback; then the composition $\beta \circ \alpha\colon \Lambda^{\cdot}_Y \to \Lambda^{\cdot}_Z$ is again a pullback.

For future induction procedure we denote by $(\mathrm{C})_{k,m,n}$ the property (C) when $\dim Z \leq k$, $\dim X \leq m$, $\dim Y \leq n$.

(EP) (Existence of pullback). Let $f\colon X \to Y$ be a morphism, and fix $\Lambda^{\cdot}_Y \in \mathcal{R}(Y)$; then there exists a $\Lambda^{\cdot}_X \in \mathcal{R}(X)$ and a pullback $\Lambda^{\cdot}_Y \to \Lambda^{\cdot}_X$.

As above we denote by $(\mathrm{EP})_{m,n}$ the property (EP) when $\dim X \leq m$, $\dim Y \leq n$.

(U) (Uniqueness of pullback). Let $f\colon X \to Y$ be a morphism, and $\alpha\colon \Lambda^{\cdot}_Y \to \Lambda^{\cdot}_X$, $\beta\colon \Lambda^{\cdot}_Y \to \Lambda^{\cdot}_X$ two pullback corresponding to f; then $\alpha = \beta$.

We denote by $(\mathrm{U})_{m,n}$ the property (U) when $\dim X \leq m$, $\dim Y \leq n$.

(F) (Filtering). If $\Lambda^{\cdot,1}_X$, $\Lambda^{\cdot,2}_X \in \mathcal{R}(X)$, there exists a third $\Lambda^{\cdot}_X \in \mathcal{R}(X)$ and two pullback $\Lambda^{\cdot,1}_X \to \Lambda^{\cdot}_X$, $\Lambda^{\cdot,2}_X \to \Lambda^{\cdot}_X$ corresponding to the identity.

We denote by $(\mathrm{F})_m$ the property (F) when $\dim X \leq m$.

As noted above, since the definition of the complexes $\Lambda^{\cdot}_X$ uses the definition and the existence of the pullback, we shall be forced to construct both at the same time. This will be done by a recursion on the dimensions of the complex spaces involved.

2.2 Definitions and statements

2.2.1 Definition of the family $\mathcal{R}(X)$

We denote by $\Lambda^{\cdot}_X(-1)$ the complex obtained by shifting the degree in $\Lambda^{\cdot}_X$: more precisely $\Lambda^{p}_X(-1) = \Lambda^{p-1}_X$.

We define the family $\mathcal{R}(X)$ by induction on $n = \dim X$. If $\dim X = 0$, $\mathcal{R}(X)$ contains only the complex $\mathbb{C}^{\cdot}_X$ with $\mathbb{C}^0_X = \mathbb{C}_X$ and $\mathbb{C}^p_X = 0$ for $p > 0$. We suppose $\mathcal{R}(Y)$ to be known for complex spaces Y of dimension $< n$; then:

Definition 2.1 ($(\mathrm{D})_n$)**.** Let X be a complex space of dimension n. An element $\Lambda^{\cdot}_X \in \mathcal{R}(X)$ is the assignment of the following data:

i. a nowhere dense closed subspace $E \subset X$, $\mathrm{sing}\,(X) \subset E$ and a proper modification

$$\begin{array}{ccc} \tilde{E} & \xrightarrow{\;i\;} & \tilde{X} \\ {\scriptstyle q}\downarrow & & \downarrow{\scriptstyle \pi} \\ E & \xrightarrow{\;j\;} & X \end{array}$$

where $j\colon E \to X$ is the natural inclusion, $\tilde{X}$ is a smooth manifold, $\tilde{E} = \pi^{-1}(E)$ and π induces an isomorphism $\tilde{X} \setminus \tilde{E} \simeq X \setminus E$;

ii. there exist $\Lambda^{\cdot}_E \in \mathcal{R}(E)$, $\Lambda^{\cdot}_{\tilde{E}} \in \mathcal{R}(\tilde{E})$, and two pullback

$$\phi\colon \Lambda^{\cdot}_E \to \Lambda^{\cdot}_{\tilde{E}}$$
$$\psi\colon \mathcal{E}^{\cdot}_{\tilde{X}} \to \Lambda^{\cdot}_{\tilde{E}}$$

(corresponding respectively to q and i);

iii. the complex $\Lambda^{\cdot}_X$ is defined by

$$\Lambda^{\cdot}_X = \pi_* \mathcal{E}^{\cdot}_{\tilde{X}} \oplus j_* \Lambda^{\cdot}_E \oplus (j \circ q)_* \Lambda^{\cdot}_{\tilde{E}}(-1) \tag{2.2}$$

with differential given by

$$\begin{aligned} d\colon \Lambda^p_X &\to \Lambda^{p+1}_X \quad \text{with} \\ \Lambda^p_X &= \pi_* \mathcal{E}^p_{\tilde{X}} \oplus j_* \Lambda^p_E \oplus (j \circ q)_* \Lambda^{p-1}_{\tilde{E}}, \\ \Lambda^{p+1}_X &= \pi_* \mathcal{E}^{p+1}_{\tilde{X}} \oplus j_* \Lambda^{p+1}_E \oplus (j \circ q)_* \Lambda^p_{\tilde{E}}, \\ d(\omega, \sigma, \theta) &= \big(d\omega, d\sigma, d\theta + (-1)^p(\psi(\omega) - \phi(\sigma))\big) \end{aligned} \tag{2.3}$$

iv. the augmentation

$$\begin{aligned} \mathbb{C}_X &\to \Lambda^0_X \\ c &\mapsto (c, c, 0) \end{aligned}$$

makes $\Lambda^{\cdot}_X$ a resolution of $\mathbb{C}_X$;

v. there is a uniquely determined family $(X_l, h_l)_{l \in L}$ of smooth manifolds X_l and proper maps $h_l\colon X_l \to X$ such that

$$\Lambda^p_X = \bigoplus_l h_{l*} \mathcal{E}^{p-q(l)}_{X_l}$$

where $q(l)$ is a nonnegative integer; and there exist mappings $h_{lm}\colon X_l \to X_m$, commuting with h_l and h_m, such that the differential $\Lambda^p_X \to \Lambda^{p+1}_X$ is given by

$$d\left(\oplus_l \omega_l\right) = \left(\oplus_l \left[d\omega_l + \sum_m \epsilon^{(p)}_{lm} h^*_{lm} \omega_m\right]\right)$$

where $\epsilon^{(p)}_{lm}$ can take the values $0, \pm 1$.

The pullback $\phi\colon \Lambda^{\cdot}_E \to \Lambda^{\cdot}_{\tilde{E}}$ and $\psi\colon \mathcal{E}^{\cdot}_{\tilde{X}} \to \Lambda^{\cdot}_{\tilde{E}}$ in (ii) are called inner pullback of the complex $\Lambda^{\cdot}_X$.

The family $(X_l, h_l)_{l \in L}$ will be called the hypercovering of X associated to $\Lambda^{\cdot}_X$, and $q(l) = q_X(l)$ will be the rank of X_l.

Explanation 2.1. In order to understand the above definition, we must assume that we already know, by induction:

- the family of the $\Lambda^{\cdot}_Y \in \mathcal{R}(Y)$ for any space Y with $\dim Y < \dim X$ (so that we understand $\Lambda^{\cdot}_E$ and $\Lambda^{\cdot}_{\tilde{E}}$).
- For every morphism $f\colon Y \to Z$ with $\dim Y < \dim X$, $\dim Z < \dim X$ the notion of pullback $\Lambda^{\cdot}_Z \to \Lambda^{\cdot}_Y$ in $\mathcal{R}(f)$; if moreover Y is smooth of dimension $\leq \dim X$, and $\dim Z < \dim X$, the notion of pullback $\Lambda^{\cdot}_Z \to \mathcal{E}^{\cdot}_Y$.

- If $U \subset X$ is an open set, and $\tilde{U} = \pi^{-1}(U)$, an element of $\Lambda^p_X(U)$ is a triple of forms (ω, σ, θ) with $\omega \in \Gamma(\tilde{U}, \mathcal{E}^p_{\tilde{X}})$, $\sigma \in \Gamma(U \cap E, \Lambda^p_E)$, $\theta \in \Gamma(\tilde{U} \cap \tilde{E}, \Lambda^{p-1}_{\tilde{E}})$, and the differential is given by the formula (2.3).

- Also the hypercovering $(X_l)_{l \in L}$ of $\Lambda^{\cdot}_X$ comes by induction. If $(E_m)_{m \in M}$ is the hypercovering of $\Lambda^{\cdot}_E$ and $(\tilde{E}_s)_{s \in S}$ is the hypercovering of $\Lambda^{\cdot}_{\tilde{E}}$, (X_l) is the family $\left(\tilde{X}, E_m, \tilde{E}_s\right)_{m \in M, s \in S}$. The rank of $\tilde{X}$ is 0, the rank of E_m in X is the same as its rank in E; but the rank of $\tilde{E}_s$ in X is given by

$$q_X(s) = q_{\tilde{E}}(s) + 1$$

i.e. $\tilde{E}$ plays a 1-shifted role in $\Lambda^{\cdot}_X$.

To simplify the notations, in the sequel we will write $\Lambda^p_X = \bigoplus_l \mathcal{E}^{p-q(l)}_{X_l}$ instead of $\bigoplus_l h_{l*} \mathcal{E}^{p-q(l)}_{X_l}$.

2.2.2 Construction-existence theorem

Theorem 2.1 ($(\mathrm{E})_n$). *Let X be a complex space of dimension $\leq n$. Let $E \subset X$ be a nowhere dense closed subspace with* $\operatorname{sing}(X) \subset E$, *$j \colon E \to X$ the natural inclusion, and*

$$\begin{array}{ccc} \tilde{E} & \xrightarrow{i} & \tilde{X} \\ {\scriptstyle q}\downarrow & & \downarrow{\scriptstyle \pi} \\ E & \xrightarrow{j} & X \end{array}$$

be a proper desingularization. Let $\Lambda^{\cdot}_E \in \mathcal{R}(E)$. There exists $\Lambda^{\cdot}_{\tilde{E}} \in \mathcal{R}(\tilde{E})$, a pullback $\phi \colon \Lambda^{\cdot}_E \to \Lambda^{\cdot}_{\tilde{E}}$ (corresponding to q), a pullback $\psi \colon \mathcal{E}^{\cdot}_{\tilde{X}} \to \Lambda^{\cdot}_{\tilde{E}}$ (corresponding to i) with the following property: the complex

$$\Lambda^{\cdot}_X = \pi_* \mathcal{E}^{\cdot}_{\tilde{X}} \oplus j_* \Lambda^{\cdot}_E \oplus (j \circ q)_* \Lambda^{\cdot}_{\tilde{E}}(-1)$$

whose differential is by definition

$$\begin{aligned} d \colon \Lambda^p_X &\to \Lambda^{p+1}_X \quad \textit{with} \\ \Lambda^p_X &= \pi_* \mathcal{E}^p_{\tilde{X}} \oplus j_* \Lambda^p_E \oplus (j \circ q)_* \Lambda^{p-1}_{\tilde{E}}, \\ \Lambda^{p+1}_X &= \pi_* \mathcal{E}^{p+1}_{\tilde{X}} \oplus j_* \Lambda^{p+1}_E \oplus (j \circ q)_* \Lambda^p_{\tilde{E}}, \\ d(\omega, \sigma, \theta) &= \big(d\omega, d\sigma, d\theta + (-1)^p (\psi(\omega) - \phi(\sigma))\big) \end{aligned} \tag{2.4}$$

is a fine resolution of $\mathbb{C}_X$.

Notation. Throughout all the book we will write for simplicity $\mathcal{E}^{\cdot}_{\tilde{X}} \oplus \Lambda^{\cdot}_E \oplus \Lambda^{\cdot}_{\tilde{E}}(-1)$ instead of $\pi_* \mathcal{E}^{\cdot}_{\tilde{X}} \oplus j_* \Lambda^{\cdot}_E \oplus (j \circ q)_* \Lambda^{\cdot}_{\tilde{E}}(-1)$.

2.2.3 Definition of a primary pullback for irreducible spaces

Let $f\colon X \to Y$ be a morphism of irreducible complex spaces, $\dim X \leq m$, $\dim Y \leq n$, $\Lambda^p_X = \mathcal{E}^p_{\tilde{X}} \oplus \Lambda^p_E \oplus \Lambda^{p-1}_{\tilde{E}}$, $\Lambda^p_Y = \mathcal{E}^p_{\tilde{Y}} \oplus \Lambda^p_F \oplus \Lambda^{p-1}_{\tilde{F}}$, with $\Lambda^{\cdot}_X \in \mathcal{R}(X)$ and $\Lambda^{\cdot}_Y \in \mathcal{R}(Y)$. Let us consider the corresponding diagrams

$$\begin{array}{ccc} \tilde{E} & \longrightarrow & \tilde{X} \\ \downarrow & & \downarrow \\ E & \xrightarrow{j} & X \end{array} \qquad \begin{array}{ccc} \tilde{F} & \longrightarrow & \tilde{Y} \\ \downarrow & & \downarrow \\ F & \xrightarrow{k} & Y \end{array}$$

In order to define a primary pullback $\phi\colon \Lambda^{\cdot}_Y \to \Lambda^{\cdot}_X$ we proceed by double induction on (m, n), i.e. $(\mathrm{DP})_{m,n-1}$ and $(\mathrm{DP})_{m-1,n} \implies (\mathrm{DPP})_{m,n}$ (for $(\mathrm{DP})_{m,n}$ see the definition 2.4 below).

Definition 2.2 ($(\mathrm{DPP})_{m,n}$). We say that ϕ is a primary pullback (corresponding to f) if it satisfies the following properties:

P_0) ϕ is a morphism of complexes, i.e. it commutes with differentials.

P_1) Let $(X_l, h_l)_{l\in L}$, $(Y_s, g_s)_{s\in S}$ be the hypercoverings associated to $\Lambda^{\cdot}_X, \Lambda^{\cdot}_Y$, i.e.

$$\Lambda^p_X = \bigoplus_l \mathcal{E}^{p-q(l)}_{X_l}, \quad \Lambda^p_Y = \bigoplus_s \mathcal{E}^{p-q(s)}_{Y_s}$$

For every X_l there exist at most one Y_s, having the same rank q as X_l, and a commutative diagram

$$\begin{array}{ccc} X_l & \xrightarrow{f_{ls}} & Y_s \\ {\scriptstyle h_l}\downarrow & & \downarrow{\scriptstyle g_s} \\ X & \xrightarrow{f} & Y \end{array}$$

such that the composition

$$\mathcal{E}^{p-q}_{Y_s} \longrightarrow \Lambda^p_Y \xrightarrow{\phi} \Lambda^p_X \longrightarrow \mathcal{E}^{p-q}_{X_l}$$

is either identically zero for every p, or coincides with the De Rham pullback f^*_{ls}.

P_2) Let $\alpha\colon \mathcal{E}^{\cdot}_{\tilde{Y}} \to \mathcal{E}^{\cdot}_{\tilde{X}}$ be induced by ϕ. Then $\alpha \equiv 0$ if and only if $f(X) \subset F$; moreover in this case ϕ is the composition $\Lambda^{\cdot}_Y \to \Lambda^{\cdot}_F \to \Lambda^{\cdot}_X$ where $\Lambda^{\cdot}_Y \to \Lambda^{\cdot}_F$ is the projection onto the summand $\Lambda^{\cdot}_F$ and $\Lambda^{\cdot}_F \to \Lambda^{\cdot}_X$ is a pullback (inductively defined) corresponding to the induced morphism $X \to F$.

P_3) If $\alpha\colon \mathcal{E}^{\cdot}_{\tilde{Y}} \to \mathcal{E}^{\cdot}_{\tilde{X}}$ is not identically zero, then, according to P_2, $f(X) \not\subset F$; in that case we assume the following properties:

i. $f^{-1}(F) \subset E$;

ii. the morphism f extends to a morphism $\tilde{f} \colon \tilde{X} \to \tilde{Y}$ and $\alpha = \tilde{f}^*$ is the ordinary De Rham pullback;

iii. the morphism ϕ is given by

$$\begin{array}{ccccc} \mathcal{E}^{\cdot}_{\tilde{Y}} & \oplus & \Lambda^{\cdot}_{F} & \oplus & \Lambda^{\cdot}_{\tilde{F}}(-1) \\ \downarrow \alpha & \searrow \beta & \downarrow \gamma & \swarrow \delta & \downarrow \epsilon \\ \mathcal{E}^{\cdot}_{\tilde{X}} & \oplus & \Lambda^{\cdot}_{E} & \oplus & \Lambda^{\cdot}_{\tilde{E}}(-1) \end{array}$$

where $(\beta, \gamma, \delta) \colon \Lambda^{\cdot}_{Y} \to \Lambda^{\cdot}_{E}$ is a pullback corresponding to the composition $f \circ j \colon E \to Y$ (inductively defined).

Explanation 2.2. In the above definition, we must assume that we already know, by induction:

- the family of the $\Lambda^{\cdot}_{Y} \in \mathcal{R}(Y)$ for any space Y with $\dim Y \leq \dim X$;
- for every morphism $f \colon T \to Z$ with $\dim T \leq \dim X$, $\dim Z < \dim X$ the notion of pullback $\Lambda^{\cdot}_{Z} \to \Lambda^{\cdot}_{T}$;
- for every morphism $f \colon T \to Z$ with $\dim T < \dim X$, $\dim Z \leq \dim X$, the notion of pullback $\Lambda^{\cdot}_{Z} \to \Lambda^{\cdot}_{T}$.

Remark. Note that the property i): $f^{-1}(F) \subset E$, in the situation P_3, cannot be forgotten. As an important example let us consider a complex $\Lambda^{\cdot}_{X} = \mathcal{E}^{\cdot}_{\tilde{X}} \oplus \Lambda^{\cdot}_{E} \oplus \Lambda^{\cdot}_{\tilde{E}}(-1)$ as in the definition 2.1. The projection $\Lambda^{\cdot}_{X} \to \mathcal{E}^{\cdot}_{\tilde{X}}$ **is not a pullback** corresponding to the map $\pi \colon \tilde{X} \to X$, because $\pi^{-1}(E) \not\subset \emptyset$. Here $\mathcal{E}^{\cdot}_{\tilde{X}}$ must be understood as the complex attached to the trivial diagram

$$\begin{array}{ccc} \emptyset & \longrightarrow & \tilde{X} \\ \downarrow & & \downarrow \\ \emptyset & \longrightarrow & \tilde{X} \end{array}$$

2.2.4 Definition of a pullback morphism: the general case

Let X be a complex space of dimension m.

Let $X = \bigcup_i X_i$ be the decomposition of X into its irreducible components. Let $E \subset X$ be a closed subspace such that $X \setminus E$ is smooth and dense in X, and $f \colon (\tilde{X}, \tilde{E}) \to (X, E)$ be a proper desingularization. Then: $\tilde{X} = \bigsqcup_i \tilde{X}_i$, $\tilde{E} = \bigsqcup_i \tilde{E}_i$, $\tilde{E}_i = \tilde{E} \cap \tilde{X}_i$ (here $\bigsqcup$ denotes disjoint union); moreover, $f|_{\tilde{X}_i} \colon (\tilde{X}_i, \tilde{E}_i) \to (X_i, E_i)$ is a desingularization. Let $\Lambda^{\cdot}_{X} = \mathcal{E}^{\cdot}_{\tilde{X}} \oplus \Lambda^{\cdot}_{E} \oplus \Lambda^{\cdot}_{\tilde{E}}(-1) \in \mathcal{R}(X)$; let us denote by $\phi \colon \Lambda^{\cdot}_{E} \to \Lambda^{\cdot}_{\tilde{E}}$, $\psi \colon \mathcal{E}^{\cdot}_{\tilde{X}} \to \Lambda^{\cdot}_{\tilde{E}}$ the inner pullback of $\Lambda^{\cdot}_{X}$ (see definition 2.1 $(\mathrm{D})_m$).

Then $\mathcal{E}^{\cdot}_{\tilde{X}} = \bigoplus_i \mathcal{E}^{\cdot}_{\tilde{X}_i}$, $\Lambda^{\cdot}_{\tilde{E}} = \bigoplus_i \Lambda^{\cdot}_{\tilde{E}_i}$.

Let us denote by $p_i : \mathcal{E}^{\cdot}_{\tilde{X}} \to \mathcal{E}^{\cdot}_{\tilde{X}_i}$, $q_i : \Lambda^{\cdot}_{\tilde{E}} \to \Lambda^{\cdot}_{\tilde{E}_i}$ the projections.

Definition 2.3. A morphism of complexes $\zeta\colon \Lambda^{\cdot}_X = \mathcal{E}^{\cdot}_{\tilde{X}} \oplus \Lambda^{\cdot}_E \oplus \Lambda^{\cdot}_{\tilde{E}}(-1) \to \Lambda^{\cdot}_{X_i}$ corresponding to the inclusion $X_i \to X$ is called a natural pullback if $\Lambda^{\cdot}_{X_i} = \mathcal{E}^{\cdot}_{\tilde{X}_i} \oplus \Lambda^{\cdot}_{E_i} \oplus \Lambda^{\cdot,1}_{\tilde{E}_i}$ with $\Lambda^{\cdot}_{E_i} \in \mathcal{R}(E_i)$, $\Lambda^{\cdot,1}_{\tilde{E}_i} \in \mathcal{R}(\tilde{E}_i)$ and there exists a commutative diagram of pullback

$$
\begin{array}{ccccc}
 & & \Lambda^{\cdot}_E & \xrightarrow{\eta_i} & \Lambda^{\cdot}_{E_i} \\
 & & \downarrow{\scriptstyle \phi} & & \\
\mathcal{E}^{\cdot}_{\tilde{X}} & \xrightarrow{\psi} & \Lambda^{\cdot}_{\tilde{E}} & & \downarrow{\scriptstyle \phi_i} \\
\downarrow{\scriptstyle p_i} & & \downarrow{\scriptstyle q_i} & & \\
\mathcal{E}^{\cdot}_{\tilde{X}_i} & \xrightarrow{\psi_i} & \Lambda^{\cdot}_{\tilde{E}_i} & \xrightarrow{\mu_i} & \Lambda^{\cdot,1}_{\tilde{E}_i}
\end{array}
\tag{2.5}
$$

such that

$$\zeta(\omega, \sigma, \theta) = (p_i(\omega), \eta_i(\sigma), \mu_i(q_i(\theta))) \tag{2.6}$$

Let $f\colon X \to Y$ be a morphism between (reducible) complex spaces, $X = \bigcup_j X_j$, $Y = \bigcup_k Y_k$ be the respective decompositions into irreducible components. Let $\Lambda^{\cdot}_X = \mathcal{E}^{\cdot}_{\tilde{X}} \oplus \Lambda^{\cdot}_E \oplus \Lambda^{\cdot}_{\tilde{E}}(-1)$, $\Lambda^{\cdot}_Y = \mathcal{E}^{\cdot}_{\tilde{Y}} \oplus \Lambda^{\cdot}_F \oplus \Lambda^{\cdot}_{\tilde{F}}(-1)$.

For a given j two cases can occur:

(a) $f(X_j) \subset F$;

(b) $f(X_j) \not\subset F$, and there exists a unique Y_k with $f(X_j) \subset Y_k$ (because F contains, by definition, the singularities of Y).

Definition 2.4 ($(\mathrm{DP})_{m,n}$). A morphism of complexes $\phi\colon \Lambda^{\cdot}_Y \to \Lambda^{\cdot}_X$ is called a pullback corresponding to f if

- it satisfies P_0 and P_1 in definition $(\mathrm{DPP})_{m,n}$ (definition 2.2);

- (P_4) the composition $\Lambda^{\cdot}_Y \to \Lambda^{\cdot}_X \to \Lambda^{\cdot}_E$ (where the second morphism is the projection onto the summand) is a pullback corresponding to the composition $f \circ j\colon E \to Y$ (inductively defined);

- for every component X_j of X there exists:

- in case (a) ($f(X_j) \subset F$): a commutative diagram

$$\begin{array}{ccc} \Lambda^{\cdot}_{Y} & \longrightarrow & \Lambda^{\cdot}_{X} \\ \downarrow & & \downarrow \\ \Lambda^{\cdot}_{F} & & \Lambda^{\cdot}_{X_j} \\ & \searrow & \downarrow \\ & & \Lambda^{\cdot,0}_{X_j} \end{array} \tag{2.7}$$

where $\Lambda^{\cdot}_{F} \to \Lambda^{\cdot,0}_{X_j}$ is a pullback corresponding to $f|_{X_j} : X_j \to F$ (inductively defined), $\Lambda^{\cdot}_{X} \to \Lambda^{\cdot}_{X_j}$ and $\Lambda^{\cdot}_{X_j} \to \Lambda^{\cdot,0}_{X_j}$ are natural pullback;

- in case (b) ($f(X_j) \not\subset F$): a commutative diagram

$$\begin{array}{ccc} \Lambda^{\cdot}_{Y} & \longrightarrow & \Lambda^{\cdot}_{X} \\ \downarrow & & \downarrow \\ \Lambda^{\cdot}_{Y_k} & & \Lambda^{\cdot}_{X_j} \\ & \searrow & \downarrow \\ & & \Lambda^{\cdot,0}_{X_j} \end{array} \tag{2.8}$$

where $\Lambda^{\cdot}_{Y} \to \Lambda^{\cdot}_{Y_k}$, $\Lambda^{\cdot}_{X} \to \Lambda^{\cdot}_{X_j}$ and $\Lambda^{\cdot}_{X_j} \to \Lambda^{\cdot,0}_{X_j}$ are natural pullback, and $\Lambda^{\cdot}_{Y_k} \to \Lambda^{\cdot,0}_{X_j}$ is a primary pullback corresponding to $f|_{X_j} : X_j \to Y_k$.

Explanation 2.3. In the above definition, we must assume that we already know, by induction:

- for every morphism $f\colon T \to Z$ with $\dim T < \dim X$, $\dim Z \leq \dim X$, the notion of pullback $\Lambda^{\cdot}_{Z} \to \Lambda^{\cdot}_{T}$;

- for every morphism $f\colon T \to Z$ with T irreducible, $\dim T \leq \dim X$, $\dim Z < \dim X$ the notion of pullback $\Lambda^{\cdot}_{Z} \to \Lambda^{\cdot}_{T}$;

- for every morphism $f\colon T \to Z$ with T and Z irreducible, $\dim T \leq \dim X$, $\dim Z \leq \dim X$ the notion of primary pullback $\Lambda^{\cdot}_{Z} \to \Lambda^{\cdot}_{T}$.

Remark.

i. In the above definition we can take $\Lambda^{\cdot}_{X_j} = \Lambda^{\cdot,0}_{X_j}$, because by the proposition 2.1 below the composition $\Lambda^{\cdot}_{X} \to \Lambda^{\cdot}_{X_j} \to \Lambda^{\cdot,0}_{X_j}$ is a natural pullback;

ii. when X and Y are irreducible, we obtain a definition of pullback which is more general that the one in definition $(\text{DPP})_{m,n}$. Although we conjecture that every pullback between irreducible spaces is primary, the reader should keep in mind that a priori there are pullback between irreducible spaces which are not primary.

2.2.5 Existence of primary pullback (the irreducible case)

Theorem 2.2 ($(\text{EPP})_{m,n}$). *Let $f\colon X \to Y$ be a morphism between irreducible complex spaces, $\dim X = m, \dim Y = n$ and fix $\Lambda^{\cdot}_Y \in \mathcal{R}(Y)$; there exists a $\Lambda^{\cdot}_X \in \mathcal{R}(X)$ and a primary pullback $\Lambda^{\cdot}_Y \to \Lambda^{\cdot}_X$.*

Remark 2.1. In the particular case $X = Y, f = id$, $m = n$ it will follow from the proof that in order to obtain $(\text{EPP})_{m,n}$ (i.e. $(\text{EPP})_{m,m}$) we do not need the assumptions $(\text{F})_m$, $(\text{EP})_{m,n-1}$, $(\text{U})_{m,n-1}$ (i.e. $(\text{EP})_{m,m-1}$, $(\text{U})_{m,m-1}$).

As a consequence of the proof of the theorem we will obtain the following more precise statement.

Theorem 2.3. *Let $f\colon X \to Y$ be a morphism between irreducible complex spaces, $\dim X = m, \dim Y = n$ and fix $\Lambda^{\cdot}_Y = \mathcal{E}^{\cdot}_{\tilde{Y}} \oplus \Lambda^{\cdot}_F \oplus \Lambda^{\cdot}_{\tilde{F}}(-1) \in \mathcal{R}(Y)$; let E be a nowhere dense subspace of X, such that*

i. $f^{-1}(F) \subset E$*;*

ii. there are two commutative diagrams

$$\begin{array}{ccc} \tilde{F} & \longrightarrow & \tilde{Y} \\ \downarrow & & \downarrow p \\ F & \longrightarrow & Y \end{array}$$

$$\begin{array}{ccccc} \tilde{E} & \longrightarrow & \tilde{X} & \xrightarrow{\tilde{f}} & \tilde{Y} \\ \downarrow & & h \downarrow & & \downarrow p \\ E & \longrightarrow & X & \xrightarrow{f} & Y \end{array}$$

Then there exists a $\Lambda^{\cdot}_X = \mathcal{E}^{\cdot}_{\tilde{X}} \oplus \Lambda^{\cdot}_E \oplus \Lambda^{\cdot}_{\tilde{E}}(-1) \in \mathcal{R}(X)$ and a primary pullback $\Lambda^{\cdot}_Y \to \Lambda^{\cdot}_X$.

Remark 2.2. In particular, if we already know that there is a pullback

$$\Lambda^{\cdot}_Y = \mathcal{E}^{\cdot}_{\tilde{Y}} \oplus \Lambda^{\cdot}_F \oplus \Lambda^{\cdot}_{\tilde{F}}(-1) \to \Lambda^{\cdot}_X = \mathcal{E}^{\cdot}_{\tilde{X}} \oplus \Lambda^{\cdot}_E \oplus \Lambda^{\cdot}_{\tilde{E}}(-1)$$

then for any $\Lambda^{\cdot,1}_Y$ of the form $\mathcal{E}^{\cdot}_{\tilde{Y}} \oplus \Lambda^{\cdot,1}_F \oplus \Lambda^{\cdot,1}_{\tilde{F}}(-1)$ there exist a $\Lambda^{\cdot,1}_X = \mathcal{E}^{\cdot}_{\tilde{X}} \oplus \Lambda^{\cdot,1}_E \oplus \Lambda^{\cdot,1}_{\tilde{E}}(-1)$ and a pullback $\Lambda^{\cdot,1}_Y \to \Lambda^{\cdot,1}_X$.

2.2.6 Uniqueness of primary pullback (the irreducible case)

Theorem 2.4 ($(\mathrm{UP})_{m,n}$). *Let $f\colon X \to Y$ be a morphism between irreducible complex spaces, $\dim X \leq m, \dim Y \leq n$ and let $\phi_j\colon \Lambda^{\cdot}_Y \to \Lambda^{\cdot}_X$, $j = 1, 2$ be two primary pullback corresponding to f; then $\phi_1 = \phi_2$.*

2.2.7 Existence of pullback: the general case

Theorem 2.5 ($(\mathrm{EP})_{m,n}$). *Let $f\colon X \to Y$ be a morphism between complex spaces, $\dim X \leq m, \dim Y \leq n$ and fix $\Lambda^{\cdot}_Y \in \mathcal{R}(Y)$; then there exist a $\Lambda^{\cdot}_X \in \mathcal{R}(X)$ and a pullback $\Lambda^{\cdot}_Y \to \Lambda^{\cdot}_X$.*

Remark 2.3. In the particular case $X = Y, f = \mathrm{id}$, $m = n$ it follows from the proof that in order to obtain $(\mathrm{EP})_{m,n}$ (i.e. $(\mathrm{EP})_{m,m}$) we do not need the assumptions $(\mathrm{EP})_{m,n-1}$, $(\mathrm{U})_{m,n-1}$, (i.e. $(\mathrm{EP})_{m,m-1}$, $(\mathrm{U})_{m,m-1}$); moreover we need $(\mathrm{F})_m$ only for the (irreducible) X_j.

2.2.8 Uniqueness of pullback: the general case

Theorem 2.6 ($(\mathrm{U})_{m,n}$). *Let $f\colon X \to Y$ be a morphism between complex spaces, $\dim X \leq m, \dim Y \leq n$ and let $\phi_j\colon \Lambda^{\cdot}_Y \to \Lambda^{\cdot}_X, j = 1, 2$ be two pullback corresponding to f; then $\phi_1 = \phi_2$.*

2.2.9 Composition of primary pullback (the irreducible case)

Theorem 2.7 ($(\mathrm{CP})_{k,m,n}$). *Let Z, X, Y, be irreducible complex spaces, $\dim Z \leq k$, $\dim X \leq m$, $\dim Y \leq n$, $g\colon Z \to X$, $f\colon X \to Y$ two morphisms, $\psi\colon \Lambda^{\cdot}_X \to \Lambda^{\cdot}_Z$, $\phi\colon \Lambda^{\cdot}_Y \to \Lambda^{\cdot}_X$ be two primary pullback corresponding to g and f respectively. Then the composition $\psi \circ \phi\colon \Lambda^{\cdot}_Y \to \Lambda^{\cdot}_Z$ is a primary pullback corresponding to $f \circ g$.*

2.2.10 Composition of pullback: the general case

Theorem 2.8 ($(\mathrm{C})_{k,m,n}$). *Let Z, X, Y, be complex spaces, $\dim Z \leq k$, $\dim X \leq m$, $\dim Y \leq n$, $g\colon Z \to X$, $f\colon X \to Y$ two morphisms, $\psi\colon \Lambda^{\cdot}_X \to \Lambda^{\cdot}_Z$, $\phi\colon \Lambda^{\cdot}_Y \to \Lambda^{\cdot}_X$ be two pullback corresponding to g and f respectively. Then the composition $\psi \circ \phi\colon \Lambda^{\cdot}_Y \to \Lambda^{\cdot}_Z$ is a pullback corresponding to $f \circ g$.*

2.2.11 The filtration property

Theorem 2.9 ($(\mathrm{F})_m$). *Let X be a complex space, $\dim X \leq m$. If $\Lambda^{\cdot,1}_X$, $\Lambda^{\cdot,2}_X \in \mathcal{R}(X)$, there exists a third $\Lambda^{\cdot}_X \in \mathcal{R}(X)$ and two pullback $\Lambda^{\cdot,1}_X \to \Lambda^{\cdot}_X$, $\Lambda^{\cdot,2}_X \to \Lambda^{\cdot}_X$.*

Remark 2.4. If $\Lambda^{\cdot,1}_X = \mathcal{E}^{\cdot}_{\tilde{X}_1} \oplus \Lambda^{\cdot,1}_E \oplus \Lambda^{\cdot,1}_{\tilde{E}_1}(-1)$, $\Lambda^{\cdot,2}_X = \mathcal{E}^{\cdot}_{\tilde{X}_2} \oplus \Lambda^{\cdot,2}_E \oplus \Lambda^{\cdot,2}_{\tilde{E}_2}(-1)$, then it is possible to choose $\Lambda^{\cdot,3}_X = \mathcal{E}^{\cdot}_{\tilde{X}_3} \oplus \Lambda^{\cdot,3}_E \oplus \Lambda^{\cdot,3}_{\tilde{E}_3}(-1)$.

2.3 The induction procedure

Let $s, t \in \mathbb{N} \times \mathbb{N}$, $s = (m, n), t = (p, q)$. We define the following order on $\mathbb{N} \times \mathbb{N}$: $s > t$ if $\sup(m, n) > \sup(p, q)$ or $\sup(m, n) = \sup(p, q)$ and $(m, n) > (p, q)$ in the lexicographic order.

We write $(\mathrm{EP})_s$, $(\mathrm{U})_s, \ldots$ instead of $(\mathrm{EP})_{m,n}$, $(\mathrm{U})_{m,n}$.

Then we prove the following implications:

$(\mathrm{E})_p + (\mathrm{F})_q + (\mathrm{EP})_t + (\mathrm{U})_t$ for $p < n$, $q < n$, $t < (0, n) \implies (\mathrm{E})_n$

$(\mathrm{E})_p + (\mathrm{F})_q + (\mathrm{EP})_t + (\mathrm{U})_t$ for $p < n$, $q < n$, $t < (n, 0) \implies (\mathrm{F})_n$

$(\mathrm{E})_p + (\mathrm{F})_q + (\mathrm{EP})_t + (\mathrm{U})_t$ for $(0, p) \leq s$, $(q, 0) \leq s$, $t < s \implies (\mathrm{EPP})_s$ and $(\mathrm{EP})_s$

$(\mathrm{E})_p + (\mathrm{F})_q + (\mathrm{EP})_t + (\mathrm{U})_t$ for $(0, p) \leq s$, $(q, 0) \leq s$, $t < s \implies (\mathrm{UP})_s$ and $(\mathrm{U})_s$

Also the definitions $(\mathrm{D})_t$, $(\mathrm{DPP})_t$, $(\mathrm{DP})_t$, are given by induction. Their order is as follows: $(\mathrm{U})_r$ for $r < t$, $(\mathrm{D})_t$, $(\mathrm{DPP})_t$, $(\mathrm{DP})_t$, $(\mathrm{E})_t$.

Finally, $(\mathrm{CP})_{m,n,k}$ preceeds $(\mathrm{C})_{m,n,k}$, and $(\mathrm{CP})_{m,n,k}$ (resp. $(\mathrm{C})_{m,n,k}$) must be proved after $(\mathrm{DPP})_{(m,n)}$, $(\mathrm{DPP})_{(n,k)}$, $(\mathrm{DPP})_{(m,k)}$ (resp. $(\mathrm{DP})_{(m,n)}$, $(\mathrm{DP})_{(n,k)}$, $(\mathrm{DP})_{(m,k)}$). The induction scheme is

$(\mathrm{C})_{k-1,m,n} + (\mathrm{C})_{k,m-1,n} + (\mathrm{C})_{k,m,n-1} \implies (\mathrm{CP})_{k,m,n}$ and $(\mathrm{C})_{k,m,n}$.

In the course of each proof the corresponding induction assumptions will be made more explicit.

2.4 The proofs

Proposition 2.1. *We suppose that $(EP)_{m-1,m-1}$, $(F)_{m-1}$, $(U)_{m-1,m-1}$, $(E)_m$ are already proved. Let X be a complex space of dimension m, X_i an irreducible component of X.*

i. Given $\Lambda^{\cdot}_X = \mathcal{E}^{\cdot}_{\tilde{X}} \oplus \Lambda^{\cdot}_E \oplus \Lambda^{\cdot}_{\tilde{E}}(-1)$ there exists a natural pullback $\zeta\colon \Lambda^{\cdot}_X \to \Lambda^{\cdot}_{X_i}$.

ii. Let $\zeta_1\colon \Lambda^{\cdot}_X \to \Lambda^{\cdot,1}_{X_i}$, $\zeta_2\colon \Lambda^{\cdot}_X \to \Lambda^{\cdot,2}_{X_i}$ two natural pullback; then there exist $\Lambda^{\cdot,3}_{X_i}$ and pullback $\beta_1\colon \Lambda^{\cdot,1}_{X_i} \to \Lambda^{\cdot,3}_{X_i}$, $\beta_2\colon \Lambda^{\cdot,2}_{X_i} \to \Lambda^{\cdot,3}_{X_i}$ such that $\beta_1 \circ \zeta_1 = \beta_2 \circ \zeta_2$ and the composition $\Lambda^{\cdot}_X \to \Lambda^{\cdot,3}_{X_i}$ is a natural pullback.

Proof.

i. By $(\mathrm{EP})_{m-1,m-1}$ there exist pullback $\eta_i\colon \Lambda^{\cdot}_E \to \Lambda^{\cdot}_{E_i}$, $a\colon \Lambda^{\cdot}_{E_i} \to \Lambda^{\cdot,2}_{E_i}$ and $b\colon \Lambda^{\cdot}_{\tilde{E}_i} \to \Lambda^{\cdot,3}_{\tilde{E}_i}$; by $(\mathrm{F})_{m-1}$ there are pullback $c\colon \Lambda^{\cdot,2}_{\tilde{E}_i} \to \Lambda^{\cdot,1}_{\tilde{E}_i}$ and $e\colon \Lambda^{\cdot,3}_{\tilde{E}_i} \to \Lambda^{\cdot,1}_{\tilde{E}_i}$; if we apply $(\mathrm{U})_{m-1,m-1}$ to the pullback from $\Lambda^{\cdot}_E$ to $\Lambda^{\cdot,1}_{\tilde{E}_i}$

we obtain $c \circ a \circ \eta_i = e \circ b \circ q_i \circ \phi$. Putting $\phi_i = c \circ a$ and $\mu_i = e \circ b$ we obtain a commutative diagram like (2.5) and we define ζ by the formula (2.6).

ii. Let $\Lambda^{\cdot,j}_{X_i} = \mathcal{E}^{\cdot,j}_{\tilde{X}_i} \oplus \Lambda^{\cdot,j}_{E_i} \oplus \Lambda^{\cdot,j}_{\tilde{E}_i}(-1)$, $j = 1, 2$; the two pullback ζ_1 and ζ_2 differ only by the morphisms $\Lambda^{\cdot}_{E} \to \Lambda^{\cdot,j}_{E_i}$ and $\Lambda^{\cdot}_{\tilde{E}}(-1) \to \Lambda^{\cdot,j}_{\tilde{E}_i}(-1)$. Arguing as in i), we find $\Lambda^{\cdot,3}_{E_i}$ and $\Lambda^{\cdot,3}_{\tilde{E}_i}$ and commutative diagrams

$$\begin{array}{ccc} \Lambda^{\cdot,j}_{E_i} & \longrightarrow & \Lambda^{\cdot,3}_{E_i} \\ \downarrow & & \downarrow \\ \Lambda^{\cdot,j}_{\tilde{E}_i} & \longrightarrow & \Lambda^{\cdot,3}_{\tilde{E}_i} \end{array}$$

$(j = 1, 2)$ from which we easily construct the morphisms β_1 and β_2. □

Lemma 2.1. *Let $\Lambda^{\cdot}_Y \to \Lambda^{\cdot}_X$ be a pullback corresponding to $f\colon X \to Y$, $\dim X = m, \dim Y = n$. We suppose that $(EP)_{m-1,n-1}$, $(U)_{m-1,n-1}$, $(F)_q$, $(E)_q$ $q = \sup\{m, n\}$ are already proved, as well as $(EPP)_{m,n}$ and $(UP)_{m,n}$ (for morphisms between irreducible spaces). Then*

(a) If $f(X_j) \subset F$, for any natural pullback $\Lambda^{\cdot}_X \to \Lambda^{\cdot}_{X_j}$ there exists a commutative diagram (2.7).

(b) If $f(X_j) \not\subset F$ and $f(X_j) \subset Y_k$, for every natural pullback $\Lambda^{\cdot}_X \to \Lambda^{\cdot}_{X_j}$ and every natural pullback $\Lambda^{\cdot}_Y \to \Lambda^{\cdot}_{Y_k}$ there exists a commutative diagram (2.8).

Proof. We check b), leaving to the reader the proof of a), which is similar. By definition there exists a commutative diagram

$$\begin{array}{ccc} \Lambda^{\cdot}_Y & \longrightarrow & \Lambda^{\cdot}_X \\ \downarrow & & \downarrow \\ \Lambda^{\cdot,1}_{Y_k} & & \Lambda^{\cdot,1}_{X_j} \\ & \searrow & \downarrow \\ & & \Lambda^{\cdot,2}_{X_j} \end{array} \tag{2.9}$$

where $\Lambda^{\cdot}_Y \to \Lambda^{\cdot,1}_{Y_k}$, $\Lambda^{\cdot}_X \to \Lambda^{\cdot,1}_{X_j}$ and $\Lambda^{\cdot,1}_{X_j} \to \Lambda^{\cdot,2}_{X_j}$ are natural pullback, and $\Lambda^{\cdot,1}_{Y_k} \to \Lambda^{\cdot,2}_{X_j}$ is a pullback corresponding to $f|_{X_j} : X_j \to Y_k$. By proposition 2.1, (ii), we find pullback $\Lambda^{\cdot,1}_{Y_k} \to \Lambda^{\cdot,3}_{Y_k}$ and $\Lambda^{\cdot}_{Y_k} \to \Lambda^{\cdot,3}_{Y_k}$ such that their compositions $\Lambda^{\cdot}_Y \to \Lambda^{\cdot,3}_{Y_k}$ are identical, and are natural pullback. It is clear that for our purposes we can replace $\Lambda^{\cdot}_{Y_k}$ by $\Lambda^{\cdot,3}_{Y_k}$ hence we can suppose from the beginning that $\Lambda^{\cdot}_Y \to \Lambda^{\cdot}_{Y_k}$ decomposes through $\Lambda^{\cdot}_Y \to \Lambda^{\cdot,1}_{Y_k} \to \Lambda^{\cdot}_{Y_k}$; the same argument shows that we can suppose that the morphism $\Lambda^{\cdot}_X \to \Lambda^{\cdot}_{X_j}$ decomposes through $\Lambda^{\cdot}_Y \to \Lambda^{\cdot,1}_{X_j} \to \Lambda^{\cdot}_{X_j}$. We apply (EPP)$_{m,n}$ to $f|_{X_j} : X_j \to Y_k$ and we get a pullback $\Lambda^{\cdot}_{Y_k} \to \Lambda^{\cdot,4}_{X_j}$; moreover by the remark 2.2 we can suppose $\Lambda^{\cdot,4}_{X_j} = \mathcal{E}^{\cdot}_{\tilde{X}_i} \oplus \Lambda^{\cdot,4}_{E_i} \oplus \Lambda^{\cdot,4}_{\tilde{E}_i}(-1)$; by (F)$_m$ and remark 2.4 (for X_j) there are natural pullback $\Lambda^{\cdot,2}_{X_j} \to \Lambda^{\cdot,0}_{X_j}$ and $\Lambda^{\cdot,4}_{X_j} \to \Lambda^{\cdot,0}_{X_j}$; finally we apply (UP)$_{m,n}$ to $f|_{X_j} : X_j \to Y_k$: the two compositions $\Lambda^{\cdot,1}_{Y_k} \to \Lambda^{\cdot,2}_{X_j} \to \Lambda^{\cdot,0}_{X_j}$ and $\Lambda^{\cdot,1}_{Y_k} \to \Lambda^{\cdot}_{Y_k} \to \Lambda^{\cdot,4}_{X_j} \to \Lambda^{\cdot,0}_{X_j}$ coincide. This completes the proof. □

Corollary 2.1.

i. If $\Lambda^{\cdot}_Y \to \Lambda^{\cdot}_X$ is a pullback, and $\Lambda^{\cdot}_X \to \Lambda^{\cdot}_{X_j}$ is a natural pullback, the composition $\Lambda^{\cdot}_Y \to \Lambda^{\cdot}_{X_j}$ is a pullback.

ii. Conversely let $\Lambda^{\cdot}_Y \to \Lambda^{\cdot}_X$ be a morphism of complexes satisfying (P_0), (P_1), (P_4) in definition 2.4, and suppose that for every irreducible component X_j of X there exists a natural pullback $\Lambda^{\cdot}_X \to \Lambda^{\cdot}_{X_j}$ such that the composition $\Lambda^{\cdot}_Y \to \Lambda^{\cdot}_{X_j}$ is a pullback; then $\Lambda^{\cdot}_Y \to \Lambda^{\cdot}_X$ is a pullback.

2.4.1 Proof of theorem 2.7: composition of primary pullback (the irreducible case)

The proof is by triple induction on (k, m, n) where $\dim Z \leq k, \dim X \leq m, \dim Y \leq n$. More precisely we prove
(C)$_{k-1,m,n}$ and (C)$_{k,m-1,n}$ and (C)$_{k,m,n-1}$ $\Longrightarrow$ (CP)$_{k,m,n}$ for primary pullback.

It is obvious that $\psi \circ \phi$ commutes to differentials and satisfies the property (P$_1$) in the definition 2.2. Let $\Lambda^{\cdot}_Z = \mathcal{E}^{\cdot}_{\tilde{Z}} \oplus \Lambda^{\cdot}_G \oplus \Lambda^{\cdot}_{\tilde{G}}(-1)$, $\Lambda^{\cdot}_X = \mathcal{E}^{\cdot}_{\tilde{X}} \oplus \Lambda^{\cdot}_E \oplus \Lambda^{\cdot}_{\tilde{E}}(-1)$, $\Lambda^{\cdot}_Y = \mathcal{E}^{\cdot}_{\tilde{Y}} \oplus \Lambda^{\cdot}_F \oplus \Lambda^{\cdot}_{\tilde{F}}(-1)$, $\alpha \colon \mathcal{E}^{\cdot}_{\tilde{Y}} \to \mathcal{E}^{\cdot}_{\tilde{X}}$, $\alpha' \colon \mathcal{E}^{\cdot}_{\tilde{X}} \to \mathcal{E}^{\cdot}_{\tilde{Z}}$, be induced by ϕ and ψ respectively.

1) Case $\alpha \neq 0, \alpha' \neq 0$. Then we find a commutative diagram

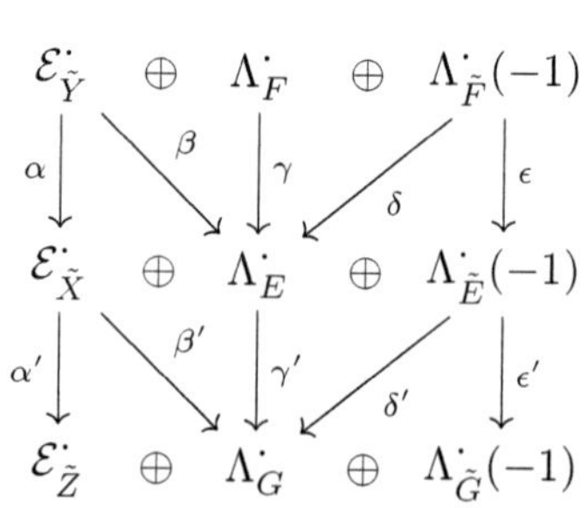

Since $\alpha \neq 0, \alpha' \neq 0$, f and g extend respectively to $\tilde{f}: \tilde{X} \to \tilde{Y}$ and $\tilde{g}: \tilde{Z} \to \tilde{X}$ so that $f \circ g$ extends to $\tilde{f} \circ \tilde{g}: \tilde{Z} \to \tilde{Y}$, and $\alpha' \circ \alpha$ is the ordinary De Rham pullback $(\tilde{f} \circ \tilde{g})^*$. Hence it remains to check that $\psi \circ \phi$ satisfies (P_3) (i) (iii) in definition 2.2.

We check (P_3) (i). Since $\alpha \neq 0$, $\alpha' \neq 0$, we see that ϕ and ψ satisfy (P_3) (i); hence $f^{-1}(F) \subset E$, $g^{-1}(E) \subset G$ and finally $(f \circ g)^{-1}(F) \subset G$.

The morphism $\Lambda^{\cdot}_Y \to \Lambda^{\cdot}_G$ is the composition $\Lambda^{\cdot}_Y \to \Lambda^{\cdot}_X \to \Lambda^{\cdot}_G$; by ($P_3$) (iii) applied to ψ, $\Lambda^{\cdot}_X \to \Lambda^{\cdot}_G$ is a pullback; since $\dim G \leq k-1$, it follows by $(C)_{k-1,m,n}$ that $\Lambda^{\cdot}_Y \to \Lambda^{\cdot}_G$ is a pullback, which is (P_3) (iii).

2) Case $\alpha = 0$. In this case $\alpha' \circ \alpha = 0$ so we must check P_2 for $\psi \circ \phi$. $\alpha = 0$ implies $f(X) \subset F$, hence $(f \circ g)(Z) \subset F$; the morphism $\Lambda^{\cdot}_Y \to \Lambda^{\cdot}_X$ decomposes through $\Lambda^{\cdot}_Y \to \Lambda^{\cdot}_F \to \Lambda^{\cdot}_X$. The composition $\Lambda^{\cdot}_F \to \Lambda^{\cdot}_X \to \Lambda^{\cdot}_Z$ is a pullback by $(C)_{k,m,n-1}$, so that $\Lambda^{\cdot}_Y \to \Lambda^{\cdot}_Z$ decomposes through $\Lambda^{\cdot}_Y \to \Lambda^{\cdot}_F \to \Lambda^{\cdot}_Z$, which is exactly P_2.

3) Case $\alpha' = 0, \alpha \neq 0$. In this case $g(Z) \subset E$ and $\Lambda^{\cdot}_X \to \Lambda^{\cdot}_Z$ decomposes through $\Lambda^{\cdot}_X \to \Lambda^{\cdot}_E \to \Lambda^{\cdot}_Z$, where $\Lambda^{\cdot}_E \to \Lambda^{\cdot}_Z$ is a pullback. The composite mapping $\Lambda^{\cdot}_Y \to \Lambda^{\cdot}_X \to \Lambda^{\cdot}_E$ is a pullback by (P_2) ; finally the composition $\Lambda^{\cdot}_Y \to \Lambda^{\cdot}_E \to \Lambda^{\cdot}_Z$ is a pullback by $(C)_{k,m-1,n}$.

2.4.2 Proof of theorem 2.8: composition of pullback (the general case)

The proof is by triple induction on (k, m, n) where $\dim Z \leq k, \dim X \leq m, \dim Y \leq n$. More precisely we prove
$(C)_{k-1,m,n}$ and $(C)_{k,m-1,n}$ and $(C)_{k,m,n-1} \implies (C)_{k,m,n}$.

We already know $(CP)_{k,m,n}$ is true for primary pullback between irreducible spaces. It is obvious that $\psi \circ \phi$ commutes to differentials and satisfies the property (P_1) in the definition 2.4. Let $\Lambda^{\cdot}_Z = \mathcal{E}^{\cdot}_{\tilde{Z}} \oplus \Lambda^{\cdot}_G \oplus \Lambda^{\cdot}_{\tilde{G}}(-1)$, $\Lambda^{\cdot}_X = \mathcal{E}^{\cdot}_{\tilde{X}} \oplus \Lambda^{\cdot}_E \oplus \Lambda^{\cdot}_{\tilde{E}}(-1)$, $\Lambda^{\cdot}_Y = \mathcal{E}^{\cdot}_{\tilde{Y}} \oplus \Lambda^{\cdot}_F \oplus \Lambda^{\cdot}_{\tilde{F}}(-1)$; the composition $\Lambda^{\cdot}_X \to \Lambda^{\cdot}_Z \to \Lambda^{\cdot}_G$ is a pullback, hence the composite mapping $\Lambda^{\cdot}_Y \to \Lambda^{\cdot}_G$ is a pullback by $(C)_{k-1,m,n}$.

- Let Z_l be an irreducible component of Z; if $g(Z_l) \subset E$ there exists a commutative diagram

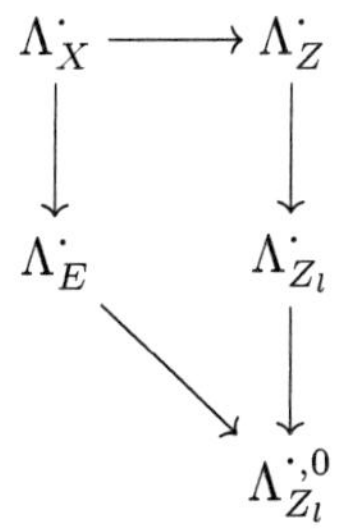

The composition $\Lambda^{\cdot}_{Y} \to \Lambda^{\cdot}_{X} \to \Lambda^{\cdot}_{E}$ is a pullback because $\Lambda^{\cdot}_{Y} \to \Lambda^{\cdot}_{X}$ is, hence the composition $\Lambda^{\cdot}_{Y} \to \Lambda^{\cdot}_{E} \to \Lambda^{\cdot,0}_{Z_l}$ is a pullback by $(\mathrm{C})_{k,m-1,n}$. If $g(f(Z_l)) \not\subset F$, by lemma 2.1, b) (applied to the spaces Z_l and Y) there exists a commutative diagram

$$\begin{array}{ccc}
\Lambda^{\cdot}_{Y} & \longrightarrow & \Lambda^{\cdot,0}_{Z_l} \\
\downarrow & & \downarrow \\
\Lambda^{\cdot}_{Y_k} & & \Lambda^{\cdot,1}_{Z_l} \\
& \searrow & \downarrow \\
& & \Lambda^{\cdot,2}_{Z_l}
\end{array}$$

where $\Lambda^{\cdot}_{Y_k} \to \Lambda^{\cdot,2}_{Z_l}$ is a primary pullback. So we obtain a commutative diagram

$$\begin{array}{ccc}
\Lambda^{\cdot}_{Y} & \longrightarrow & \Lambda^{\cdot}_{Z} \\
\downarrow & & \downarrow \\
\Lambda^{\cdot}_{Y_k} & \longrightarrow & \Lambda^{\cdot,2}_{Z_l}
\end{array}$$

- if $g(f(Z_l)) \subset F$ arguing in the same way we find a commutative diagram

$$\begin{array}{ccc}
\Lambda^{\cdot}_{Y} & \longrightarrow & \Lambda^{\cdot}_{Z} \\
\downarrow & & \downarrow \\
\Lambda^{\cdot}_{F} & \longrightarrow & \Lambda^{\cdot,3}_{Z_l}
\end{array}$$

These last two diagrams imply that $\Lambda^{\cdot}_{Y} \to \Lambda^{\cdot}_{Z}$ is a pullback.

- if $g(Z_l) \not\subset E$ there exists a unique irreducible component X_j of X with $g(Z_l) \subset X_j$; for simplicity we suppose that there is a unique irreducible component Y_k of Y with $f(X_j) \subset Y_k$ (we leave to the reader the case $f(X_j) \subset F$). According to the definition 2.4 there is a commutative diagram

$$\begin{array}{ccc}
\Lambda^{\cdot}_{Y} & \longrightarrow & \Lambda^{\cdot}_{X} \\
\downarrow & & \downarrow \\
\Lambda^{\cdot}_{Y_k} & & \Lambda^{\cdot}_{X_j} \\
& \searrow & \downarrow \\
& & \Lambda^{\cdot,0}_{X_j}
\end{array}$$

and according to the lemma 2.1 another commutative diagram

$$\begin{array}{ccc}
\Lambda^{\cdot}_{X} & \longrightarrow & \Lambda^{\cdot}_{Z} \\
\downarrow & & \downarrow \\
\Lambda^{\cdot,0}_{X_j} & & \Lambda^{\cdot}_{Z_l} \\
 & \searrow & \downarrow \\
 & & \Lambda^{\cdot,0}_{Z_l}
\end{array}$$

The composite morphism $\Lambda^{\cdot}_{Y_k} \to \Lambda^{\cdot,0}_{X_j} \to \Lambda^{\cdot,0}_{Z_l}$ is the composition of two primary pullback between irreducible spaces, hence it is a pullback by $(\mathrm{CP})_{m,n}$. If $f(g(Z_l)) \not\subset F$ the proof is finished. If $f(g(Z_l)) \subset F$, then by definition 2.2 (primary pullback for irreducible spaces) the morphism $\Lambda^{\cdot}_{Y_k} \to \Lambda^{\cdot}_{Z_l}$ decomposes as $\Lambda^{\cdot}_{Y_k} \to \Lambda^{\cdot}_{F_k} \to \Lambda^{\cdot,0}_{Z_l}$ ($F_k = F \cap Y_k$). Taking into account the commutative diagram

$$\begin{array}{ccc}
\Lambda^{\cdot}_{Y} & \longrightarrow & \Lambda^{\cdot}_{Y_k} \\
\downarrow & & \downarrow \\
\Lambda^{\cdot}_{F} & \longrightarrow & \Lambda^{\cdot}_{F_k}
\end{array}$$

we finally get by composition the diagram

$$\begin{array}{ccc}
\Lambda^{\cdot}_{Y} & \longrightarrow & \Lambda^{\cdot}_{Z} \\
\downarrow & & \downarrow \\
\Lambda^{\cdot}_{F} & \longrightarrow & \Lambda^{\cdot,0}_{Z_l}
\end{array}$$

which concludes the proof.

2.4.3 Proof of theorem 2.1: construction-existence

We suppose that $(\mathrm{EP})_{n-1,n-1}$, $(\mathrm{U})_{n-1,n-1}$, $(\mathrm{F})_{n-1}$, $(\mathrm{E})_{n-1}$ are already proved.

Let us establish first the following

Lemma 2.2. *Let $f\colon X \to Y$ be a morphism of complex spaces, with Y smooth, $\dim X = r \le n-1$, $\dim Y \le n$. We suppose that $(C)_{k,m,l}$, $(EP)_{m,l}$, $(F)_m$, $(U)_{m,l}$, $(E)_m$ are already proved for $k < n$, $m < n$, $l < n$.*

i. There exists a $\Lambda^{\cdot}_{X} \in \mathcal{R}(X)$ and a pullback $\psi\colon \mathcal{E}^{\cdot}_{Y} \to \Lambda^{\cdot}_{X}$ (corresponding to f).

ii. Moreover if $\psi_1 : \mathcal{E}^{\cdot}_Y \to \Lambda^{\cdot}_X$ *is another pullback corresponding to* f *then* $\psi = \psi_1$.

The proof of the lemma is by induction on r, the case $r = 0$ being trivial. First we check (i). Let $E = \operatorname{sing} X$. Since $\dim E < r$ there is a $\Lambda^{\cdot}_E \in \mathcal{R}(E)$ and a pullback $\rho : \mathcal{E}^{\cdot}_Y \to \Lambda^{\cdot}_E$. Let us take a desingularization $\pi : \tilde{X} \to X$ with exceptional space $\tilde{E} = \pi^{-1}(E)$.

By $(\mathrm{E})_{n-1}$ we construct $\Lambda^{\cdot}_X = \mathcal{E}^{\cdot}_{\tilde{X}} \oplus \Lambda^{\cdot}_E \oplus \Lambda^{\cdot}_{\tilde{E}}(-1)$. Let us consider the commutative diagram of pullback

$$\begin{array}{ccc} \mathcal{E}^{\cdot}_Y & \xrightarrow{\rho} & \Lambda^{\cdot}_E \\ {\scriptstyle\alpha}\downarrow & & \downarrow{\scriptstyle\phi} \\ \mathcal{E}^{\cdot}_{\tilde{X}} & \xrightarrow{\theta} & \Lambda^{\cdot}_{\tilde{E}} \end{array}$$

where $\alpha = (f \circ \pi)^*$ is the ordinary De Rham pullback, and ϕ, θ are the inner pullback of $\Lambda^{\cdot}_X$; using (ii) in smaller dimension (replace X by E) we see that $\phi \circ \rho = \theta \circ \alpha$.

Then we define $\psi : \mathcal{E}^{\cdot}_Y \to \Lambda^{\cdot}_X$ by $\psi(\omega) = ((f \circ \pi)^*(\omega), \rho(\omega), 0)$, which is a pullback (left to the reader).

We check the uniqueness (ii); we remark that in our situation any pullback $\psi_1 : \mathcal{E}^{\cdot}_Y \to \Lambda^{\cdot}_X$ is given by

$$\mathcal{E}^{\cdot}_Y \xrightarrow{(\alpha_1, \beta_1, 0)} \mathcal{E}^{\cdot}_{\tilde{X}} \oplus \Lambda^{\cdot}_E \oplus \Lambda^{\cdot}_{\tilde{E}}(-1)$$

where $\alpha_1 = (f \circ \pi)^*$ and $\beta_1 : \Lambda^{\cdot}_Y \to \Lambda^{\cdot}_E$ is a pullback corresponding to the composition $f \circ j : E \to Y$ (inductively defined). Hence we immediately get $\alpha_1 = \alpha$, and $\beta_1 = \rho$ follows from $(\mathrm{U})_{n-1,n-1}$.

Let us go back to the construction-existence theorem. First we recall the proposition 1.1 of the chapter 1.

Proposition 2.2. *Let* $U \subset X$ *be an open neighborhood of a point* $x \in E$.

i. Let $\eta_1, \eta_2 \in H^k\left(\pi^{-1}(U), \mathbb{C}\right)$ *be two cohomology classes whose restrictions to* $H^k\left(\pi^{-1}(U) \cap \tilde{E}, \mathbb{C}\right)$ *coincide. There exists an open neighborhood* $V \subset U$ *of* x *such that the restrictions of* η_1, η_2 *to* $H^k\left(\pi^{-1}(V), \mathbb{C}\right)$ *coincide.*

ii. Let $\theta \in H^k\left(\pi^{-1}(U) \cap \tilde{E}, \mathbb{C}\right)$. *There exists an open neighborhood* $V \subset U$ *of* x *and* $\eta \in H^k\left(\pi^{-1}(V), \mathbb{C}\right)$ *inducing* $\theta|_{\pi^{-1}(V) \cap \tilde{E}}$.

By the above lemma there is a pullback $\rho : \mathcal{E}^{\cdot}_{\tilde{X}} \to \Lambda^{\cdot,1}_{\tilde{E}}$; by $(\mathrm{EP})_{n-1,n-1}$ there is a pullback $\Lambda^{\cdot}_E \to \Lambda^{\cdot,2}_{\tilde{E}}$; by $(\mathrm{F})_{n-1}$ and $(\mathrm{EP})_{n-1,n-1}$ there exist $\Lambda^{\cdot}_{\tilde{E}}$ and morphisms $\Lambda^{\cdot,1}_{\tilde{E}} \to \Lambda^{\cdot}_{\tilde{E}}$ and $\Lambda^{\cdot,2}_{\tilde{E}} \to \Lambda^{\cdot}_{\tilde{E}}$; by composition we obtain $\phi : \Lambda^{\cdot}_E \to \Lambda^{\cdot}_{\tilde{E}}$ (corresponding to q), and $\psi : \mathcal{E}^{\cdot}_{\tilde{X}} \to \Lambda^{\cdot}_{\tilde{E}}$ (corresponding to i). Let us prove that the complex $\Lambda^{\cdot}_X = \mathcal{E}^{\cdot}_{\tilde{X}} \oplus \Lambda^{\cdot}_E \oplus \Lambda^{\cdot}_{\tilde{E}}(-1)$ is a resolution of $\mathbb{C}_X$. Let $x \in X$, U a neighbhorhood of x in X and $(\omega, \sigma, \theta) \in \Lambda^p_X(U)$, i.e. $\omega \in \mathcal{E}^p_{\tilde{X}}\left(\pi^{-1}(U)\right)$, $\sigma \in \Lambda^p_E(U \cap E)$, $\theta \in \Lambda^{p-1}_{\tilde{E}}\left(\pi^{-1}(U) \cap \tilde{E}\right)$, with $\big(d\omega, d\sigma, d\theta + (-1)^p(\psi(\omega) - \phi(\sigma))\big) = (0,0,0)$. Then $d\sigma = 0$ implies by induction on the dimension of X that $\sigma = d\sigma'$ (after possibly shrinking U). Then we have

$$d\omega = 0, \quad d(\theta - (-1)^p\phi(\sigma')) = -(-1)^p\psi(\omega)$$

which implies that ω gives a cohomology class in $H^p\left(\pi^{-1}(U), \mathbb{C}\right)$ whose restriction to $H^p\left(\pi^{-1}(U) \cap \tilde{E}, \mathbb{C}\right)$ is zero; by proposition 2.2 we can write (after possibly shrinking U) $\omega = d\omega'$ where $\omega' \in \mathcal{E}^{p-1}_{\tilde{X}}\left(\pi^{-1}(U)\right)$; it follows

$$d\big[\theta + (-1)^p\psi(\omega') - (-1)^p\phi(\sigma')\big] = 0$$

thus $\theta + (-1)^p\psi(\omega') - (-1)^p\phi(\sigma')$ gives a class in $H^{p-1}\left(\pi^{-1}(U) \cap \tilde{E}, \mathbb{C}\right)$. By proposition 2.2 we can write (again after possibly shrinking U)

$$\theta + (-1)^p\psi(\omega') - (-1)^p\phi(\sigma') = (-1)^{p-1}\psi(\omega'') + d\theta'$$

where $\omega'' \in \mathcal{E}^{p-1}_{\tilde{X}}\left(\pi^{-1}(U)\right)$, $d\omega'' = 0$, and $\theta' \in \Lambda^{p-2}_{\tilde{E}}\left(\pi^{-1}(U) \cap \tilde{E}\right)$ (here we suppose of course $p \geq 2$: the case $p \leq 1$ needs minor modifications). As a consequence

$$(\omega, \sigma, \theta) = d(\omega' + \omega'', \sigma', \theta')$$

Finally, for every p, $\Lambda^p_X = \mathcal{E}^p_{\tilde{X}} \oplus \Lambda^p_E \oplus \Lambda^{p-1}_{\tilde{E}}$ is a fine sheaf; in fact it is a direct sum of direct images of fine sheaves (we use again induction on the dimension).

2.4.4 Proof of theorem 2.2: existence of primary pullback (the irreducible case)

We suppose that $(\mathrm{EP})_{m-1,n}$, $(\mathrm{EP})_{m,n-1}$, $(\mathrm{EP})_{m-1,m}$, $(\mathrm{U})_{m-1,n}$, $(\mathrm{F})_m$, $(\mathrm{E})_q$, $q = \sup\{m, n\}$ are already proved.

Let $\Lambda^p_Y = \mathcal{E}^p_{\tilde{Y}} \oplus \Lambda^p_F \oplus \Lambda^{p-1}_{\tilde{F}}$ be defined by a desingularization

$$\begin{array}{ccc} \tilde{F} & \longrightarrow & \tilde{Y} \\ \downarrow & & \downarrow{\scriptstyle p} \\ F & \longrightarrow & Y \end{array}$$

with inner pullback given by $u_1 \colon \mathcal{E}^{\cdot}_{\tilde{Y}} \to \Lambda^{\cdot}_{\tilde{F}}$ and $p_1 \colon \Lambda^{\cdot}_F \to \Lambda^{\cdot}_{\tilde{F}}$. If $f(X) \subset F$, by $(\mathrm{EP})_{m,n-1}$ there exists a pullback $\Lambda^{\cdot}_F \to \Lambda^{\cdot}_X$; hence we obtain the required pullback as composition $\Lambda^{\cdot}_Y \to \Lambda^{\cdot}_F \to \Lambda^{\cdot}_X$, where $\Lambda^{\cdot}_Y \to \Lambda^{\cdot}_F$ is the projection onto the summand.

From now on we suppose $f(X) \not\subset F$. Since X is irreducible, the subspace $E \equiv f^{-1}(F) \cup \operatorname{sing} X$ is nowhere dense in X.

Lemma 2.3. *We can construct two commutative diagrams*

$$\begin{array}{ccccc} \tilde{E} & \longrightarrow & \tilde{X} & \xrightarrow{\tilde{f}} & \tilde{Y} \\ \downarrow & & \downarrow{\scriptstyle h} & & \downarrow{\scriptstyle p} \\ E & \longrightarrow & X & \xrightarrow{f} & Y \end{array}$$

where h is a desingularization of X.

Proof of the lemma. By Hironaka Chow lemma (part I, chapter 7) the modification $f\colon \tilde{Y} \to Y$ is dominated by the composition $g\colon Y_1 \to Y$ of a locally finite sequence of blowing-up; following step by step the sequence of blowing-up, and lifting them by lemma 7.1 of part I, chapter 7, we can extend g to $u\colon X_1 \to X$: i.e. we obtain a commutative diagram

$$\begin{array}{ccc} X_1 & \xrightarrow{h_1} & Y_1 \\ {\scriptstyle u}\downarrow & & \downarrow{\scriptstyle g} \\ X & \xrightarrow{f} & Y \end{array}$$

and, by composition, a diagram

$$\begin{array}{ccc} X_1 & \xrightarrow{f_1} & \tilde{Y} \\ {\scriptstyle u}\downarrow & & \downarrow{\scriptstyle p} \\ X & \xrightarrow{f} & Y \end{array}$$

We take as $\tilde{X}$ a desingularization of X_1. This proves the lemma. □

Now, $\dim \tilde{F} < n, \dim \tilde{f}^{-1}(\tilde{F}) < m$, so that by $(\mathrm{EP})_{m-1,m}$ and $(\mathrm{EP})_{m-1,n-1}$ there exist two pullback $u_2\colon \mathcal{E}^{\cdot}_{\tilde{X}} \to \Lambda^{\cdot}_{\tilde{f}^{-1}(\tilde{F})}$ and $p_2\colon \Lambda^{\cdot}_{\tilde{F}} \to \Lambda^{\cdot}_{\tilde{f}^{-1}(\tilde{F})}$ (here we need also $(\mathrm{F})_{m-1}$ in order to insure the same target $\Lambda^{\cdot}_{\tilde{f}^{-1}(\tilde{F})}$).

By $(\mathrm{U})_{m-1,n}$

$$p_2 \circ u_1 = u_2 \circ \tilde{f}^*$$

From the trivial proper modification $\mathrm{id}\colon (\tilde{X}, \tilde{f}^{-1}(\tilde{F})) \to (\tilde{X}, \tilde{f}^{-1}(\tilde{F}))$ we build the complex $\Lambda^{\cdot}_{\tilde{X}} \in \mathcal{R}(\tilde{X})$:

$$\Lambda^p_{\tilde{X}} = \mathcal{E}^p_{\tilde{X}} \oplus \Lambda^p_{\tilde{f}^{-1}(\tilde{F})} \oplus \Lambda^{p-1}_{\tilde{f}^{-1}(\tilde{F})}$$

with inner pullback given by $u_2\colon \mathcal{E}^p_{\tilde{X}} \to \Lambda^p_{\tilde{f}^{-1}(\tilde{F})}$ and $\mathrm{id}\colon \Lambda^p_{\tilde{f}^{-1}(\tilde{F})} \to \Lambda^p_{\tilde{f}^{-1}(\tilde{F})}$. Let us define the morphism

$$\psi\colon \Lambda^p_{\tilde{Y}} = \mathcal{E}^p_{\tilde{Y}} \oplus \Lambda^p_{\tilde{F}} \oplus \Lambda^{p-1}_{\tilde{F}} \to \Lambda^p_{\tilde{X}} = \mathcal{E}^p_{\tilde{X}} \oplus \Lambda^p_{\tilde{f}^{-1}(\tilde{F})} \oplus \Lambda^{p-1}_{\tilde{f}^{-1}(\tilde{F})}$$

$$\psi(\omega, \sigma, \theta) = (\tilde{f}^*\omega, (p_2 \circ p_1)(\sigma), p_2(\theta))$$

We check that ψ is a primary pullback (see definition 2.2) corresponding to the morphism $f \circ h \colon \tilde{X} \to Y$; in fact it commutes with the differential because of $p_2 \circ u_1 = u_2 \circ \tilde{f}^*$; the condition P_1 on the hypercoverings of $\tilde{X}$ and Y is clearly satisfied; the condition $(f \circ h)^{-1}(F) \subset \tilde{f}^{-1}(\tilde{F})$ is true because of $f \circ h = p \circ \tilde{f}$; $f \circ h$ extends trivially to $\tilde{f} \colon \tilde{X} \to Y$; finally the composition $\Lambda_Y^p \to \Lambda_{\tilde{X}}^p \to \Lambda_{\tilde{f}^{-1}(\tilde{F})}^p$ coincides with $p_2 \circ p_1$, therefore is a pullback.

Next we define

$$t \colon \mathcal{E}_{\tilde{X}}^p \to \Lambda_{\tilde{X}}^p = \mathcal{E}_{\tilde{X}}^p \oplus \Lambda_{\tilde{f}^{-1}(\tilde{F})}^p \oplus \Lambda_{\tilde{f}^{-1}(\tilde{F})}^{p-1}$$
$$t(\rho) = (\rho, u_2(\rho), 0)$$

which is also a pullback corresponding to the identity id: $\tilde{X} \to \tilde{X}$ (left to the reader). Now we use induction: by $(\mathrm{EP})_{m-1,n}$ there is a pullback $w \colon \Lambda_Y^{\cdot} \to \Lambda_E^{\cdot}$ corresponding to the composite morphism $E \to X \to Y$: again by $(\mathrm{EP})_{m-1,m}$ there is a pullback $z \colon \Lambda_{\tilde{X}}^{\cdot} \to \Lambda_{\tilde{E}}^{\cdot}$ corresponding to the embedding $\tilde{E} \to \tilde{X}$; by $(\mathrm{EP})_{m-1,m-1}$ there is a a pullback $q_1 \colon \Lambda_E^{\cdot} \to \Lambda_{\tilde{E}}^{\cdot,1}$ corresponding to the morphism $\tilde{E} \to E$. Using $(\mathrm{F})_{m-1}$ we can suppose $\Lambda_{\tilde{E}}^{\cdot,1} = \Lambda_{\tilde{E}}^{\cdot}$. Because of $(\mathrm{U})_{m-1,n}$ the two pullback $z \circ \psi, q_1 \circ w \colon \Lambda_Y^{\cdot} \to \Lambda_{\tilde{E}}^{\cdot}$ coincide:

$$z \circ \psi = q_1 \circ w$$

We define

$$\Lambda_X^p = \mathcal{E}_{\tilde{X}}^p \oplus \Lambda_E^p \oplus \Lambda_{\tilde{E}}^{p-1}$$

where the inner pullback are defined by $z \circ t \colon \mathcal{E}_{\tilde{X}}^{\cdot} \to \Lambda_{\tilde{E}}^{\cdot}$ and $q_1 \colon \Lambda_E^{\cdot} \to \Lambda_{\tilde{E}}^{\cdot}$. Finally we define

$$\phi \colon \Lambda_Y^p = \mathcal{E}_{\tilde{Y}}^p \oplus \Lambda_F^p \oplus \Lambda_{\tilde{F}}^{p-1} \to \Lambda_X^p = \mathcal{E}_{\tilde{X}}^p \oplus \Lambda_E^p \oplus \Lambda_{\tilde{E}}^{p-1}$$
$$\phi(\omega, \sigma, \theta) = \big(\tilde{f}^*\omega, w(\omega, \sigma, \theta), z(0, p_2(\theta), 0)\big)$$

Here we notice that $p_2(\theta) \in \Lambda_{\tilde{f}^{-1}(\tilde{F})}^{p-1}$, $(0, p_2(\theta), 0) \in \Lambda_X^{p-1}$ so that $z(0, p_2(\theta), 0) \in \Lambda_{\tilde{E}}^{p-1}$. In order to prove that ϕ is a primary pullback, the only non trivial property is that it commutes with differentials. Let us check it.

$$d(\omega, \sigma, \theta) = \big(d\omega, d\sigma, d\theta + (-1)^p(u_1(\omega) - p_1(\sigma))\big)$$
$$\phi(d(\omega, \sigma, \theta)) = \big(\tilde{f}^* d\omega, w(d(\omega, \sigma, \theta)), z(0, p_2(d\theta + (-1)^p(u_1(\omega) - p_1(\sigma)), 0)\big)$$

On the other hand

$$d(\phi(\omega, \sigma, \theta)) = d\big(\tilde{f}^*\omega, w(\omega, \sigma, \theta), z(0, p_2(\theta), 0)\big) =$$
$$\Big(d\tilde{f}^*\omega, d(w(\omega, \sigma, \theta)), d(z(0, p_2(d\theta), 0)) + (-1)^p\big[(z \circ t)(\tilde{f}^*\omega) - q_1(w(\omega, \sigma, \theta))\big]\Big)$$

Since $\tilde{f}^* d\omega = d\tilde{f}^*\omega$, $w(d(\omega, \sigma, \theta)) = d(w(\omega, \sigma, \theta)$, it remains to check the equality of the third components. We must be careful about signs: $d(z(0, p_2(d\theta), 0))$ in the above formula(s) is a differential of a $(p-1)$-form. Recalling that

$$\psi(\omega, \sigma, \theta) = \big(\tilde{f}^*\omega, (p_2 \circ p_1)(\sigma), p_2(\theta)\big)$$

and

$$t(\tilde{f}^*\omega) = (\tilde{f}^*\omega, u_2(\tilde{f}^*\omega), 0) = \left(\tilde{f}^*\omega, (p_2 \circ u_1)(\omega), 0\right)$$

we obtain

$$d\big(z(0, p_2(d\theta), 0)\big) + (-1)^p\Big[(z \circ t)\big(\tilde{f}^*\omega\big) - q_1\big(w(\omega, \sigma, \theta)\big)\Big] =$$
$$= z\big(0, dp_2(\theta), (-1)^p p_2(\theta)\big) + (-1)^p\Big[(z \circ t)\big(\tilde{f}^*\omega\big) - (z \circ \psi)(\omega, \sigma, \theta)\Big] =$$
$$= z\Big\{\big(0, dp_2(\theta), (-1)^p p_2(\theta)\big) + (-1)^p\big[\tilde{f}^*\omega - \psi(\omega, \sigma, \theta)\big]\Big\} =$$
$$= z\Big(0, dp_2(\theta) + (-1)^p\big[(p_2 \circ u_1)(\omega) - (p_2 \circ p_1)(\sigma)\big], 0\Big)$$

which gives the result.

It is clear that the above proof implies the more precise statement theorem 2.3 and remarks 2.1 and 2.2.

2.4.5 Proof of theorem 2.4: uniqueness of the primary pullback (the irreducible case)

We suppose that $(\mathrm{EP})_{m-1,n}$, $(\mathrm{EP})_{m,n-1}$, $(\mathrm{EP})_{n-1,n-1}$, $(\mathrm{U})_{m-1,n}$, $(\mathrm{U})_{m,n-1}$, $(\mathrm{U})_{n-1,n-1}$, $(\mathrm{F})_m$, $(\mathrm{E})_q$, $q = \sup\{m, n\}$ are already proved.

Let $\Lambda^{\cdot}_Y = \mathcal{E}^{\cdot}_{\tilde{Y}} \oplus \Lambda^{\cdot}_F \oplus \Lambda^{\cdot}_{\tilde{F}}(-1)$, $\Lambda^{\cdot}_X = \mathcal{E}^{\cdot}_{\tilde{X}} \oplus \Lambda^{\cdot}_E \oplus \Lambda^{\cdot}_{\tilde{E}}(-1)$.

We suppose first that $f(X) \subset F$, so that both ϕ_1 and ϕ_2 decompose through the projection $\Lambda^{\cdot}_Y \to \Lambda^{\cdot}_F$; the result follows if we apply $(\mathrm{U})_{m,n-1}$ to the induced morphism $X \to F$.

So we assume $f(X) \not\subset F$.

According to (P_3) in definition 2.2

i. $f^{-1}(F) \subset E$;

ii. the morphism f extends to a morphism $\tilde{f} \colon \tilde{X} \to \tilde{Y}$;

iii. the morphism ϕ_j is given by

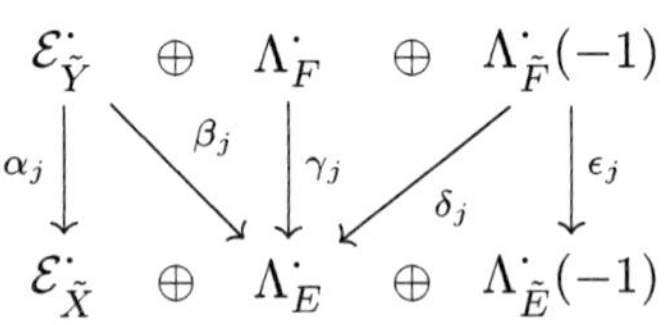

(2.10)

where $\mu_j = (\beta_j, \gamma_j, \delta_j) \colon \Lambda^{\cdot}_Y \to \Lambda^{\cdot}_E$ is a pullback corresponding to the composition $E \to X \to Y$, and $\alpha_j = \tilde{f}^*$ is the ordinary De Rham pullback.

It follows $\alpha_1 = \tilde{f}^* = \alpha_2$; by $(\mathrm{U})_{m-1,n}$ $\mu_1 = \mu_2$.

The conclusion will follow from the next lemma.

Lemma 2.4. *Let $\phi_j \colon \Lambda^{\cdot}_Y \to \Lambda^{\cdot}_X$, $j = 1, 2$, two primary pullback given by (2.10), and suppose $\alpha_1 = \alpha_2, \beta_1 = \beta_2, \gamma_1 = \gamma_2, \delta_1 = \delta_2$. Then also $\epsilon_1 = \epsilon_2$.*

In order to prove the lemma we need the following two propositions

Proposition 2.3. *Let $\phi\colon \mathcal{E}^{\cdot}_X \to \Lambda^{\cdot}_Z$ be a pullback corresponding to a morphism $f\colon Z \to X$, X being smooth, and let $(Z_s, g_s)_{s\in S}$ be the hypercovering associated to $\Lambda^{\cdot}_Z$. For every Z_s of rank 0 in Z the pullback ϕ induces the De Rham pullback $f_s^*\colon \mathcal{E}^{\cdot}_X \to \mathcal{E}^{\cdot}_{Z_s}$.*

The proof of proposition 2.3 is by induction on the dimension of Z. Let $\Lambda^{\cdot}_Z = \mathcal{E}^{\cdot}_{\tilde{Z}} \oplus \Lambda^{\cdot}_G \oplus \Lambda^{\cdot}_{\tilde{G}}(-1)$. If Z_s appears in the hypercovering of G, we conclude by induction applied to the pullback $\mathcal{E}^{\cdot}_X \to \Lambda^{\cdot}_G$. Otherwise, because of its rank, Z_s cannot appear in the hypercovering of $\tilde{G}$ so it must be a connected component of the desingularization $\tilde{Z}$, so the result is an immediate consequence of the definition of pullback.

Proposition 2.4. *Let $\Lambda^{\cdot}_X \in \mathcal{R}(X)$ and $(X_l, h_l)_{l\in L}$ be the associated hypercovering. For every X_l of positive rank $r > 0$ in X there exists X_m of rank $r-1$ and an inner differential inducing a De Rham pullback $f^*_{lm}\colon \mathcal{E}^p_{X_m} \to \mathcal{E}^p_{X_l}$.*

Proof of proposition 2.4. Here, an inner differential is one of the pullback appearing in the construction of the differential $d\colon \Lambda^{\cdot}_X \to \Lambda^{\cdot}_X$; more precisely it is one of the following:

- the pullback $\mathcal{E}^{\cdot}_{\tilde{X}} \to \Lambda^{\cdot}_{\tilde{E}}$;
- the pullback $\Lambda^{\cdot}_E \to \Lambda^{\cdot}_{\tilde{E}}$;
- an inner differential in $\Lambda^{\cdot}_E$ (inductively defined);
- an inner differential in $\Lambda^{\cdot}_{\tilde{E}}$ (inductively defined).

The proof is by induction on $m = \dim X$. Since $r > 0$, X_l appears in the hypercovering of E or in the hypercovering of $\tilde{E}$. If X_l belongs to the hypercovering of E, we conclude by induction. If X_l belongs to the hypercovering of $\tilde{E}$, then the rank s of X_l in $\tilde{E}$ is $r-1$; if $r > 1$, then $s > 0$ and we conclude by induction ($\tilde{E}$ in the place of X); if $r = 1$, then $s = 0$, and we apply the proposition 2.3 to the inner pullback $\mathcal{E}^{\cdot}_{\tilde{X}} \to \Lambda^{\cdot}_{\tilde{E}}$. □

Proof of lemma 2.4. By the hypothesis we define $\alpha = \alpha_1 = \alpha_2$, $\beta = \beta_1 = \beta_2$, $\gamma = \gamma_1 = \gamma_2$, $\delta = \delta_1 = \delta_2$.

Let u, p be the inner pullback in $\Lambda^{\cdot}_X$, and v, q those in $\Lambda^{\cdot}_Y$.

Then $\phi \circ d = d \circ \phi$ implies for $\epsilon = \epsilon_1$ or $\epsilon = \epsilon_2$:

$$\epsilon(v(\omega)) = u(f^*(\omega)) - (p \circ \phi)(\omega, 0, 0) \tag{2.11}$$

$$(\epsilon \circ d - d \circ \epsilon)(\gamma) = (-1)^p (p \circ \phi)(0, 0, \gamma) \tag{2.12}$$

Let (X_l) and (Y_s) be the hypercoverings corresponding to $\Lambda^{\cdot}_X$ and $\Lambda^{\cdot}_Y$; for every X_l appearing in $\tilde{E}$ there is at most a Y_s appearing in $\tilde{F}$ such that the morphism $\epsilon_{1,ls}\colon \mathcal{E}^p_{Y_s} \to \mathcal{E}^p_{X_l}$ induced a by ϵ_1 is a De Rham pullback; otherwise

$\epsilon_{1,ls} = 0$; the same for $\epsilon_{2,ls}$. Since $\epsilon_{1,ls}$ and $\epsilon_{2,ls}$ completely determine ϵ_1 and ϵ_2, it will be enough to prove that $\epsilon_{1,ls} = 0$ if and only if $\epsilon_{2,ls} = 0$; more precisely we check that (2.11) and (2.12) completely determine whether ϵ_{ls} is zero or nonzero.

Let $s \geq 0$ be the rank of X_l in $\tilde{E}$, so that the rank of X_l in X is $r = s+1 > 0$. We proceed by induction on s.

Let first s be 0. We apply proposition 2.3 to the inner pullback $v\colon \mathcal{E}^{\cdot}_{\tilde{Y}} \to \Lambda^{\cdot}_{\tilde{F}}$; hence for every s such that Y_s appears in $\tilde{F}$ with rank 0 there is $\omega \in \mathcal{E}^{0}_{\tilde{Y}}$ such that $v(\omega) \neq 0 \in \mathcal{E}^{0}_{Y_s}$. it follows that $\epsilon_{ls} = 0$ or $\neq 0$ according to the second member of (2.11) , which is the same for ϵ_1 and ϵ_2.

We consider now the case $s > 0$. Let us fix l and s; by proposition 2.4 there exist Y_k of rank $s - 1 \geq 0$ in $\tilde{F}$ and an inner pullback inducing a De Rham pullback $a_{sk}\colon \mathcal{E}^{\cdot}_{Y_k} \to \mathcal{E}^{\cdot}_{Y_s}$.

Let $\theta \in \Lambda^{\cdot}_{\tilde{F}}$ be defined as $\theta = (0, \ldots, \theta_k, \ldots, 0)$; then $(d\theta)_l = \pm a_{sk}(\theta_k)$. Moreover $\epsilon(\theta) = (\epsilon_{tk}(\theta_k))_t$ where $\epsilon_{tk}\colon \mathcal{E}^{p}_{Y_k} \to \mathcal{E}^{p}_{X_t}$ involves only components Y_k and X_t of rank $< s$. hence by induction $\epsilon_{1,tk} = \epsilon_{2,tk}$ and $d\epsilon_1(\theta) = d\epsilon_2(\theta)$. From (2.12) it follows that $\epsilon_{sl}(a_{sk}(\theta_k)$ is the same for ϵ_1 and ϵ_2. Taking any $\theta_k \neq 0 \in \mathcal{E}^{0}_{Y_k}$ we obtain $a_{sk}(\theta_k) \neq 0 \in \mathcal{E}^{0}_{Y_s}$ so that $\epsilon_{1,ls} = 0$ if and only if $\epsilon_{2,ls} = 0$. □

2.4.6 Proof of theorem 2.5: existence of pullback (the general case)

We suppose that $(\mathrm{EP})_{m-1,n}$, $(\mathrm{EP})_{m-1,m-1}$, $(\mathrm{U})_{m-1,n}$, $(\mathrm{F})_m$, $(\mathrm{E})_q$, $q = \sup\{m,n\}$ are already proved. Moreover we suppose $(\mathrm{EPP})_{m,n}$ true for morphisms between irreducible spaces.

First we deal with the case X irreducible. There exists an irreducible component Y_k of Y such that $f(X) \subset Y_k$; let $\Lambda^{\cdot}_Y \to \Lambda^{\cdot}_{Y_k}$ be a natural pullback; by $(\mathrm{EPP})_{m,n}$ there exists a pullback $\Lambda^{\cdot}_{Y_k} \to \Lambda^{\cdot}_X$. By composition we get the required pullback.

Let us consider the case where X is reducible; let $X = \bigcup_j X_j$ be the decomposition of X into its irreducible components. By the above case for every j there exists a pullback $\phi_j\colon \Lambda^{\cdot}_Y \to \Lambda^{\cdot}_{X_j}$. Let $\Lambda^{\cdot}_{X_j} = \mathcal{E}^{\cdot}_{\tilde{X}_j} \oplus \Lambda^{\cdot}_{E_j} \oplus \Lambda^{\cdot}_{\tilde{E}_j}(-1)$ and denote by $\rho_j\colon \Lambda^{\cdot}_{E_j} \to \Lambda^{\cdot}_{\tilde{E}_j}$ and $\eta_j\colon \mathcal{E}^{\cdot}_{\tilde{X}_j} \to \Lambda^{\cdot}_{\tilde{E}_j}$ the inner pullback. Then $\mathrm{sing}(X_j) \subset E_j$; using $(\mathrm{F})_m$ and $(\mathrm{E})_m$ we can enlarge E_j in order to obtain $\mathrm{sing}(X) \cap X_j \subset E_j$, so that if we define $E = \bigcup_j E_j$ we obtain $\mathrm{sing}(X) \subset E$. Let moreover $\tilde{X} = \bigsqcup_j \tilde{X}_j$, $\tilde{E} = \bigsqcup_j \tilde{E}_j$.

By $(\mathrm{EP})_{m-1,n}$ and $(\mathrm{EP})_{m-1,m-1}$ there are pullback $\beta\colon \Lambda^{\cdot}_Y \to \Lambda^{\cdot}_E$ and $\mu_j\colon \Lambda^{\cdot}_E \to \Lambda^{\cdot,1}_{E_j}$; if we apply $(\mathrm{F})_{m-1}$ to E_j we find two pullback $\Lambda^{\cdot}_{E_j} \to \Lambda^{\cdot,0}_{E_j}, \Lambda^{\cdot,1}_{E_j} \to \Lambda^{\cdot,0}_{E_j}$; by $(\mathrm{E})_m$ there exists $\Lambda^{\cdot,0}_{X_j} = \mathcal{E}^{\cdot}_{\tilde{X}_j} \oplus \Lambda^{\cdot,0}_{E_j} \oplus \Lambda^{\cdot,0}_{\tilde{E}_j}(-1)$ and a pullback $\theta_j\colon \Lambda^{\cdot}_{X_j} \to \Lambda^{\cdot,0}_{X_j}$. Finally, after replacing $\Lambda^{\cdot}_{E_j}$ by $\Lambda^{\cdot,0}_{E_j}$, $\Lambda^{\cdot}_{X_j}$ by

$\Lambda^{\cdot,0}_{X_j}$ and ϕ_j by $\theta_j \circ \phi_j$ we can assume that $\mu_j \colon \Lambda^{\cdot}_E \to \Lambda^{\cdot}_{E_j}$. Let ϕ_j be given by

$$\begin{aligned} \Lambda^p_Y \to \Lambda^p_{X_j} &= \mathcal{E}^p_{\tilde{X}_j} \oplus \Lambda^p_{E_j} \oplus \Lambda^{p-1}_{\tilde{E}_j} \\ \omega &\mapsto \big(\alpha_j(\omega), \beta_j(\omega), \theta_j(\omega)\big) \end{aligned}$$

Recall that by the definition of pullback, $\beta_j \colon \Lambda^{\cdot}_Y \to \Lambda^{\cdot}_{E_j}$ is a pullback. By $(\mathrm{U})_{m-1,n}$

$$\beta_j = \mu_j \circ \beta$$

We can now define

$$\Lambda^{\cdot}_X = \mathcal{E}^{\cdot}_{\tilde{X}} \oplus \Lambda^{\cdot}_E \oplus \Lambda^{\cdot}_{\tilde{E}}(-1)$$

with inner pullback given by

$$\xi \colon \Lambda^{\cdot}_E \to \Lambda^{\cdot}_{\tilde{E}} = \oplus_j \Lambda^{\cdot}_{\tilde{E}_j}$$

$$\sigma \to \oplus_j \rho_j(\mu_j(\sigma))$$

and

$$\lambda \colon \mathcal{E}^{\cdot}_{\tilde{X}} = \oplus_j \mathcal{E}^{\cdot}_{\tilde{X}_j} \to \Lambda^{\cdot}_{\tilde{E}} = \oplus_j \Lambda^{\cdot}_{\tilde{E}_j}$$

$$(\tau_j) \to \oplus_j \eta_j(\tau_j)$$

Finally we define the morphism

$$\begin{aligned} \phi \colon \Lambda^p_Y \to \Lambda^p_X &= \oplus_j \mathcal{E}^p_{\tilde{X}_j} \oplus \Lambda^p_E \oplus \oplus_j \Lambda^{p-1}_{\tilde{E}_j} \\ \omega &\mapsto \big(\oplus_j\, \alpha_j(\omega), \beta(\omega), \oplus_j \theta_j(\omega)\big) \end{aligned}$$

We check that ϕ commutes with differentials. Since ϕ_j commutes with differentials, we get for $\omega \in \Lambda^p_Y$:

$$\begin{aligned} d\alpha_j(\omega) &= \alpha_j(d\omega) \\ d\beta_j(\omega) &= \beta_j(d\omega) \\ d\theta_j(\omega) &= \theta_j(d\omega) - (-1)^p\big[\eta_j(\alpha_j(\omega)) - \rho_j(\beta_j(\omega))\big] \end{aligned}$$

Using the equality $\rho_j \circ \beta_j = \rho_j \circ \mu_j \circ \beta$ a simple computation shows that ϕ commutes to d. It is clear by construction that ϕ satisfies all the properties in the definition 2.4 so it is the required pullback.

As to remark 2.3, it is clear from the proof that if $X = Y, f = \mathrm{id}$, $m = n$ we do not need the assumptions $(\mathrm{EP})_{m,m-1}$, $(\mathrm{U})_{m,m-1}$, because for every j, $f(X_j) \not\subset E$; moreover we need $(\mathrm{F})_m$ only for the (irreducible) X_j.

2.4.7 Proof of theorem 2.6: uniqueness of the pullback (the general case)

We suppose that $(\mathrm{U})_{m-1,n}$, $(\mathrm{F})_m$, $(\mathrm{E})_q$, $q = \sup\{m,n\}$ are already proved. Moreover we suppose $(\mathrm{UP})_{m,n}$ true for morphisms between irreducible spaces.

Let $f\colon X \to Y$ be a morphism between (reducible) complex spaces, $X = \bigcup_j X_j$, $Y = \bigcup_k Y_k$ be the respective decompositions into irreducible components. Let $\Lambda^{\cdot}_X = \mathcal{E}^{\cdot}_{\tilde{X}} \oplus \Lambda^{\cdot}_E \oplus \Lambda^{\cdot}_{\tilde{E}}(-1)$, $\Lambda^{\cdot}_Y = \mathcal{E}^{\cdot}_{\tilde{Y}} \oplus \Lambda^{\cdot}_F \oplus \Lambda^{\cdot}_{\tilde{F}}(-1)$. Let $\phi_i\colon \Lambda^{\cdot}_Y \to \Lambda^{\cdot}_X$, $i = 1,2$ be two pullback. Let u, p be the inner pullback in $\Lambda^{\cdot}_X$, and v, q those in $\Lambda^{\cdot}_Y$.

For $(\omega, \sigma, \theta) \in \Lambda^p_Y$, we have

$$\begin{aligned}
\phi_i(\omega,0,0) &= \big(\alpha_i(\omega), \beta_i(\omega), 0\big)\\
\phi_i(0,\sigma,0) &= \big(\rho_i(\sigma), \gamma_i(\sigma), \eta_i(\sigma)\big)\\
\phi_i(0,0,\theta) &= \big(0, \delta_i(\theta), \epsilon_i(\theta)\big)
\end{aligned}$$

In fact the morphism $\mathcal{E}^p_{\tilde{Y}} \to \Lambda^{p-1}_{\tilde{E}}$ induced by ϕ_i is identically zero because the components of $\tilde{Y}$ have rank 0 in Y, while the components of the hypercovering of $\Lambda^{\cdot}_X$ belonging to $\Lambda^{\cdot}_{\tilde{E}}$ have strictly positive rank in X; for an analogous reason the morphism $\Lambda^{p-1}_{\tilde{F}} \to \mathcal{E}^p_{\tilde{X}}$ is zero.

By (P_4) in the definition 2.4, and $(\mathrm{U})_{m-1,n}$, we obtain $\beta_1 = \beta_2 = \beta, \gamma_1 = \gamma_2 = \gamma, \delta_1 = \delta_2 = \delta$.

Moreover it is clear that $\alpha_i = 0$ on $\tilde{X}_j$ in case $f(X_j) \subset F$, while α_i is an ordinary De Rham pullback on $\tilde{X}_j$ if $f(X_j) \not\subset F$; it follows $\alpha_1 = \alpha_2 = \alpha$.

For any irreducible component X_j of X we consider a commutative diagram like (2.8) (or (2.7): left to the reader) for ϕ_1. Because of lemma 2.1 we have for ϕ_2 an analogous diagram, with the same $\Lambda^{\cdot,0}_{X_j}$ and $\Lambda^{\cdot}_{Y_k}$, and same natural pullback $\Lambda^{\cdot}_X \to \Lambda^{\cdot,0}_{X_j}$ and $\Lambda^{\cdot}_Y \to \Lambda^{\cdot}_{Y_k}$.

By $(\mathrm{UP})_{m,n}$ applied to the morphism of irreducible spaces $X_j \to Y_k$ we can suppose that the diagrams (2.8) for ϕ_1 and ϕ_2 coincide, so that the same is true for the composite morphisms $\pi_i\colon \Lambda^{\cdot}_Y \to \Lambda^{\cdot}_{Y_k} \to \Lambda^{\cdot,0}_{X_j}$, i.e. $\pi_1 = \pi_2$. Since by construction π_i and ϕ_i induce the same morphisms $\Lambda^p_{\tilde{F}} \to \mathcal{E}^p_{\tilde{X}_j}$ for every j, we conclude that also $\rho_1 = \rho_2$.

Since ϕ_i commutes to the differentials we can deduce

$$\epsilon_i(v(\omega)) = u(\alpha(\omega)) - p(\beta(\omega)) \tag{2.13}$$

$$(\epsilon_i \circ d - d \circ \epsilon_i)(\theta) = (-1)^p(p(\delta(\theta)) \tag{2.14}$$

$$(\eta_i \circ d - d \circ \eta_i)(\sigma) = (-1)^p \epsilon_i(q(\sigma)) + (-1)^p\big[u(d\rho(\sigma)) - q(d\gamma(\sigma))\big] \tag{2.15}$$

Arguing as in the proof of lemma 2.4, by (2.13) and (2.14) we obtain $\epsilon_1 = \epsilon_2$; then using (2.15) (instead of (2.12)) again an argument similar to that in the proof of the lemma 2.4 shows that $\eta_1 = \eta_2$ (the reader should note that η_1 and η_2 do not involve $\tilde{Y}$).

2.4.8 Proof of theorem 2.9: filtering

We suppose that $(\mathrm{EP})_{m-1,m-1}$, $(\mathrm{U})_{m-1,m-1}$, $(\mathrm{F})_{m-1}$, $(\mathrm{E})_m$, are already proved.

Let $\Lambda^{\cdot,1}_X = \mathcal{E}^{\cdot}_{\tilde{X}_1} \oplus \Lambda^{\cdot,1}_E \oplus \Lambda^{\cdot,1}_{\tilde{E}}(-1)$, $\Lambda^{\cdot,2}_X = \mathcal{E}^{\cdot}_{\tilde{X}_2} \oplus \Lambda^{\cdot,2}_F \oplus \Lambda^{\cdot,2}_{\tilde{F}}(-1)$.

First we consider the case $\tilde{X}_1 = \tilde{X}_2 = \tilde{X}, E = F, \tilde{E} = \tilde{F}$. By $(\mathrm{F})_{m-1}$ there are pullback $\Lambda^{\cdot,1}_E \to \Lambda^{\cdot}_E$, $\Lambda^{\cdot,2}_E \to \Lambda^{\cdot}_E$, $\Lambda^{\cdot,1}_{\tilde{E}} \to \Lambda^{\cdot}_{\tilde{E}}$, $\Lambda^{\cdot,2}_{\tilde{E}} \to \Lambda^{\cdot}_{\tilde{E}}$. Using $(\mathrm{EP})_{m-1,m-1}$, $(\mathrm{U})_{m-1,m-1}$ and again $(\mathrm{F})_{m-1}$ we find a pullback $\Lambda^{\cdot}_E \to \Lambda^{\cdot}_{\tilde{E}}$ such that the following two diagrams for $j = 1, 2$ are commutative

$$\begin{array}{ccc} \Lambda^{\cdot,j}_E & \longrightarrow & \Lambda^{\cdot}_E \\ \downarrow & & \downarrow \\ \Lambda^{\cdot,j}_{\tilde{E}} & \longrightarrow & \Lambda^{\cdot}_{\tilde{E}} \end{array}$$

where the left vertical arrows come from the inner differentials in $\Lambda^{\cdot,j}_X$. Finally by $(\mathrm{E})_m$ we find $\Lambda^{\cdot}_X = \mathcal{E}^{\cdot}_{\tilde{X}} \oplus \Lambda^{\cdot}_E \oplus \Lambda^{\cdot}_{\tilde{E}}(-1)$ and obvious pullback $\Lambda^{\cdot,j}_X \to \Lambda^{\cdot}_X$.

Lemma 2.5. *Under the assumptions of the theorem, let* $\Lambda^{\cdot}_X = \mathcal{E}^{\cdot}_{\tilde{X}} \oplus \Lambda^{\cdot}_E \oplus \Lambda^{\cdot}_{\tilde{E}}(-1)$*, and let* G *be a nowhere dense closed subspace of* X *with* $E \subset G$*. Then there exists* $\Lambda^{\cdot,0}_X = \mathcal{E}^{\cdot}_{\tilde{X}} \oplus \Lambda^{\cdot}_G \oplus \Lambda^{\cdot}_{\tilde{G}}(-1) \in \mathcal{R}(X)$ *and a pullback* $\Lambda^{\cdot}_X \to \Lambda^{\cdot,0}_X$ *corresponding to the identity.*

It is now clear how to prove $(\mathrm{F})_m$: we take $G = E \cup F$. By the lemma we find $\Lambda^{\cdot,0}_X = \mathcal{E}^{\cdot}_{\tilde{X}} \oplus \Lambda^{\cdot}_G \oplus \Lambda^{\cdot}_{\tilde{G}}(-1)$ and two pullback $\Lambda^{\cdot,j}_X \to \Lambda^{\cdot,0}_X$, $j = 1, 2$.

Sketch of the proof of the lemma. In the case where X is irreducible, the lemma is a consequence of the remark 2.1. In the general case it follows from the remark 2.3, because meanwhile $(\mathrm{F})_m$ has been already proved for each irreducible component of X. □

2.5 Kähler hypercoverings

We recall that in part I, chapter 7 we have defined a (B)-Kähler space as a compact complex space X dominated by a compact Kähler manifold, or, equivalently, there is a modification $f\colon M \to X$. A (B)-Kähler manifold is a smooth (B)-Kähler space (it is not necessarily a Kähler manifold). For a (B)-Kähler space X, we can deal with complexes $\Lambda^{\cdot}_X$ whose associated hypercovering is formed by Kähler manifolds. More precisely:

Definition 2.5. The hypercovering $(X_l)_{l \in L}$ associated to $\Lambda^{\cdot}_X$ is called a Kähler hypercovering if each X_l is a Kähler manifold.

Theorem 2.10.

i. *Let X be a (B)-Kähler space. Then there exists a $\Lambda^{\cdot}_X \in \mathcal{R}(X)$ with a Kähler hypercovering.*

ii. *Let $f\colon X \to Y$ be a morphism of (B)-Kähler spaces, and fix $\Lambda^{\cdot}_Y \in \mathcal{R}(Y)$ with a Kähler hypercovering. Then there exists a $\Lambda^{\cdot}_X \in \mathcal{R}(X)$ with a Kähler hypercovering and a pullback $\Lambda^{\cdot}_Y \to \Lambda^{\cdot}_X$.*

iii. *If X is a (B)-Kähler space and $\Lambda^{\cdot,1}_X$, $\Lambda^{\cdot,2}_X \in \mathcal{R}(X)$ admit a Kähler hypercovering, there exists a third $\Lambda^{\cdot}_X \in \mathcal{R}(X)$ with a Kähler hypercovering and two pullback $\Lambda^{\cdot,1}_X \to \Lambda^{\cdot}_X$, $\Lambda^{\cdot,2}_X \to \Lambda^{\cdot}_X$ corresponding to the identity.*

This is easily seen by induction on $\dim X$. For example, let us check (i). If X is (B)-Kähler, there exist diagrams (2.1) where $\tilde{X}$ is Kähler, and $E \subset X$, $\tilde{E} \subset \tilde{X}$ (as subspaces of a (B)-Kähler space) are (B)-Kähler. Let $\Lambda^{\cdot}_X = \mathcal{E}^{\cdot}_{\tilde{X}} \oplus \Lambda^{\cdot}_E \oplus \Lambda^{\cdot}_{\tilde{E}}(-1)$ be obtained from such a diagram (2.1). By induction the hypercoverings of $\Lambda^{\cdot}_E$ and $\Lambda^{\cdot}_{\tilde{E}}$ can be taken Kähler. Hence the hypercovering of $\Lambda^{\cdot}_X$, which is the union of the hypercoverings of $\Lambda^{\cdot}_E$, $\Lambda^{\cdot}_{\tilde{E}}$, plus $\tilde{X}$, is Kähler.

2.6 Chains and homology

For any real analytic space M we denote by $\mathcal{S}_{M,k}$ the sheaf of germs of subanalytic k-chains on M (with coefficients in $\mathbb{C}$), and $\partial\colon \mathcal{S}_{M,k} \to \mathcal{S}_{M,k-1}$ the boundary operator (see part I, chapter 7 , section 7.8)

For any complex space X and any $\Lambda^{\cdot}_X = \mathcal{E}^{\cdot}_{\tilde{X}} \oplus \Lambda^{\cdot}_E \oplus \Lambda^{\cdot}_{\tilde{E}}(-1)$ (corresponding to the diagram (2.1)) we introduce **the co-complex of sheaves of chains** $(\mathcal{C}_{X,\cdot}, \partial)$ **dual to** $\Lambda^{\cdot}_X$, and for every morphism $f\colon X \to Y$ of complex spaces and every pullback $\alpha\colon \Lambda^{\cdot}_Y \to \Lambda^{\cdot}_X$ a pushdown $\alpha_*\colon f_*\mathcal{C}_{X,\cdot} \to \mathcal{C}_{Y,\cdot}$ satisfying the pullback-pushdown formula

$$\langle \alpha(\omega), \gamma \rangle = \langle \omega, \alpha_* \gamma \rangle \tag{2.16}$$

Inductively we define

$$\mathcal{C}_{X,k} = \pi_* \mathcal{S}_{\tilde{X},k} \oplus j_* \mathcal{C}_{E,k} \oplus (j \circ q)_* \mathcal{C}_{\tilde{E},k-1}$$

which we simply write

$$\mathcal{C}_{X,k} = \mathcal{S}_{\tilde{X},k} \oplus \mathcal{C}_{E,k} \oplus \mathcal{C}_{\tilde{E},k-1}$$

and the boundary operator as

$$\begin{aligned} \partial\colon \mathcal{C}_{X,k} = \mathcal{S}_{\tilde{X},k} \oplus \mathcal{C}_{E,k} \oplus \mathcal{C}_{\tilde{E},k-1} &\to \mathcal{C}_{X,k-1} = \mathcal{S}_{\tilde{X},k-1} \oplus \mathcal{C}_{E,k-1} \oplus \mathcal{C}_{\tilde{E},k-2} \\ \partial(\alpha, \beta, \gamma) &= \big(\partial\alpha + (-1)^k \psi_* \gamma, \partial\beta - (-1)^k \phi_* \gamma, \partial\gamma\big) \end{aligned} \tag{2.17}$$

where ψ_* and ϕ_* are the pushdown dual to the inner pullback of $\Lambda^{\cdot}_X$. It is clear that $\partial\partial = 0$, so that we obtain a co-complex $(\mathcal{C}_{X,\cdot}, \partial)$.

For an open set U of X let $C_k(U) = \Gamma_c(U, \mathcal{C}_{X,k})$ be the $\mathbb{C}$-module of the chains with compact support in U. Then

$$C_k(U) = S_{\tilde{X},k}(\tilde{U}) \oplus C_{E,k}(U \cap E) \oplus C_{\tilde{E},k-1}(\tilde{U} \cap \tilde{E})$$

where $S_{\tilde{X},k}(\tilde{U})$, $C_{E,k}(U \cap E)$, $C_{\tilde{E},k-1}(\tilde{U} \cap \tilde{E})$ are the respective chains with compact support (inductively defined for E and $\tilde{E}$).

Theorem 2.11. *The homology of the complex of global sections $(C.(X), \partial)$ is the Borel-Moore homology with compact supports of X:*

$$H^c_k(X, \mathbb{C}) = \frac{\ker\{\,\partial\colon C_k(X) \to C_{k-1}(X)\,\}}{\partial C_{k+1}(X)} \tag{2.18}$$

Proof. The presheaf on X defined by $U \mapsto \hom_{\mathbb{C}}(C_k(U), \mathbb{C})$ is a flabby sheaf, which we denote by $\mathcal{C}^k_X$. The boundaries ∂ give by transposition differentials $\delta\colon \mathcal{C}^k_X \to \mathcal{C}^{k+1}_X$ such that $\delta \circ \delta = 0$, hence $(\mathcal{C}^{\cdot}_X, \delta)$ is a complex. It is enough to prove that $(\mathcal{C}^{\cdot}_X, \delta)$ is a resolution of $\mathbb{C}_X$.

The sheaf $\mathcal{C}^k_X$ decomposes as

$$\mathcal{C}^k_X = \mathcal{S}^k_{\tilde{X}} \oplus \mathcal{C}^k_E \oplus \mathcal{C}^{k-1}_{\tilde{E}} \tag{2.19}$$

and by part I, chapter 7, theorem 7.22, $\mathcal{S}^{\cdot}_{\tilde{X}}$ is a resolution of constant sheaf on $\tilde{X}$, and by induction on the dimensions, $\mathcal{C}^{\cdot}_E$ and $\mathcal{C}^{\cdot}_{\tilde{E}}$ are resolutions of the constant sheaf on E and $\tilde{E}$ respectively.

The augmentation $\mathbb{C}_X \to \mathcal{C}^0_X$ is given by $c \mapsto (c, c, 0)$.

The duality between $(\omega, \sigma, \theta) \in \mathcal{C}^k_X$ and $(\alpha, \beta, \gamma) \in \mathcal{C}_{X,k}$ is given by

$$\langle\,(\omega, \sigma, \theta), (\alpha, \beta, \gamma)\,\rangle = \langle\,\omega, \alpha\,\rangle + \langle\,\sigma, \beta\,\rangle + \langle\,\theta, \gamma\,\rangle \tag{2.20}$$

and δ is defined by

$$\langle\,\delta(\omega, \sigma, \theta), (\alpha, \beta, \gamma)\,\rangle = \langle\,(\omega, \sigma, \theta), \partial(\alpha, \beta, \gamma)\,\rangle \tag{2.21}$$

By (2.18), (2.20), (2.21) and using the pullback-pushdown formula (2.16)

$$\langle\,\psi(\omega), \gamma\,\rangle = \langle\,\omega, \psi_*\gamma\,\rangle, \quad \langle\,\phi(\sigma), \gamma\,\rangle = \langle\,\sigma, \phi_*\gamma\,\rangle$$

we easily obtain

$$\delta(\omega, \sigma, \theta) = \big(\delta\omega, \delta\sigma, \delta\theta + (-1)^k(\psi(\omega) - \phi(\sigma))\big) \tag{2.22}$$

The above formula looks exactly like the formula for d in (2.3). Hence, *mutatis mutandis*, the proof that $(\mathcal{C}^{\cdot}_X, \delta)$ is a resolution of $\mathbb{C}_X$ follows almost word by word that of theorem 2.1. □

2.7 Integration of forms on chains

There is a duality between forms in Λ_X^k and chains in $\mathcal{C}_{X,k}$, given by integration: for $(\alpha,\beta,\gamma) \in \mathcal{C}_{X,k}$ and $(\omega,\sigma,\theta) \in \Lambda_X^k$ we put

$$\int_{(\alpha,\beta,\gamma)} (\omega,\sigma,\theta) = \int_\alpha \omega + \int_\beta \sigma + \int_\gamma \theta \tag{2.23}$$

where the $\int_\alpha \omega$ is the usual integral of a form on a subanalytic chain on the manifold $\tilde{X}$, and $\int_\beta \sigma$, $\int_\gamma \theta$ are defined by induction on the dimensions of complex spaces.

Lemma 2.6 (Stokes theorem). *Let $\lambda \in \Gamma(X,\Lambda_X^k)$ and $\eta \in \Gamma(X,\mathcal{C}_{X,k})$ such that λ, or η, has compact support. Then*

$$\int_\eta d\lambda = \int_{\partial\eta} \lambda \tag{2.24}$$

Proof. Let $\lambda = (\omega,\sigma,\theta)$, $\eta = (\alpha,\beta,\gamma)$; we must prove:

$$\int_{(\alpha,\beta,\gamma)} d(\omega,\sigma,\theta) = \int_{\partial(\alpha,\beta,\gamma)} (\omega,\sigma,\theta) \tag{2.25}$$

In fact by (2.3)

$$\int_{(\alpha,\beta,\gamma)} d(\omega,\sigma,\theta) = \int_\alpha d\omega + \int_\beta d\sigma + \int_\gamma (d\theta + (-1)^k(\psi(\omega) - \phi(\sigma)))$$

We suppose by induction the Stokes theorem true for E and $\tilde{E}$; using the Stokes theorem for the manifold $\tilde{X}$, we obtain

$$\int_\alpha d\omega = \int_{\partial\alpha} \omega, \quad \int_\beta d\sigma = \int_{\partial\beta} \sigma, \quad \int_\gamma d\theta = \int_{\partial\gamma} \theta$$

on the other hand

$$\int_{\partial(\alpha,\beta,\gamma)} (\omega,\sigma,\theta) = \int_{\partial\alpha} \omega + (-1)^k \int_{\psi_*\gamma} \omega + \int_{\partial\beta} \sigma - (-1)^k \int_{\phi_*\gamma} \sigma + \int_{\partial\gamma} \theta =$$
$$= \int_{\partial\alpha} \omega + (-1)^k \int_\gamma \psi(\omega) + \int_{\partial\beta} \sigma - (-1)^k \int_\gamma \phi(\sigma) + \int_{\partial\gamma} \theta$$

which implies (2.25). □

As a consequence we obtain the following analogous of one of the classical theorems of De Rham.

Theorem 2.12. *Let X be a complex space. Then the integration formula (2.23) induces the natural duality between the cohomology $H^k(X,\mathbb{C})$ and the homology $H_k^c(X,\mathbb{C})$.*

2.8 The complex of Grauert and Grothendieck

A complex of differential forms on any complex (or real analytic) space X, which we denote by $D^{\cdot}_X$, goes back to H. Grauert, and A. Grothendieck. The definition of $D^{\cdot}_X$ is local. Let U be an open set of X which embeds as a subspace into an open set $V \subset \mathbb{C}^N$. Let $i\colon U^* \to V$ be the restriction of the embedding to the set U^* of the regular points of U. Let $\mathcal{N}^p \subset \mathcal{E}^p_V$ be the subsheaf of the $\omega \in \mathcal{E}^p_V$ such that $i^*\omega = 0$. Then by definition

$$D^p_U = (\mathcal{E}^p_V / \mathcal{N}^p)|_U$$

and a differential d is induced because of the inclusion $d(\mathcal{N}^p) \subset \mathcal{N}^{p+1}$.

$D^{\cdot}_X$ is a complex of fine sheaves, provided with an augmentation $\mathbb{C}_X \to D^{\cdot}_X$; for $p > 2\dim X$, $D^p_X = 0$. The restriction $D^{\cdot}_X|_U$ to the open subset U of smooth points of X is the ordinary De Rham complex $\mathcal{E}^{\cdot}_U$. If $f\colon X \to Y$ is a morphism of complex spaces, the pullback $f^*\colon D^{\cdot}_Y \to D^{\cdot}_X$ is defined; it commutes to differentials, and is functorial.

Unfortunately, the complex $\left(D^{\cdot}_X, d\right)$ is not exact in general (see [GK] and [BH]), hence it is not a resolution of $\mathbb{C}_X$.

It is easy to see that the complex $D^{\cdot}_X$ is in a natural way a subcomplex of $\Lambda^{\cdot}_X$. More precisely for any $\Lambda^{\cdot}_X \in \mathcal{R}(X)$ there is an injective morphism of complexes

$$\eta_X : D^{\cdot}_X \to \Lambda^{\cdot}_X$$

such that for every pullback $\alpha\colon \Lambda^{\cdot}_Y \to \Lambda^{\cdot}_X$ corresponding to a morphism $f\colon X \to Y$ the following diagram

$$\begin{array}{ccc} D^{\cdot}_Y & \xrightarrow{\eta_Y} & \Lambda^{\cdot}_Y \\ {\scriptstyle f^*}\big\downarrow & & \big\downarrow{\scriptstyle \alpha} \\ D^{\cdot}_X & \xrightarrow{\eta_X} & \Lambda^{\cdot}_X \end{array} \tag{2.26}$$

commutes.

With the notations of definition 2.1: $\Lambda^{\cdot}_X = \pi_*\mathcal{E}^{\cdot}_{\tilde{X}} \oplus j_*\Lambda^{\cdot}_E \oplus (j \circ q)_*\Lambda^{\cdot}_{\tilde{E}}(-1)$ we define

$$\eta_X(\omega) = \left(\pi^*(\omega), \eta_E(j^*(\omega)), 0\right)$$

where η_E has been defined by induction, and $\pi^*(\omega) \in D^{\cdot}_{\tilde{X}} = \mathcal{E}^{\cdot}_{\tilde{X}}$. In other words, looking at the hypercovering $(X_l, h_l)_{l\in L}$, we have $\eta_X(\omega)_l = h_l{}^*(\omega)$ if $q(l) = 0$ and $\eta_X(\omega)_l = 0$ otherwise. Hence $\eta_X(\omega)$ lives on the 0-skeleton of the hypercovering, i.e. on the spaces X_l with $q(l) = 0$. Again by induction we prove $\psi\left(\pi^*(\omega)\right) = \phi\left(\eta_E(j^*(\omega))\right)$ which implies that η_X commutes to differentials in $D^{\cdot}_X$ and $\Lambda^{\cdot}_X$. Since π^* is injective, η_X is injective too.

Chapter 3

Mixed Hodge structures on compact spaces

3.1 Introduction

Let X be a complex space, $\Lambda^{\cdot}_X = \mathcal{E}^{\cdot}_{\tilde{X}} \oplus \Lambda^{\cdot}_E \oplus \Lambda^{\cdot}_{\tilde{E}}(-1)$ an element of $\mathcal{R}(X)$ as in chapter 2. In the present chapter we introduce three filtrations on $\Lambda^{\cdot}_X$: the weight filtration W, the Hodge filtration F and its conjugate $\bar{F}$. They are defined by induction on the dimensions of the spaces; they are supposed to be already defined on $\Lambda^{\cdot}_E$ and $\Lambda^{\cdot}_{\tilde{E}}$, on the other hand they are known for the De Rham complex $\mathcal{E}^{\cdot}_{\tilde{X}}$. So we define the filtrations on $\Lambda^{\cdot}_X$ as direct sums (up a shift on $\Lambda^{\cdot}_{\tilde{E}}$ in the case of the weight filtration). The above filtrations induce corresponding filtrations on the complex of global sections $\Gamma(X, \Lambda^{\cdot}_X)$ and on the cohomology $H^k(X, \mathbb{C})$.

For the complex $\Gamma(X, \Lambda^{\cdot}_X)$ and its filtration W, we construct the corresponding spectral sequence denoted by $E_r^{m,k}$. Then, under the assumption that the hypercovering associated to $\Lambda^{\cdot}_X$ is formed by compact Kähler manifolds, we prove:

1. the spectral sequence degenerates at level 2: $d_r = 0$, hence $E_r^{m,k} = E_2^{m,k}$, for $r \geq 2$.

2. the $E_2^{m,k}$ carry a pure Hodge structure, and are isomorphic to the graded quotients $\frac{W_m H^k(X,\mathbb{C})}{W_{m-1} H^k(X,\mathbb{C})}$ of the cohomology $H^k(X, \mathbb{C})$ with respect to the weight filtrations.

The main consequence is that $H^k(X, \mathbb{C})$ acquires a mixed Hodge structure.

3.2 Filtration by the degree: the weight filtration

Let M be a complex manifold. The **weight filtration** on the De Rham complex $\mathcal{E}^{\cdot}_M$ is the trivial increasing filtration

$$W_m\mathcal{E}^k_M = \mathcal{E}^k_M \quad \text{for } m \geq 0, W_m\mathcal{E}^k_M = 0 \quad \text{for } m < 0$$

Let X be a complex space, $\Lambda^{\cdot}_X$ an element of $\mathcal{R}(X)$. We define an increasing **weight filtration** W_m on $\Lambda^{\cdot}_X$ with the following properties:

i. $W_m\Lambda^k_X = \Lambda^k_X \quad \text{for } m \geq 0$.

ii. $W_m\Lambda^k_X = 0$ for $m < -k$.

iii. $(\Lambda^{\cdot}_X, d)$ is a filtered complex, namely

$$d(W_m\Lambda^k_X) \subset W_m\Lambda^{k+1}_X$$

iv. If X is a complex manifold and $\mathcal{E}^{\cdot}_X$ is the usual De Rham complex, $W_m\mathcal{E}^k_X = \mathcal{E}^k_X$ for $m \geq 0$, $W_m\mathcal{E}^k_X = 0$ for $m < 0$.

v. If $f\colon X \to Y$ is a morphism of complex spaces any pullback $\Phi\colon \Lambda^{\cdot}_Y \to \Lambda^{\cdot}_X$ in $\mathcal{R}(f)$ is a strict morphism of filtered complexes (see part I, chapter 1 for definitions).

vi. If $\Lambda^k_X = \mathcal{E}^k_{\tilde{X}} \oplus \Lambda^k_E \oplus \Lambda^{k-1}_{\tilde{E}}$ one has

$$W_m\Lambda^k_X = W_m\mathcal{E}^k_{\tilde{X}} \oplus W_m\Lambda^k_E \oplus \; W_{m+1}\Lambda^{k-1}_{\tilde{E}} \tag{3.1}$$

Moreover, if $(X_l, h_l)_{l\in L}$ is the hypercovering of X associated to Λ^k_X, so that

$$\Lambda^k_X = \bigoplus_{l\in L}(h_l)_*\mathcal{E}^{k-q(l)}_{X_l}$$

then

$$W_m\Lambda^k_X = \bigoplus_{\{l\in L: -q(l)\leq m\}} (h_l)_*\mathcal{E}^{k-q(l)}_{X_l} \tag{3.2}$$

so that $W_m\Lambda^k_X$ contains only forms of degree $\leq k+m$ on the various spaces X_l of the hypercovering associated to Λ^k_X.

Proof of the statements on W_m. We define $W_m\Lambda^k_X$ by recursion on $\dim X$ using the equation (3.1) and the definition of W_m on the standard De Rham complex. It is then obvious that (i) and (ii) hold. Moreover the equation (3.2) is also true: one has

$$W_m\Lambda^k_X = \bigoplus_{l\in L} W_{m+q_X(l)}(h_l)_*\mathcal{E}^{k-q_X(l)}_{X_l} \tag{3.3}$$

In fact, one of the X_l, let say X_a, is $\tilde{X}$ itself, with $q_X(a) = 0$, so that for $l = a$, $W_m\mathcal{E}^k_{\tilde{X}} = W_{m+q_X(a)}(h_a)_*\mathcal{E}^{k-q_X(a)}_{X_a}$ appears in (3.3) exactly as in (3.1). If X_l is in the hypercovering of $\Lambda^{\cdot}_E$, we have $q_X(l) = q_E(l)$ and (3.3) holds for $W_m\Lambda^{\cdot}_E$ by recursion on the dimension.

Finally, if X_l is in the hypercovering of $\tilde{E}$ one has $q_X(l) = q_{\tilde{E}}(l) + 1$, so by recursion on the dimension,

$$W_{m+1}\Lambda^k_{\tilde{E}} = \bigoplus_{l\in L} W_{m+1+q_{\tilde{E}}(l)}(h_l)_*\mathcal{E}^{k-1-q_{\tilde{E}}(l)}_{\tilde{E}} = \bigoplus_{l\in L} W_{m+q_X(l)}(h_l)_*\mathcal{E}^{k-q_X(l)}_{X_l}$$

This proves (3.3).

But $W_{m+q_X(l)}(h_l)_*\mathcal{E}^{k-q_X(l)}_{X_l}$ is equal to $(h_l)_*\mathcal{E}^{k-q_X(l)}_{X_l}$ for $m \geq -q_X(l)$ and is 0 for $m < -q_X(l)$. This, together with (3.3) proves (3.2).

We prove now the assertion (v) concerning morphisms $f\colon X \to Y$. Let $\Phi\colon \Lambda^{\cdot}_Y \to \Lambda^{\cdot}_X$ in $\mathcal{R}(f)$, and $(X_l, h_l)_{l\in L}$, $(Y_s, g_s)_{s\in S}$ be the hypercoverings associated to $\Lambda^{\cdot}_X, \Lambda^{\cdot}_Y$, i.e.

$$\Lambda^k_X = \bigoplus_l \mathcal{E}^{k-q(l)}_{X_l}, \quad \Lambda^k_Y = \bigoplus_s \mathcal{E}^{k-q(s)}_{Y_s}$$

We know by chapter 2 that for every X_l there exist at most one Y_s, having the same rank q as X_l, and a commutative diagram

$$\begin{array}{ccc} X_l & \xrightarrow{f_{ls}} & Y_s \\ {\scriptstyle h_l}\downarrow & & \downarrow{\scriptstyle g_s} \\ X & \xrightarrow{f} & Y \end{array}$$

such that the composition

$$\mathcal{E}^{k-q}_{Y_s} \longrightarrow \Lambda^k_Y \xrightarrow{\Phi} \Lambda^k_X \longrightarrow \mathcal{E}^{k-q}_{X_l}$$

is either identically zero for every k, or coincides with the De Rham pullback f^*_{ls}. But a standard De Rham pullback preserves the degree of a form or kills the form. This implies that

$$\operatorname{im}\Phi \cap W_m\Lambda^{\cdot}_X = \Phi\big(W_m(\Lambda^{\cdot}_Y)\big)$$

so that Φ is a strict morphism of filtered complexes.

Finally we prove that d respects the filtration. This is a consequence of the formula for d:

$$d(\omega,\sigma,\theta) = \big(d\omega, d\sigma, d\theta + (-1)^k(\psi(\omega) - \phi(\sigma))\big)$$

If $(\omega,\sigma,\theta) \in W_m\Lambda^k_X$, an induction on the dimension shows that $d\sigma \in W_m\Lambda^{k+1}_E$ because $\dim E < \dim X$, and analogously $d\theta \in W_{m+1}\Lambda^k_{\tilde{E}}$, because $\dim\tilde{E} < \dim X$; the inner pullback ψ and ϕ preserve the filtration. Finally we obtain $d(\omega,\sigma,\theta) \in W_m\Lambda^{k+1}_X$. □

From equation (3.3) we deduce immediately:

Lemma 3.1. *If (X_l) is the hypercovering associated to $\Lambda^{\cdot}_X$, the graded complex with respect the filtration W_m is given by*

$$\frac{W_m\Lambda^{\cdot}_X}{W_{m-1}\Lambda^{\cdot}_X} = \bigoplus_{\{l\in L:-q(l)=m\}} (h_l)_* \mathcal{E}^{k-q(l)}_{X_l} \tag{3.4}$$

In other words, the graded complex is exactly the subspace of forms of Λ^k_X whose nonzero components have degree $k+m$ on each X_l.

3.3 The weight filtration in cohomology

The filtration W_m induces a filtration, still denoted W, on the complex of global sections:

$$W_m\Gamma(X,\Lambda^k_X) = \Gamma(X, W_m\Lambda^k_X) \tag{3.5}$$

Since the sheaves $\frac{W_m\Lambda^{\cdot}_X}{W_{m-1}\Lambda^{\cdot}_X}$ are fine, we have the natural isomorphism of complexes

$$\frac{\Gamma(X,W_m\Lambda^{\cdot}_X)}{\Gamma(X,W_{m-1}\Lambda^{\cdot}_X)} = \Gamma\left(X, \frac{W_m\Lambda^{\cdot}_X}{W_{m-1}\Lambda^{\cdot}_X}\right) \tag{3.6}$$

The filtration W induces a filtration on the cohomology:

$$W_mH^k(X,\mathbb{C}) = \frac{\ker\{\,d\colon \Gamma(X,W_m\Lambda^k_X)\to\Gamma(X,W_m\Lambda^{k+1}_X)\,\}}{d\Gamma(X,\Lambda^{k-1}_X)\cap\Gamma(X,W_m\Lambda^k_X)}$$

Any pullback $\Phi\colon \Lambda^{\cdot}_Y \to \Lambda^{\cdot}_X$ corresponding to a morphism $f\colon X\to Y$ induces the standard pullback $f^*\colon H^k(Y,\mathbb{C})\to H^k(X,\mathbb{C})$ on cohomology so that f^* becomes a morphism of filtered spaces for the weight filtrations W.

3.4 The action of d on the filtered complexes

Let $(X_l,h_l)_{l\in L}$ be the hypercovering of X associated to $\Lambda^{\cdot}_X$, and let $\omega=\oplus_l\omega_l$ be an element of Λ^k_X; we denote

$$P_m\omega = \oplus_{\{-q(l)=m\}}\omega_l \tag{3.7}$$

the projection of ω on the quotient $\frac{W_m\Lambda^k_X}{W_{m-1}\Lambda^k_X}$, according to (3.4). We also denote

$$Q_m\omega = \oplus_{\{-q(l)\le m\}}\omega_l$$

the projection on $W_m \Lambda_X^k$, so that

$$Q_m = \sum_{j \leq m} P_j \tag{3.8}$$

The differential $d\omega$ is given (see chapter 2) by:

$$\begin{aligned} d\omega &= \oplus_l (d\omega)_l \\ (d\omega)_l &= \left(d\omega_l + \textstyle\sum_s \epsilon_{ls}^{(k)} h_{ls}^* \omega_s \right) \end{aligned} \tag{3.9}$$

where $\epsilon_{ls}^{(k)}$ can take the values $0, \pm 1$ and the h_{ls}^* are the inner pullback of the complex $\Lambda_X^{\cdot}$; moreover $q(s) = q(l) - 1$.

For further references and to fix completely the notations, we prove the following

Lemma 3.2.

i. *There exists a linear operator*

$$\Phi_m^{(k)} : \frac{W_{m+1}\Lambda_X^k}{W_m \Lambda_X^k} \to \frac{W_m \Lambda_X^k}{W_{m-1}\Lambda_X^k}$$

or

$$\Phi_m^{(k)} : \oplus_{\{l: -q(l) = m+1\}} (h_l)_* \mathcal{E}_{X_l}^{k-q(l)} \to \oplus_{\{b: -q(b) = m\}} (h_b)_* \mathcal{E}_{X_b}^{k-q(b)}$$

such that for all $\pi \in \Lambda_X^k$ one has

$$P_m(d\pi) = d(P_m \pi) + \Phi_m^{(k)} P_{m+1} \pi \tag{3.10}$$

where in $d(P_m \pi)$ the differential is calculated component by component of $P_m \pi$.

ii. *One has*

$$Q_m(d\pi) = d(Q_m \pi) + \sum_{j \leq m} \Phi_j^{(k)} (P_{j+1} \pi)$$

where again in $d(Q_m \pi)$ the differential is calculated component by component of $Q_m \pi$.

iii. *In particular, if $\pi \in W_m \Lambda_X^k$*

$$P_m(d\pi) = d(P_m \pi) \tag{3.11}$$

where again the differential in the second member is calculated component by component.

iv. *If $\pi \in W_m \Lambda_X^k$ and $d\pi = 0$ then all the components of $P_m \pi$ are d-closed for the usual De Rham complexes to which they belong.*

v. *One has*

$$\Phi_{m-1}^{(k+1)} \circ \Phi_m^{(k)} = 0 \tag{3.12}$$

$$d\Phi_m^{(k)} + \Phi_m^{(k+1)} d = 0 \tag{3.13}$$

where, in (3.13), d *is calculated component by component on* $\frac{W_m \Lambda_X^k}{W_{m-1}\Lambda_X^k}$.

Proof.

i. To calculate $P_m(d\pi)$ we need to calculate $(d\pi)_l$ for l such that $-q(l) = m$ (see (3.7) defining P_m). This is done using the equation (3.9)

$$(d\pi)_l = \left(d\pi_l + \sum_s \epsilon_{ls}^{(k)} h_{ls}^* \pi_s\right)$$

where the sum is on the s such that $-q(s) = q(l) + 1 = m + 1$, so that the π_s are components of $P_{m+1}\pi$. If $\omega = \oplus_l \omega_l \in W_{m+1}\Lambda_X^k$ we define

$$\Phi_m^{(k)} \omega = \sum_s \epsilon_{ls}^{(k)} h_{ls}^* \omega_s \tag{3.14}$$

so that (3.10) is a consequence of the definition.

ii. Using (3.8), we have

$$Q_m(d\pi) = Q_{m-1}(d\pi) + P_m(d\pi) = Q_{m-1}(d\pi) + d(P_m\pi) + \Phi_m^{(k)}(P_{m-1}\pi)$$

so we obtain (ii) by induction on m.

iii. If $\pi \in W_m\Lambda_X^k$, $P_{m+1}\pi = 0$ and (3.11) is a consequence of (3.10) .

iv. is then evident from (iii).

v. We consider a form $\pi \in W_m\Lambda_X^k$ so that $\pi = P_m\pi$; then $d\pi \in W_m\Lambda_X^{k+1}$; from (3.10) and (3.11) we obtain

$$P_m(d\pi) = d(P_m\pi) \tag{3.15}$$

$$P_{m-1}(d\pi) = \Phi_m^{(k-1)} P_m\pi = \Phi_{m-1}^{(k)}\pi \tag{3.16}$$

$$P_j(d\pi) = 0 \quad (j \leq m-2) \tag{3.17}$$

From $dd\pi = 0$, using (3.15), (3.16), (3.17), we obtain:

$$0 = P_m(dd\pi) = d(P_m d\pi) = dd(P_m\pi)$$

$$\begin{aligned} 0 = P_{m-1}(dd\pi) &= dP_{m-1}(d\pi) + \Phi_{m-1}^{(k+1)} P_m(d\pi) = \\ &= d\Phi_{m-1}^{(k)}\pi + \Phi_{m-1}^{(k+1)} dP_m\pi = (d\Phi_{m-1}^{(k)} + \Phi_{m-1}^{(k+1)} d)\pi \end{aligned}$$

$$0 = P_{m-2}(dd\pi) = dP_{m-1}(d\pi) + \Phi_{m-2}^{(k+1)} P_{m-1}\pi = \Phi_{m-2}^{(k+1)} \Phi_{m-1}^{(k)}\pi;$$

the above formulas are correct for any form $\pi = P_m\pi$ in $\frac{W_m\Lambda_X^k}{W_{m-1}\Lambda_X^k}$, so that $\Phi_{m-2}^{(k+1)} \circ \Phi_{m-1}^{(k)} = 0$ and $d\Phi_{m-1}^{(k)} + \Phi_{m-1}^{(k+1)} d = 0$ (in the last equation d is calculated component by component). Hence (up to a shift $m-1 \to m$) we have proved (3.12) and (3.13). □

3.5 The first term of the spectral sequence

For the complex $\Gamma(X, \Lambda_X^{\cdot})$ and its filtration W_m given by (3.5), we can construct the corresponding spectral sequence denoted by $E_r^{m,k}$ with first term

$$E_1^{m,k} = H^k\left(X, \frac{W_m \Lambda_X^{\cdot}}{W_{m-1}\Lambda_X^{\cdot}}\right) = H^k\left(\frac{\Gamma(X, W_m\Lambda_X^{\cdot})}{\Gamma(X, W_{m-1}\Lambda_X^{\cdot})}\right)$$

where the second equality follows from (3.6).

Then $E_1^{m,k}$ appears as the cohomology of degree k of the complex

$$\cdots E_0^{m,k} = \frac{\Gamma(X, W_m\Lambda_X^k)}{\Gamma(X, W_{m-1}\Lambda_X^k)} \longrightarrow E_0^{m,k+1} = \frac{\Gamma(X, W_m\Lambda_X^{k+1})}{\Gamma(X, W_{m-1}\Lambda_X^{k+1})} \cdots \tag{3.18}$$

whose differential is induced by d. By lemma 3.1 it is also the cohomology of degree k of the complex

$$\cdots \bigoplus_{\{-q(l)=m\}} \Gamma(X_l, \mathcal{E}_{X_l}^{k-q(l)}) \longrightarrow \bigoplus_{\{-q(l)=m\}} \Gamma(X_l, \mathcal{E}_{X_l}^{k+1-q(l)}) \cdots \tag{3.19}$$

where the differentials are the direct sums of the differentials on each X_l. We can compute $E_1^{m,k}$ and the differential

$$d_1 \colon E_1^{m,k} \to E_1^{m-1,k+1}$$

as follows.

Lemma 3.3. *Let $(X_l, h_l)_{l\in L}$ be the hypercovering associated to $\Lambda_X^{\cdot}$.*

i. The first term of the spectral sequence is

$$E_1^{m,k} = \bigoplus_{\{-q(l)=m\}} H^{k-q(l)}(X_l, \mathbb{C})$$

ii. The differential $d_1 \colon E_1^{m,k} \to E_1^{m-1,k+1}$ is constructed as follows. Let $[\pi] = [\pi_l]$ be an element of $E_1^{m,k}$, with $\pi_l \in \Gamma(X_l, \mathcal{E}_{X_l}^{k-q(l)})$, $d\pi_l = 0, -q(l) = m$. We define an element $\pi' \in \Gamma(X, W_m\Lambda_X^k)$ by

$$\begin{cases} \pi_l' = \pi_l & \text{if } -q(l) = m \\ \pi_l' = 0 & \text{if } -q(l) \neq m \end{cases} \tag{3.20}$$

so that $P_m\pi' = (\pi_l)$. Then $d\pi' \in \Gamma(X, \Lambda_X^{k+1})$ defines an element of $E_1^{m-1,k+1}$ which is $d_1[\pi]$ and in fact

$$d_1[\pi] = \left[\Phi_{m-1}^{(k)}(\pi_l)\right] \tag{3.21}$$

iii. *If* $f\colon X \to Y$ *is a morphism, a pullback* $\Lambda_Y^{\cdot} \to \Lambda_X^{\cdot}$ *in* $\mathcal{R}(f)$ *induces a natural morphism*

$$E_1^{m,k}(Y) \to E_1^{m,k}(X)$$

(with an obvious meaning of the notations).

Proof. The proof is a specification of the general facts on spectral sequences (see part I, chapter 1); computing the cohomology in (3.19) gives (i).

To prove (ii), we use lemma 3.2 (formula (3.11)), so that

$$P_m(d\pi') = d(P_m\pi') = 0 \tag{3.22}$$

because $P_m\pi'$ is formed with the closed forms π_l. Then

$$P_{m-1}(d\pi') = \Phi_{m-1}^{(k)}(P_m\pi') \tag{3.23}$$

$$P_j(d\pi') = 0 \quad \text{for } j \le m-1 \tag{3.24}$$

So, in fact $d\pi' \in \Gamma\big(X, W_{m-1}\Lambda_X^{k+1}\big)$, and $\pi' \in Z_1^{m,k}$, $d\pi' \in Z_1^{m-1,k+1}$ and one can define the class $[d\pi']_1$ in $E_1^{m-1,k+1}$. This class is exactly $d_1[\pi']$ and in view of (3.22) (3.23) (3.24) one has also

$$d_1[\pi] = \Big[\Phi_{m-1}^{(k)}(\pi_l)\Big] \in \bigoplus_{\{-q(b)=m-1\}} H^{k+1-q(b)}(X_b, \mathbb{C})$$

the last space being $E_1^{m-1,k+1}$. □

3.6 The second term of the spectral sequence

We know that

$$E_2^{m,k} = \frac{\ker\Big\{d_1\colon E_1^{m,k} \to E_1^{m-1,k+1}\Big\}}{\operatorname{im}\Big\{d_1\colon E_1^{m+1,k-1} \to E_1^{m,k}\Big\}} \tag{3.25}$$

We can calculate the differential

$$d_2\colon E_2^{m,k} \to E_2^{m-2,k+1}$$

in the following manner.

Lemma 3.4. *Let* $[\pi]_1 = \oplus[\pi_l] \in E_1^{m,k}$ *be an element in* $\ker d_1$, *where* $\pi_l \in \Gamma\big(X_l, \mathcal{E}_{X_l}^{k-q(l)}\big)$, $d\pi_l = 0$, *and* $-q(l) = m$.

i. *There exist forms* $\omega_b \in \Gamma\big(X_b, \mathcal{E}_{X_b}^{k-q(b)}\big)$ *with*

$$\left(\Phi_{m-1}^{(k)}(\pi)\right)_b = d\omega_b \tag{3.26}$$

for $-q(b) = m-1$.

ii. *Let* $\omega = \oplus\omega_b$*; we define* $\pi'' \in \Gamma\big(X, W_m\Lambda_X^k\big)$ *by*

$$\begin{cases} P_m\pi'' = \pi \\ P_{m-1}\pi'' = \omega \\ P_j\pi'' = 0 \quad \text{for } j \neq m, m-1 \end{cases} \tag{3.27}$$

Then $d\pi'' \in \Gamma(X, W_{m-2}\Lambda_X^{k+1})$ *and*

$$\begin{cases} P_{m-2}(d\pi'') = \Phi_{m-2}^{(k)}(\omega) \\ P_j(d\pi'') = 0 \quad \text{for } j \neq m-2 \end{cases} \tag{3.28}$$

We have $d\pi'' \in Z_2^{m-2,k+1}$ *and this induces an element* $[d\pi'']_2 \in E_2^{m-2,k+1}$ *which is the* d_2 *of the class of* $[\pi]$ *in* $E_1^{m,k}$:

$$[d\pi'']_2 = d_2[\pi]_1 \tag{3.29}$$

iii. *Moreover* $\Phi_{m-2}^{(k)}(\omega) \in \ker\left\{ d_1 \colon E_1^{m-2,k+1} \to E_1^{m-3,k+1} \right\}$ *and its class in the quotient* (3.25) *for* $E_1^{m-2,k+1}$ *is also* $d_2[\pi]_1$.

Proof.

i. We start with $[\pi]_1 \in E_1^{m,k} = \bigoplus_{\{-q(l)=m\}} H^{k-q(l)}(X_l, \mathbb{C})$ and define π' as in (3.20).

Now, we write that $d_1[\pi]_1$ is 0 in $E_1^{m-1,k+1}$. Using (3.21), this means that $\Phi_{m-1}^{(k)}(\pi)_b$ is an exact differential on X_b for $-q(b) = m-1$ so that (3.26) holds.

ii. The form π defines a form $\pi'' \in \Gamma\big(X, W_m\Lambda_X^k\big)$ by the prescription of (3.27). The differential d_2 will be the differential d acting on π'' which can be calculated, using lemma 3.2, formula (3.10):

$$P_{m-2}(d\pi'') = d(P_{m-2}\pi) + \Phi_{m-2}^{(k)}(P_{m-1}\pi'') = \Phi_{m-2}^{(k)}(\omega)$$

and

$$P_j(d\pi'') = d(P_j\pi) + \Phi_j^{(k)}(P_{j+1}\pi'') = 0 \quad (j \neq m-2)$$

By definition, $d\pi'' \in Z_2^{m-2,k+1}$ and has a class $[d\pi'']_2 \in E_2^{m-2,k+1}$ which is the d_2 of the class of $[\pi]_1$.

iii. $\Phi^{(k)}_{m-2}(\omega)$ defines also an element in the kernel of $d_1 : E_1^{m-2,k+1} \to E_1^{m-3,k+2}$ whose image in $E_2^{m-2,k+1}$ will be also $d_2[\pi]_1$.

In fact, let $\Phi^{(k)}_{m-2}(\omega)_c$ be a component with $-q(c) = m-2$. This is a closed form in X_c, because by (3.13)

$$d(\Phi^{(k)}_{m-2}(\omega)_c) = -\left(\Phi^{(k+1)}_{m-2}(d\omega)\right)_c$$

and by (3.26) this is a component of

$$-\Phi^{(k+1)}_{m-2}\Phi^{(k)}_{m-1}(\pi)$$

which is 0 by (3.12), hence

$$\Phi^{(k)}_{m-2}(\omega) \in \bigoplus_{\{-q(c)=m-2\}} H^{k+1-q(c)}(X_c, \mathbb{C}) = E_1^{m-2,k+1}$$

We now compute d_1 of the above element. We define $\omega' \in \Gamma(X, W_{m-2}\Lambda_X^{k+1})$ by

$$\begin{cases} P_{m-2}\omega' = \Phi^{(k)}_{m-2}(\omega) \\ P_j\omega' = 0 \quad \text{for } j \neq m-2 \end{cases}$$

Using lemma 3.2 again, we see that $d\omega' = 0$, so that ω' induces an element of $E_1^{m-2,k+1}$ which is in the kernel of d_1. □

3.7 Computation of d_r

Here we construct explicitly the differential $d_r : E_r^{m,k} \to E_r^{m-r,k+1}$ of the spectral sequence associated to the filtration W_m on the complex $\Gamma(X, \Lambda_X^{\cdot})$. We consider an element $[\pi]_r \in E_r^{m,k}$. This means that we start with a certain element $[\pi^{(1)}]_1 \in E_1^{m,k}$ with $[\pi^{(1)}]_1 \in \ker d_1$ so that, by definition, $[\pi^{(1)}]_1$ induces an element $[\pi^{(2)}]_2 \in E_2^{m,k}$ (which is the cohomology of the complex $(E_1^{m,\cdot} d_1)$) and $[\pi^{(2)}]_2$ is in $\ker d_2, \ldots$, until we arrive at an element $[\pi^{(r-1)}]_{r-1} \in E_{r-1}^{m,k}$ which is in $\ker d_{r-1}$ and induces the element $[\pi]_r \in E_r^{m,k}$. So, let $[\pi^{(1)}]_1 \in \ker d_1$. This means, according to lemma 3.4, the following facts.

i. We have

$$[\pi^{(1)}]_1 = \oplus_{\{-q(l)=m\}}[\pi_l], \quad [\pi_l] \in H^{k+m}(X_l, \mathbb{C})$$

so that π_l is a d-closed form on X_l;

ii. there exist forms $\omega_b^{(1)} \in \Gamma(X_b, \mathcal{E}_{X_b}^{k+m-1})$ for $b \in L$ with $-q(b) = m-1$ such that

$$\left(\Phi^{(k)}_{m-1}(\pi^{(1)})\right)_b = d\omega_b^{(1)}$$

iii. we define $\pi^{(2)} \in \Gamma\big(X, W_m\Lambda_X^k\big)$ by the conditions of the formula (3.27)

$$\begin{cases} P_m\pi^{(2)} = \pi \\ P_{m-1}\pi^{(2)} = \omega^{(1)} \\ P_j\pi^{(2)} = 0 \quad \text{for } j \neq m, m-1 \end{cases}$$

The form $\pi^{(2)}$ is in $Z_2^{m,k}$ and so it induces an element $[\pi^{(2)}]_2 \in E_2^{m,k}$ and

$$d_2[\pi^{(2)}]_2 = [d\pi^{(2)}]_2$$

In fact $d_2[\pi^{(2)}]_2$ is induced by $\Phi_{m-2}^{(k)}(\omega^{(1)})$ (see formulas (3.28) and (3.29)).

iv. if we assume that $d_2[\pi^{(2)}]_2 = 0$, there exist $\omega_c^{(2)} \in \Gamma\big(X_c, \mathcal{E}_{X_c}^{k+m-2}\big)$ for the indices $c \in L$ with $-q(c) = m-2$ and

$$(\Phi_{m-2}^{(k)}(\omega^{(1)}))_c = d\omega_c^{(2)}$$

and we construct $\pi^{(3)} \in \Gamma\big(X, W_m\Lambda_X^k\big)$ as follows:

$$\begin{cases} P_m\pi^{(3)} = \pi \\ P_{m-1}\pi^{(3)} = \omega^{(1)} \\ P_{m-2}\pi^{(3)} = \omega^{(2)} \\ P_j\pi^{(3)} = 0 \quad \text{for } j \neq m, m-1, m-2 \end{cases}$$

and $\pi^{(3)} \in Z_3^{m,k}$ and so induces an element $[\pi^{(3)}]_3 \in E_3^{m,k}$ which we assume to be in $\ker d_3, \ldots$

v. we arrive at $\pi^{(r)} \in \Gamma\big(X, W_m\Lambda_X^k\big)$ such that

$$\begin{cases} P_m\pi^{(r)} = \pi \\ P_{m-1}\pi^{(r)} = \omega^{(1)} \\ P_{m-2}\pi^{(r)} = \omega^{(2)} \\ \ldots \\ P_{m-(r-1)}\pi^{(r)} = \omega^{(r-1)} \\ P_j\pi^{(r)} = 0 \quad \text{for } j < m-r \text{ or } j > m \end{cases}$$

$\pi^{(r)} \in Z_r^{m,k}$ and induces an element $[\pi^{(r)}]_r \in E_r^{m,k}$.

Let us consider $d\pi^{(r)}$. By construction it will satisfy

$$\begin{cases} P_j(d\pi^{(r)}) = 0 \quad \text{for } j > m-r \\ P_{m-r}d\pi^{(r)} = \Phi_{m-r}^{(k)}(\omega^{(r-1)}) \\ P_j(d\pi^{(r)}) = 0 \quad \text{for } j < m-r, j > m \end{cases}$$

Then one defines

$$d_r[\pi^{(r)}]_r = [d\pi^{(r)}]_r$$

because $d\pi^{(r)}$ is in $Z_r^{m-r,k+1}$ and so induces an element in $E_r^{m-r,k+1}$.

3.8 The filtrations F^p and $\bar{F}^q$

In this section, we define the filtration by the types on the complexes $\Lambda^{\cdot}_X$.
Let X be a complex space, $\Lambda^{\cdot}_X$ an element of $\mathcal{R}(X)$. We define a decreasing filtration (**the Hodge filtration, or the filtration by the types**) F^p on $\Lambda^{\cdot}_X$, by recursion on the dimension, by the formula

$$F^p\Lambda^k_X = F^p\mathcal{E}^k_{\tilde{X}} \oplus F^p\Lambda^k_E \oplus F^p\Lambda^{k-1}_{\tilde{E}} \tag{3.30}$$

with the following properties.

i. If X is a complex manifold and $\mathcal{E}^{\cdot}_X$ is the usual De Rham complex, F^p is the standard Hodge filtration on $\mathcal{E}^{\cdot}_X$ (see part I, chapter 2)

ii. F^p defines a decreasing filtration and a filtered complex for d

$$\cdots \subset F^{p+1}\Lambda^k_X \subset F^p\Lambda^k_X \subset \cdots$$
$$d(F^p\Lambda^k_X) \subset F^p\Lambda^{k+1}_X$$

iii. If $(X_l, h_l)_{l\in L}$ is the hypercovering associated to $\Lambda^{\cdot}_X$, one has

$$F^p\Lambda^k_X = \bigoplus_l (h_l)_* F^p\mathcal{E}^{k-q(l)}_{X_l} \tag{3.31}$$

iv. If $f\colon X \to Y$ is a morphism of complex spaces any pullback $\Phi\colon \Lambda^{\cdot}_Y \to \Lambda^{\cdot}_X$ in $\mathcal{R}(f)$ is a strict morphism of filtered complexes for $F^p\Lambda^{\cdot}_Y$ and $F^p\Lambda^{\cdot}_X$ respectively.

v. The conjugate filtration $\bar{F}^q$ is defined as

$$\bar{F}^q\Lambda^k_X = \overline{F^q\Lambda^k_X}$$

vi. $F^p\Lambda^k_X$ and $\bar{F}^p\Lambda^k_X$ are zero for $p > k$, and are equal to Λ^k_X for $p \le 0$.

Proof. The definition (3.30) together with the requirement that F^p be the Hodge filtration on the standard De Rham complex proves that $F^p\Lambda^{\cdot}_X$ is uniquely defined and proves also (3.31). The properties (i) and (vi) are trivial. Moreover (iv) is evident because in the hypercovering language, Φ is a family of standard De Rham pullback which are all strict morphisms for the standard Hodge filtrations. Finally, the formula for the d

$$d(\omega, \sigma, \theta) = \big(d\omega, d\sigma, d\theta + (-1)^k(\psi(\omega) - \phi(\sigma))\big)$$

where ψ, ϕ are the inner pullbacks of $\Lambda^{\cdot}_X$, proves that d preserves F^p , using a recursion on the dimension. □

The filtration F^p (as well as $\bar{F}^q$) induces a filtration on the subspace $W_m\Lambda^k_X$ by

$$F^pW_m\Lambda^k_X = F^p\Lambda^k_X \cap W_m\Lambda^k_X \tag{3.32}$$

F^p, $\bar{F}^q$ induce filtrations on the complex $\Gamma(X, \Lambda^{\cdot}_X)$ and also filtrations on the cohomology $H^k(X, \mathbb{C})$:

$$F^pH^k(X, \mathbb{C}) = \frac{\ker\{\, d\colon \Gamma(X, F^p\Lambda^k_X) \to \Gamma(X, F^p\Lambda^{k+1}_X)\,\}}{d\Gamma(X, \Lambda^{k-1}_X) \cap \Gamma(X, F^p\Lambda^k_X)} \tag{3.33}$$

3.9 Pure Hodge structures on the spectral sequence

In this section, we suppose that X is a (B)-Kähler space. Let us recall from part I, chapter 7 that this means that there exists a modification $X' \to X$ where X' is a compact Kähler manifold. Let us recall from chapter 2, theorem 2.10, that we can work with complexes $\Lambda^{\cdot}_X$ whose associated hypercoverings are Kähler, i.e. are formed by compact Kähler manifolds.

Let X be a (B)-Kähler space, and let us consider $\Lambda^{\cdot}_X$ in $\mathcal{R}(X)$ with an associated Kähler hypercovering (X_l, h_l).

Lemma 3.5. *The filtrations F^p, $\bar{F}^q$ induce on $E_1^{m,k}$ a pure Hodge structure of weight $k+m$*

$$E_1^{m,k} = \bigoplus_{\{p+q=k+m\}} (E_1^{m,k}(X))^{p,q}$$

where

$$(E_1^{m,k}(X))^{p,q} = \bigoplus_{\{-q(l)=m\}} H^{p,q}(X_l)$$

and $H^{p,q}(X_l)$ are the Dolbeault groups of X_l (see part I, chapter 5). Moreover the differential $d_1\colon E_1^{m,k} \to E_1^{m-1,k+1}$ is a morphism of pure Hodge structures of weight $k+m$

$$d_1\colon E_1^{m,k}(X)^{p,q} \to E_1^{m-1,k+1}(X)^{p,q}$$

and so is a strict morphism for the filtrations F^p (or $\bar{F}^q$) on $E_1^{m,k}$. Finally if $f\colon X \to Y$ is a morphism, a pullback $\Lambda^{\cdot}_Y \to \Lambda^{\cdot}_X$ in $\mathcal{R}(f)$ induces a morphism of pure Hodge structures $E_1^{m,k}(Y) \to E_1^{m,k}(X)$, hence it is strict.

Proof. We know by lemma 3.3 that $E_1^{m,k}$ is the cohomology of the graded complex of global sections of

$$\frac{W_m\Lambda_X^{\cdot}}{W_{m-1}\Lambda_X^{\cdot}} = \bigoplus_{\{l\in L:-q(l)=m\}} (h_l)_*\mathcal{E}_{X_l}^{k-q(l)} \tag{3.34}$$

so that

$$E_1^{m,k} = \bigoplus_{\{l\in L:-q(l)=m\}} H^{k-q(l)}(X_l,\mathbb{C})$$

Moreover, F^p and $\bar{F}^q$ induce filtrations on the sheaf (3.34) which are the standard Hodge filtrations on the De Rham complex of each X_l:

$$\frac{W_m\Lambda_X^{\cdot}}{W_{m-1}\Lambda_X^{\cdot}} = \bigoplus_{\{p+q=k+m,-q(l)=m\}} (h_l)_*\mathcal{E}_{X_l}^{p,q}$$

So, taking the cohomology of the quotient complex, we see that the filtrations F^p, $\bar{F}^q$ induce the Hodge filtration on the cohomology of each X_l. Because X_l is a compact Kähler manifold, this induced filtration gives the usual Hodge decomposition

$$H^{k+m}(X_l,\mathbb{C}) = \bigoplus_{p+q=k+m} H^{p,q}(X_l) \tag{3.35}$$

Then we can write

$$E_1^{m,k} = \bigoplus_{\{p+q=k+m\}} (E_1^{m,k})^{p,q}$$

where

$$(E_1^{m,k})^{p,q} = \bigoplus_{\{l\in L:-q(l)=m\}} H^{p,q}(X_l)$$

The differential d_1 has been constructed in lemma 3.3, formulas (3.20) and (3.21). Take an element $[\pi] = \oplus[\pi_l]$ of $E_1^{m,k}(X)$, with $\pi_l \in \Gamma(X_l, \mathcal{E}_{X_l}^{k-q(l)})$, $d\pi_l = 0$, $-q(l) = m$, hence $[\pi_l] \in H^{k+m}(X_l,\mathbb{C})$. We decompose this cohomology class $[\pi_l]$ in pure types according to (3.35):

$$[\pi_l] = \oplus_{\{p+q=k+m\}}[\pi_l^{p,q}]$$

where $\pi_l^{p,q}$ is a d-closed form of type (p,q). Then according to lemma 3.3, formula (3.21):

$$d_1\left(\oplus[\pi_l]\right) = \left[\Phi_{m-1}^{(k)}\left(\oplus_{\{p+q=m,-q(l)=m\}}\pi_l'^{(p,q)}\right)\right]$$

where $\pi_l'^{(p,q)}$ is defined as in (3.20):

$$\begin{cases} \pi_l'^{(p,q)} = \pi_l^{(p,q)} & \text{if } -q(l)=m \\ \pi_l'^{(p,q)} = 0 & \text{if } -q(l)\neq m \end{cases}$$

But by definition (3.14) $\Phi^{(k)}_{m-1}$ is a combination of standard De Rham pullback and so it transforms a form of type (p,q) into a form of type (p,q) so that

$$d_1 \colon (E_1^{m,k})^{p,q} \to (E_1^{m-1,k+1})^{p,q}$$

which means that that d_1 is a morphism of pure Hodge structures. □

Theorem 3.1. *The term $E_2^{m,k}$ of the spectral sequence associated to the filtration W_m on $\Gamma(X, \Lambda^{\cdot}_X)$ carries a pure Hodge structure of weight $k+m$ for the natural filtrations induced by $E_1^{m,k}$:*

$$E_2^{m,k} = \bigoplus_{p+q=k+m} (E_2^{m,k})^{p,q} \tag{3.36}$$

where

$$(E_2^{m,k})^{p,q} = \frac{\ker\left\{ d_1 \colon (E_1^{m,k})^{p,q} \to (E_1^{m-1,k+1})^{p,q} \right\}}{\operatorname{im}\left\{ d_1 \colon (E_1^{m+1,k-1})^{p,q} \to (E_1^{m,k})^{p,q} \right\}}$$

Moreover the differential

$$d_r \colon E_r^{m,k} \to E_r^{m-r,k+1}$$

is identically zero for $r \geq 2$, that is, the spectral sequence degenerates at the level 2, in particular

$$E_r^{m,k} = E_2^{m,k} \quad \text{for } r \geq 2$$

If $f \colon X \to Y$ is a morphism, a pullback $\Lambda^{\cdot}_Y \to \Lambda^{\cdot}_X$ in $\mathcal{R}(f)$ induces a morphism of pure Hodge structures

$$E_r^{m,k}(Y) \to E_r^{m,k}(X)$$

which is strict.

Proof. $E_2^{m,k}$ is the cohomology of the complex $\left(E_1^{m,\cdot}, d_1\right)$ so that $E_2^{m,k}$ is bigraded by

$$(E_2^{m,k})^{p,q} = \frac{\ker\left\{ d_1 \colon (E_1^{m,k})^{p,q} \to (E_1^{m-1,k+1})^{p,q} \right\}}{\operatorname{im}\left\{ d_1 \colon (E_1^{m+1,k-1})^{p,q} \to (E_1^{m,k})^{p,q} \right\}}$$

and (3.36) holds by proposition 1.4 of part I, chapter 1.

Next we show that $d_r = 0$ for $r \geq 2$. We proceed by induction on r, the case $r = 2$ being included in the proof. We assume that $d_2 = \cdots = d_{r-1} = 0$ which implies that $E_r^{m,k} = E_2^{m,k}$, and that the natural bigrading of $E_r^{m,k}$

$$(E_r^{m,k})^{p,q} = \frac{\ker\left\{ d_{r-1} \colon (E_{r-1}^{m,k})^{p,q} \to (E_{r-1}^{m-r+1,k+1})^{p,q} \right\}}{\operatorname{im}\left\{ d_{r-1} \colon (E_{r-1}^{m+r-1,k-1})^{p,q} \to (E_{r-1}^{m,k})^{p,q} \right\}}$$

coincides with $(E_2^{m,k})^{p,q}$.

We want to compute $d_r(\alpha^{p,q})$ for an element $\alpha^{p,q}$ in $(E_r^{m,k})^{p,q}$. In fact, we start with $[\pi^{(1)}]_1^{p,q}$ in $(E_1^{m,k})^{p,q}$, $[\pi^{(1)}]_1^{p,q} = \bigoplus_{\{-q(l)=m\}}[\pi_l^{(1)(p,q)}]$

$$[\pi_l^{(1)(p,q)}] \in H^{p,q}(X_l) \quad p+q=m, -q(l)=m$$

$\pi_l^{p,q}$ is a d-closed form on X_l and $d_1[\pi^{(1)}]_1^{p,q}=0$. So we know that the (p,q)-forms $\Phi_{m-1}^{(k)}(\pi^{(1)})_b$ are exact (see lemma 3.4). Applying theorem 5.11 of part I, chapter 5 to the compact Kähler manifolds X_l we deduce that $\Phi_{m-1}^{(k)}(\pi^{(1)})_{b_1}$ are the differentials of forms of type $(p,q-1)$:

$$\Phi_{m-1}^{(k)}(\pi^{(1)})_{b_1} = d\omega_{b_1}^{(1)(p,q-1)} \quad (-q(b_1)=m-1) \tag{3.37}$$

Continuing the construction, finally we reach $\pi^{(r)}$ such that

$$\begin{aligned}
P_m\pi^{(r)} &= (\pi_l^{(1)(p,q)}) \quad (-q(l)=m)\\
P_{m-1}\pi^{(r)} &= (\omega_{b_1}^{(1)(p,q-1)}) \quad (-q(b_1)=m-1)\\
&\cdots\\
P_{m-(r-1)}\pi^{(r)} &= (\omega_{b_{r-1}}^{(r-1)(p,q-(r-1))}) \quad (-q(b_{r-1})=m-(r-1))\\
P_j\pi^{(r)} &= 0 \quad l<m-(r-1)
\end{aligned}$$

with the types as indicated. It follows $d_r(\alpha^{p,q}) = d\pi^{(r)} \in (E_r^{m,k})^{p,q-r+1}$. On the other hand, we can repeat the above procedure taking in (3.37), instead of a form $\omega_{b_1}^{(1)(p,q-1)}$, a form $\omega_{b_1}^{(1)(p-1,q)}$ such that

$$\Phi_{m-1}^{(k)}(\pi^{(1)})_{b_1} = d\omega_{b_1}^{(1)(p-1,q)} \quad (-q(b_1)=m-1)$$

and then obtain $\pi'^{(r)}$ with

$$\begin{aligned}
P_m\pi'^{(r)} &= (\pi_l^{(1)(p,q)}) \quad (-q(l)=m)\\
P_{m-1}\pi'^{(r)} &= (\omega_{b_1}^{(1)(p-1,q)}) \quad (-q(b_1)=m-1)\\
&\cdots\\
P_{m-(r-1)}\pi'^{(r)} &= (\omega_{b_{r-1}}^{(r-1)(p-(r-1),q)}) \quad (-q(b_{r-1})=m-(r-1)\\
P_j\pi'^{(r)} &= 0 \quad l<m-(r-1)
\end{aligned}$$

Again we find that $d_r(\alpha^{p,q}) = d\pi'^{(r)} \in (E_r^{m,k})^{p-r+1,q}$.

Since $(E_r^{m,k})^{p,q-r+1} \cap (E_r^{m,k})^{p-r+1,q} = (0)$ for $r \geq 2$, we conclude $d_r(\alpha^{p,q}) = 0$ for $r \geq 2$.

The assertion about the pullback is left to the reader. □

3.10 The Hodge filtrations on $E_r^{m,k}$

We recall that we are supposing that every X_l is a Kähler manifold.

On each term $E_r^{m,k} = E_r^{m,k}(X)$ of the spectral sequence of the filtration W_m there are two kinds of filtrations:

- The *direct filtrations* $F_1, \bar{F}_1$, induced by the filtrations F and $\bar{F}$ of the complex $\Gamma(X, \Lambda_X^{\cdot})$; precisely here $E_r^{m,k}$ must be considered as a quotient of the subspace $Z_r^{m,k} \subset \Gamma(X, W_m \Lambda_X^k)$.
- The *recursive filtrations* $F_2, \bar{F}_2$ induced recursively on $E_r^{m,k}$ considered as the cohomology of the complex $(E_{r-1}^{m,k}, d_{r-1})$.

It is clear that on $E_0^{m,k}$ and $E_1^{m,k}$ the filtration F_1 (resp. $\bar{F}_1$) is identical to the filtration F_2 (resp. $\bar{F}_2$). Moreover it is easy to prove that $F_1 \subset F_2$, $\bar{F}_1 \subset \bar{F}_2$.

It is clear also that the filtrations F_1 and $\bar{F}_1$, as well as F_2 and $\bar{F}_2$, are conjugate.

Lemma 3.6. *We suppose that X is a (B)-Kähler space, and all the manifolds X_l of the hypercovering of $\Lambda_X^{\cdot}$ are compact Kähler manifolds. Then the morphisms*

$$d_0 \colon E_0^{m,k} \to E_0^{m,k+1}$$

are strict for the filtrations F and $\bar{F}$.

In fact by (3.18) and (3.19) $E_0^{m,k} = \bigoplus_{\{l \in L : -q(l)=m\}} \Gamma(X_l, \mathcal{E}_{X_l}^{k-q(l)})$ and d_0 is the direct sum of the differentials $d \colon \Gamma(X_l, \mathcal{E}_{X_l}^{k-q(l)}) \to \Gamma(X_l, \mathcal{E}_{X_l}^{k+1-q(l)})$ on each X_l, so it is strict by part I, chapter 5.

Theorem 3.2. *We suppose that X is a (B)-Kähler space, and all the manifolds X_l of the hypercovering of $\Lambda_X^{\cdot}$ are compact Kähler manifolds. On $E_r^{m,k}$ the filtration F_1 (resp. $\bar{F}_1$) is identical to the filtration F_2 (resp. $\bar{F}_2$).*

Proof. Because F_1 and $\bar{F}_1$, F_2 and $\bar{F}_2$, are conjugate, it is sufficient to treat the case of F_1 and F_2. We already know $F_1 \subset F_2$, so we need to show $F_2 \subset F_1$. We shall denote by F the filtrations F_1, F_2, when they coincide.

We proceed by induction on r. Let us suppose that the conclusions are true for $s \leq r$ and let us prove it for $r+1$.

For simplicity we write $L^k = \Gamma(X, \Lambda_X^k)$.

Let $\alpha \in F_2^a E_{r+1}^{m,k}$; there is a representative of α in $F^a E_r^{m,k} \cap \ker d_r$, i.e. an element $x \in F^a Z_r^{m,k} \subset F^a W_m L^k$ with $d_r[x]_r = 0$; we have $dx \in W_{m-r} L^{k+1} \cap F^a W_m L^{k+1}$ so that

$$dx \in F^a W_{m-r} L^{k+1} \tag{3.38}$$

$d_r[x]_r = 0$ gives

$$dx = dx_1 + z_1,\ x_1 \in Z_{r-1}^{m-1,k},\ z_1 \in Z_{r-1}^{m-r-1,k+1} \tag{3.39}$$

Then we get successively by theorem 3.1: $d_{r-1}[x_1]_{r-1} = 0$ so that

$$dx_1 = dx_2 + z_2,\ x_2 \in Z_{r-2}^{m-2,k},\ z_2 \in Z_{r-2}^{m-r-1,k+1} \tag{3.40}$$

$$\dots$$

$$dx_{r-2} = dx_{r-1} + z_{r-1},\ x_{r-1} \in Z_1^{m-r+1,k},\ z_{r-1} \in Z_1^{m-r-1,k+1} \tag{3.41}$$

From the above equalities we find that

$$dx_{r-1} = dx \mod Z_0^{m-r-1,k+1} = W_{m-r-1}L^{k+1} \tag{3.42}$$

Since $x_{r-1} \in Z_1^{m-r+1,k}$, we can compute $d_1[x_{r-1}]_1$. It follows from (3.42) and (3.38) that $d_1[x_{r-1}]_1 \in F^a E_1^{m-r,k}$. But d_1 is a morphism of pure Hodge structures (lemma 3.5), hence it is strict for F. Therefore there exists $x'_{r-1} \in F^a Z_1^{m-r+1,k}$ with $d_1[x'_{r-1}]_1 = d_1[x_{r-1}]_1$, that is

$$dx_{r-1} = dx'_{r-1} + dy + z',\ y \in Z_0^{m-r,k},\ z' \in Z_0^{m-r-1,k+1} \tag{3.43}$$

Note that

$$dx'_{r-1} \in F^a W_{m-r+1} L^{k+1} \cap W_{m-r} L^{k+1} \subset F^a W_{m-r} L^{k+1} \tag{3.44}$$

We obtain an element $[y]_0 \in E_0^{m-r,k}$, and from (3.42), (3.43), (3.44) we get $d_0[y]_0 \in F^a E_0^{m-r,k}$. Since d_0 is strict by lemma 3.6, we find $y' \in F^a W_{m-r} L^k$ with $d_0[y]_0 = d_0[y']_0$ or

$$dy = dy' + z'',\ z'' \in W_{m-r-1} L^{k+1} \tag{3.45}$$

From (3.38)...(3.45) we obtain

$$x - x'_{r-1} - y' \in F^a W_m L^k$$

and

$$d(x - x'_{r-1} - y') \in W_{m-r-1} L^{k+1}$$

so that $x - x'_{r-1} - y' \in F^a Z_{r+1}^{m,k}$ and its class in $E_{r+1}^{m,k}$ is α. This proves the theorem for $r+1$. □

It follows from the theorem 3.1 that there are isomorphisms

$$E_2^{m,k} \simeq \frac{W_m H^k(X,\mathbb{C})}{W_{m-1} H^k(X,\mathbb{C})} \tag{3.46}$$

The cohomology space $H^k(X,\mathbb{C})$ carries a Hodge filtration (3.33), which we name for the moment F_c to avoid confusion. Precisely

$$F_c^p H^k(X,\mathbb{C}) = F^p H^k(X, \Lambda_X^{\cdot}) \tag{3.47}$$

Theorem 3.3. *Under the same assumptions as in theorem 3.2, let F_d be the direct (or the recursive) filtration on $E_2^{m,k}$, and F_c be the filtration induced on $E_2^{m,k}$, under the isomorphism (3.46), by the filtration F_c given by (3.47). Then $F_d = F_c$.*

Proof. As in the proof of the previous theorem, let $L^k = \Gamma(X, \Lambda_X^k)$. The isomorphism (3.46) means the following. Let $x \in Z_2^{m,k}$; then there exists $x' \in W_m L^k$ with $dx' = 0$ and

$$x' = x + d\tilde{x} + z,\ \tilde{x} \in Z_1^{m-1,k-1},\ z \in Z_1^{m-1,k}$$

or

$$x' \equiv x + z$$

where the symbol $\equiv$ means cohomologous.

Let $\alpha \in F_c^p E_2^{m,k}$; there exists $x \in F^p W_m L^k$ with $dx = 0$, inducing α in $E_2^{m,k} \simeq \frac{W_m H^k(X,\mathbb{C})}{W_{m-1} H^k(X,\mathbb{C})}$. It is clear that $x \in Z_2^{m,k}$, so that $x \in F^p W_m L^k \cap Z_2^{m,k} = F^p Z_2^{m,k}$. Hence $\alpha \in F_d^p E_2^{m,k}$. This proves $F_c \subset F_d$.

Conversely, if $\alpha \in F_d^p E_2^{m,k}$ there exists $x \in F^p Z_2^{m,k}$ inducing α. By the isomorphism (3.46) there exists $x' \in W_m L^k$ with $dx' = 0$ and

$$x' \equiv x + z, \quad z \in Z_1^{m-1,k} \tag{3.48}$$

Let $[z]_1 \in E_1^{m-1,k}$ be the class of z. By (3.48) we have

$$dz = -dx \in F^p W_m L^{k+1}$$

and since $dz \in W_{m-2} L^{k+1}$ we obtain

$$dz \in W_{m-2} L^{k+1} \cap F^p W_m L^{k+1} \subset F^p W_{m-2} L^{k+1}$$

Thus $d_1[z]_1 \in F^p E_1^{m-2,k+1}$. Since d_1 is a strict morphism for F, we find $x_1 \in F^p Z_1^{m-1,k}$ with $d_1[z]_1 = d_1[x]_1$, or $d_1[z - x_1]_1 = 0$, that is, $z - x_1$ gives a class in $E_2^{m-1,k}$. By the isomorphism (3.46) for $m-1$, there is $x'' \in W_{m-1} L^k$, $dx'' = 0$, with $x'' \equiv z - x_1 - z_1, \quad z_1 \in Z_1^{m-2,k}$, or

$$z \equiv x'' + x_1 + z_1$$

and by (3.48)

$$x' \equiv x + x'' + x_1 + z_1$$

We remark that the cohomology class of x'' belongs to $W_{m-1} H^k$ (where $H^k = H^k(X, \mathbb{C})$), so that it vanishes in the quotient (3.46); thus we write

$$x' \equiv (x + x_1) + z_1 \mod W_{m-1} H^k$$

We note that $x+x_1 \in F^pW_mL^k$. We proceed as above, finding that $d_1[z_1]_1 \in F^pE_1^{m-3,k+1}$ and we obtain

$$x' \equiv (x+x_1+x_2)+z_2 \mod W_{m-1}H^k, \quad x_2 \in F^pZ_1^{m-2,k}, \; z_2 \in Z_1^{m-3,k}$$

Going on, because $W_sL^{\cdot} = 0$ for $s \ll 0$ we finally write

$$x' \equiv x+x_1+x_2+\cdots+x_s \mod W_{m-1}H^k$$

with $x_j \in F^pZ_1^{m-j,k}$ so that $x' \in F^pW_mL^k$ and x', through W_mH^k, induces α. This proves $F_d \subset F_c$. □

3.11 The mixed Hodge structure on the cohomology

Theorem 3.4. *We suppose that X is a (B)-Kähler space, and all the manifolds X_l of the hypercovering of $\Lambda^{\cdot}_X$ are compact Kähler manifolds. We provide the cohomology $H^k(X,\mathbb{C})$ with the weight filtration W shifted by $-k$, the Hodge filtration F induced by the complex of global sections of $\Lambda^{\cdot}_X$, and the filtration $\bar{F}$ conjugate to F. Then $H^k(X,\mathbb{C})$ carries a mixed Hodge structure. More precisely, the graded quotients $\frac{W_mH^k(X,\mathbb{C})}{W_{m-1}H^k(X,\mathbb{C})}$ are isomorphic to $E_2^{m,k}(X)$ and thus have a pure Hodge structure of weight $m+k$:*

$$\frac{W_mH^k(X,\mathbb{C})}{W_{m-1}H^k(X,\mathbb{C})} = \bigoplus_{\{p+q=k+m\}} A_m^{k,p,q}(X) \tag{3.49}$$

The filtration induced by F on $E_2^{m,k}(X)$ coincides with the direct and the recursive filtration. The $\frac{W_mH^k(X,\mathbb{C})}{W_{m-1}H^k(X,\mathbb{C})}$ are zero for $m<-k$ and $m>0$.

(The shift of W by $-k$ is needed to normalize (3.49). In the shifted filtration $W'_m = W_{m-k}$ the formula becomes

$$\frac{W'_mH^k(X,\mathbb{C})}{W'_{m-1}H^k(X,\mathbb{C})} = \bigoplus_{\{p+q=m\}} A_{m-k}^{k,p,q}$$

that is, the quotient $\frac{W'_mH^k(X,\mathbb{C})}{W'_{m-1}H^k(X,\mathbb{C})}$ has weight m, as expected).

The proof follows immediately from theorems 3.1, 3.2 and 3.3.

Theorem 3.5 (Functoriality of the mixed Hodge structures). *Let X, Y be (B)-Kähler complex spaces, $f\colon X \to Y$ a morphism, and $\Phi\colon \Lambda^{\cdot}_Y \to \Lambda^{\cdot}_X$ a pull-back in $\mathcal{R}(f)$. We suppose that the hypercoverings associated to $\Lambda^{\cdot}_Y$ and $\Lambda^{\cdot}_X$ are Kähler. Then Φ induces for $r \geq 1$ a morphism of pure Hodge structures*

$$E_r^{m,k}(Y) \to E_r^{m,k}(X)$$

Moreover Φ induces on cohomology the natural pullback $f^*: H^k(Y,\mathbb{C}) \to H^k(X,\mathbb{C})$ and f^* is a morphism of mixed Hodge structures on cohomology, hence it is a strict morphism for the filtrations W_m and F^p.

Proof. It is clear by construction that the morphism

$$E_r^{m,k}(Y) \to E_r^{m,k}(X)$$

respects the types, hence is a morphism of pure Hodge structures. We know that Φ induces on cohomology the natural pullback $f^*: H^k(Y,\mathbb{C}) \to H^k(X,\mathbb{C})$, hence f^* respects both the filtrations W_m and F^p. □

Theorem 3.6 (Uniqueness of the mixed Hodge structures). *Let X be a (B)-Kähler complex space, and let $\Lambda_X^{\cdot,1}$, $\Lambda_X^{\cdot,2} \in \mathcal{R}(X)$, such that the associated hypercoverings are Kähler. Then $\Lambda_X^{\cdot,1}$ and $\Lambda_X^{\cdot,2}$ induce identical mixed Hodge structures on the cohomology of X.*

Proof. By the property of filtering for (B)-Kähler spaces (see chapter 2, theorem 2.10) there exists a third $\Lambda_X^{\cdot} \in \mathcal{R}(X)$, whose associated hypercovering is Kähler, and two pullback $\Phi_1 : \Lambda_X^{\cdot,1} \to \Lambda_X^{\cdot}$, $\Phi_2 : \Lambda_X^{\cdot,2} \to \Lambda_X^{\cdot}$ corresponding to the identity. We conclude by the previous theorem and the fact that a morphism of mixed Hodge structures which is an isomorphism of vector spaces (in our case: the identity) is an isomorphism of mixed Hodge structures. □

Remark. Let X be a (B)-Kähler manifold, which is not a Kähler manifold (for example: a Moishezon manifold, an algebraic manifold which are not projective). Then the hypercovering associated to the De Rham complex $\mathcal{E}_X^{\cdot}$ consists of the single manifold X, hence is not Kähler. Thus we cannot use the De Rham complex to detect the mixed Hodge structure of X.

3.12 The Mayer-Vietoris sequence

We consider a (B)-Kähler space X and a complex $\Lambda_X^{\cdot} = \mathcal{E}_{\tilde{X}}^{\cdot} \oplus \Lambda_E^{\cdot} \oplus \Lambda_{\tilde{E}}^{\cdot}(-1)$ in $\mathcal{R}(X)$, whose associated hypercovering is Kähler. Let $(\tilde{X}, \tilde{E}) \to (X, E)$ be the modification corresponding to $\Lambda_X^{\cdot}$. We define morphisms of complexes

$$\Psi: \theta \in \Lambda_{\tilde{E}}^{k-1} \to (0,0,\theta) \in \Lambda_X^k$$
$$\Phi: (\omega,\sigma,\theta) \in \Lambda_X^k \to (\omega,\sigma) \in \mathcal{E}_{\tilde{X}}^k \oplus \Lambda_E^k$$
$$\Theta: (\omega,\sigma) \in \mathcal{E}_{\tilde{X}}^k \oplus \Lambda_E^k \to (-1)^k(\psi(\omega) - \phi(\sigma)) \in \Lambda_{\tilde{E}}^k$$

where ψ, ϕ are the inner pullback of $\Lambda_X^{\cdot}$. The above morphisms induce morphisms in cohomologies. Then

Theorem 3.7. *The morphisms Ψ, Φ, Θ induce in cohomology the Mayer Vietoris sequence:*

$$\begin{aligned} \cdots \longrightarrow H^k(X,\mathbb{C}) \xrightarrow{\ \Phi\ } H^k(\tilde{X},\mathbb{C}) \oplus H^k(E,\mathbb{C}) \xrightarrow{\ \Theta\ } \\ \longrightarrow H^k(\tilde{E},\mathbb{C}) \xrightarrow{\ \Psi\ } H^{k+1}(X,\mathbb{C}) \longrightarrow \cdots \end{aligned} \tag{3.50}$$

The morphisms Φ and Θ are morphisms of filtered spaces for the filtration W and Φ is a strict morphism for the shifted filtrations $W_mH^k(\tilde{E},\mathbb{C}) \to W_{m-1}H^{k+1}(X,\mathbb{C})$.

The proof is almost evident. The fact that Φ, Θ respect the filtration W comes from the fact that they do so at level of complexes (as to Θ, we recall that the pullback respect the degrees of the forms). The fact that Φ is a strict morphism for the shifted filtrations is obvious by the definition.

Theorem 3.8. *Let X be a (B)-Kähler space, and $\Lambda^{\cdot}_X = \mathcal{E}^{\cdot}_{\tilde{X}} \oplus \Lambda^{\cdot}_E \oplus \Lambda^{\cdot}_{\tilde{E}}(-1)$ a complex in $\mathcal{R}(X)$ whose associated hypercovering is Kähler. Let $(\tilde{X},\tilde{E}) \to (X,E)$ be the modification corresponding to $\Lambda^{\cdot}_X$. Let us consider the Mayer - Vietoris sequence* (3.50).

i. The morphisms Φ and Θ are morphisms of mixed Hodge structures for the filtrations W_m and F^p induced by the complexes $\Lambda^{\cdot}_X$, $\mathcal{E}^{\cdot}_{\tilde{X}}$, $\Lambda^{\cdot}_E$, $\Lambda^{\cdot}_{\tilde{E}}$ on the corresponding cohomologies.

ii. Ψ is a morphism of mixed Hodge structures with a shift of the filtration W_m by -1, i.e.

$$\Psi : W_mH^k(\tilde{E},\mathbb{C}) \to W_{m-1}H^{k+1}(X,\mathbb{C}) \tag{3.51}$$

and, up to this shift, Ψ induces a morphism of mixed Hodge structures.

iii. For any $k, m \leq 0, p, q$ with $p + q = k + m$, Φ, Ψ, Θ induce an exact sequence

$$\begin{aligned} \cdots \longrightarrow A^{k,p,q}_m(X) \xrightarrow{\ \Phi\ } A^{k,p,q}_m(\tilde{X}) \oplus A^{k,p,q}_m(E) \xrightarrow{\ \Theta\ } \\ \longrightarrow A^{k,p,q}_m(\tilde{E}) \xrightarrow{\ \Psi\ } A^{k+1,p,q}_{m-1}(X) \longrightarrow \cdots \end{aligned}$$

the spaces $A^{k,p,q}_m$ being defined as in (3.49) *for the various spaces X, $\tilde{X}$, E, $\tilde{E}$ using the mixed Hodge structures induced by the complexes $\Lambda^{\cdot}_X$, $\mathcal{E}^{\cdot}_{\tilde{X}}$, $\Lambda^{\cdot}_E$, $\Lambda^{\cdot}_{\tilde{E}}$. In particular,*

$$A^{k,p,q}_m(\tilde{X}) = \begin{cases} H^{p,q}(\tilde{X}) & \text{if } m = 0 \\ 0 & \text{if } m < 0 \end{cases}$$

Proof.

i. Φ and Θ are combinations of pullback and so they induce morphisms for the filtration W_m and F^p on cohomologies, and as a consequence, they define morphisms of mixed Hodge structures.

ii. On the other hand, also Ψ is a linear combination of pullback, but a manifold X_c in the hypercovering of $\Lambda^{\cdot}_{\tilde{E}}$ with $-q_{\tilde{E}}(c) = m$, becomes a manifold in the hypercovering of $\Lambda^{\cdot}_X$ with $-q_X(c) = m-1$ (and the form of cohomological degree k becomes a form of cohomological degree k+1, although the actual degree of forms remains unchanged). This proves (3.51) and then assertion (ii).

iii. is a consequence of the results of part I, chapter 1 for exact sequences of morphisms of mixed Hodge structures. □

3.13 The differential d is a strict morphism for the filtration F^p

Theorem 3.9. *Let X be a (B)-Kähler space, and $\Lambda^{\cdot}_X = \mathcal{E}^{\cdot}_{\tilde{X}} \oplus \Lambda^{\cdot}_E \oplus \Lambda^{\cdot}_{\tilde{E}}(-1)$ a complex in $\mathcal{R}(X)$ whose associated hypercovering is Kähler. The differential $d\colon \Gamma(X, \Lambda^k_X) \to \Gamma(X, \Lambda^{k+1}_X)$ is a strict morphism for the filtration F^p. Equivalently, the spectral sequence associated to filtration $F^p\Gamma(X, \Lambda^{\cdot}_X)$ on the complex $\Gamma(X, \Lambda^{\cdot}_X)$ degenerates at E_1.*

Proof. As usual, the proof is by recursion on the dimension of X . Let

$$(\omega, \sigma, \theta) \in \Gamma(\tilde{X}, \mathcal{E}^k_{\tilde{X}}) \oplus \Gamma(E, \Lambda^k_E) \oplus \Gamma(\tilde{E}, \Lambda^{k-1}_{\tilde{E}})$$

so that

$$d(\omega, \sigma, \theta) = (d\omega, d\sigma, d\theta + (-1)^k(\psi(\omega) - \phi(\sigma)) \tag{3.52}$$

where ψ and ϕ are the inner pullback of the complex $\Lambda^{\cdot}_X$. Let us assume that

$$d(\omega, \sigma, \theta) \in F^p\Gamma(X, \Lambda^{k+1}_X) \tag{3.53}$$

In particular $d\omega \in F^p\Gamma(\tilde{X}, \mathcal{E}^{k+1}_{\tilde{X}})$; since $\tilde{X}$ is a compact Kähler manifold, so that $d\colon \Gamma(\tilde{X}, \mathcal{E}^k_{\tilde{X}}) \to \Gamma(\tilde{X}, \mathcal{E}^{k+1}_{\tilde{X}})$ is strict (theorem 5.9 of part I, chapter 5), we can write

$$\omega = \omega' + \Omega, \quad \omega' \in F^p\Gamma(\tilde{X}, \mathcal{E}^k_{\tilde{X}}), \quad d\Omega = 0 \tag{3.54}$$

Moreover, again by (3.52), (3.53) we see that $d\sigma \in F^p\Gamma(E, \Lambda^{k+1}_E)$; by recursion on the dimension, we can assume that d is strict for the complex $\Gamma(E, \Lambda^{\cdot}_E)$ with its filtration F^p, so that

$$\sigma = \sigma' + \Sigma, \quad \sigma' \in F^p\Gamma(E, \Lambda^k_E), \quad d\Sigma = 0 \tag{3.55}$$

From (3.52),(3.54), (3.55) one has

$$\begin{aligned} & d\theta + (-1)^k(\psi(\omega) - \phi(\sigma)) = \\ & = d\theta + (-1)^k\big(\psi(\omega') - \phi(\sigma')\big) + (-1)^k\big(\psi(\Omega) - \phi(\Sigma)\big) \end{aligned} \tag{3.56}$$

In the above equation by hypothesis (3.53) the first member is in $F^p\Gamma\big(\tilde{E}, \Lambda^k_{\tilde{E}}\big)$ and by construction also $(\psi(\omega') - \phi(\sigma')) \in F^p\Gamma\big(\tilde{E}, \Lambda^k_{\tilde{E}}\big)$ because the pullback are morphisms of the F^p filtrations. As a consequence the form ρ defined by

$$\rho = d\theta + (-1)^k\big(\psi(\omega) - \phi(\sigma)\big) - (-1)^k\big(\psi(\omega') - \phi(\sigma')\big)$$

has the property

$$\rho \in F^p\Gamma(\tilde{E}, \Lambda^k_{\tilde{E}})$$

and using (3.56) we have also

$$\rho = d\theta + (-1)^k(\psi(\Omega) - \phi(\Sigma)) \tag{3.57}$$

so by (3.54) and (3.55) $d\rho = 0$. Hence, ρ defines an element of $H^k(\tilde{E}, \mathbb{C})$ which, by (3.57), is cohomologous to $(-1)^k(\psi(\Omega) - \phi(\Sigma))$.

Now the morphism $\Theta\colon H^k(E, \mathbb{C}) \oplus H^k(\tilde{X}, \mathbb{C}) \to H^k(\tilde{E}, \mathbb{C})$ satisfies by definition

$$\Theta(\Omega, \Sigma) = (-1)^k(\psi(\Omega) - \phi(\Sigma))$$

Moreover, Θ is a linear combination of pullback and so, is a morphism of mixed Hodge structures (see theorem 3.5). From the results of part I, chapter 1, Θ is strict for the filtration F^p. This means that $[\rho]$, which is in $F^pH^k(\tilde{E}, \mathbb{C})$, can be rewritten as

$$\begin{gathered} \rho = d\theta' + (-1)^k(\psi(\Omega') - \phi(\Sigma')) \\ \Omega' \in F^p\Gamma\big(\tilde{X}, \mathcal{E}^k_{\tilde{X}}\big),\ \Sigma' \in F^p\Gamma\big(E, \Lambda^k_E\big), \quad d\Omega' = d\Sigma' = 0 \end{gathered} \tag{3.58}$$

Then, $\rho - (-1)^k(\psi(\Omega') - \phi(\Sigma'))$ is exact and is in $F^p\Gamma\big(\tilde{E}, \Lambda^k_{\tilde{E}}\big)$. By induction on the dimension, we can assume that d is strict for the complex $\Gamma\big(\tilde{E}, \Lambda^{\cdot}_{\tilde{E}}\big)$ for the filtration F^p, so there exists a $\theta'' \in F^p\Gamma\big(\tilde{E}, \Lambda^{k-1}_{\tilde{E}}\big)$ with

$$\rho = d\theta'' + (-1)^k(\psi(\Omega') - \phi(\Sigma'))$$

Finally, one can write

$$d(\omega, \sigma, \theta) = d\big(\omega' + \Omega', \sigma' + \Sigma', \theta''\big)$$

with

$$\big(\omega' + \Omega', \sigma' + \Sigma', \theta''\big) \in F^p\Gamma\big(X, \Lambda^{k-1}_X\big)$$

which proves that d is strict for F^p. □

Under the assumptions of theorem 3.9 the differential d is strict also with respect to the filtration W_m, up to a shift by $+1$:

Theorem 3.10. *Let us suppose that all the manifolds X_l of the hypercovering associated to $\Lambda^{\cdot}_X$ are compact Kähler manifolds. Let $\omega \in \Gamma(X, \Lambda^k_X)$ such that $d\omega \in W_m\Gamma(X, \Lambda^{k+1}_X)$; there exists $\theta \in W_{m+1}\Gamma(X, \Lambda^k_X)$ with $d\theta = d\omega$.*

The proof will be given (in a more general context) in part III, chapter 4.

Part III

Mixed Hodge structures on noncompact spaces

Chapter 1

Residues and Hodge mixed structures: Leray theory

1.1 Introduction

In this chapter we explain the classical Leray theory of residues and its main applications. The context is the following. Let X be a compact complex manifold, $D \subset X$ a smooth divisor, i.e. a smooth complex hypersurface. We are interested in the differential forms ω on the open set $X \setminus D$ having logarithmic poles along D; they are called logarithmic forms. This means that in local coordinates around any $x \in X$ we can write

$$\omega = \alpha \wedge \frac{dz}{z} + \beta$$

where $z = 0$ is the local equation of D. The *Leray residue* of ω is the restriction

$$\operatorname{Res} \omega = \alpha|_D$$

which is, in fact, a global form on D. The differential $d\omega$ of a logarithmic form is logarithmic: locally we can write $d\omega = d\alpha \wedge \frac{dz}{z} + d\beta$, so that

$$\operatorname{Res} d\omega = d\alpha|_D = d \operatorname{Res} \omega$$

hence the residue commutes to the differential. The above definition has a local nature: we can define a logarithmic form on $Y \setminus D$ for any open subset $Y \subset X$. It follows that the sheaf $\mathcal{E}^k_X\langle\log D\rangle$ of the logarithmic k-forms is well defined, and $\mathcal{E}^{\cdot}_X\langle\log D\rangle$ is a complex of fine sheaves on X (the reader should take note: on X, not only on $X \setminus D$). The logarithmic forms are particular differential forms on $X \setminus D$, hence we have an inclusion

$$\mathcal{E}^{\cdot}_X\langle\log D\rangle \subset \rho_* \mathcal{E}^{\cdot}_{X\setminus D}$$

where $\rho\colon X \setminus D \hookrightarrow X$ is the natural inclusion map. A theorem due to Leray states that the above inclusion of complexes induces an isomorphism in cohomology. This means exactly the following local statements: let U be an open neighborhood in X of a point $x \in D$, isomorphic to a ball in $\mathbb{C}^N$; then

- every closed differential form on $U \setminus D$ is cohomologous to a logarithmic form;
- every logarithmic differential form on $U \setminus D$ which is exact, is the differential of a logarithmic form.

(Actually, Leray himself did prove only the first of the above statements).

The main consequence of the above result is that the cohomology of $X \setminus D$ is the cohomology of the complex of global sections $\Gamma(X, \mathcal{E}_X^{\cdot}\langle\log D\rangle)$:

$$H^k(X, \mathcal{E}_X^{\cdot}\langle\log D\rangle) \simeq H^k(X \setminus D, \mathbb{C}) \tag{1.1}$$

On the other hand, since the residue commutes to d, it can be considered as a morphism of complexes

$$\mathrm{Res}^k : \mathcal{E}_X^k\langle\log D\rangle \to \mathcal{E}_D^{k-1}$$

which passes to the cohomology spaces:

$$\underline{\mathrm{Res}}^k : H^k(X \setminus D, \mathbb{C}) = H^k(X, \mathcal{E}_X^{\cdot}\langle\log D\rangle) \to H^{k-1}(D, \mathbb{C})$$

Leray proves that the above morphism is surjective, and that its kernel is exactly the image of $H^k(X, \mathbb{C})$ by the pullback ρ^*; the induced map

$$\underline{\mathrm{Res}}^k : \frac{H^k(X \setminus D, \mathbb{C})}{\rho^* H^k(X, \mathbb{C})} \to H^{k-1}(D, \mathbb{C}) \tag{1.2}$$

is an isomorphism.

Let us suppose that X is compact and Kähler. The isomorphism (1.2) points out in particular that the quotient of $H^k(X \setminus D, \mathbb{C})$ by $\rho^* H^k(X, \mathbb{C})$ is isomorphic to the space $H^{k-1}(D, \mathbb{C})$, which lives on a compact and Kähler manifold D, so that classical Hodge theory applies to it. Hence the quotient $\frac{H^k(X \setminus D, \mathbb{C})}{\rho^* H^k(X, \mathbb{C})}$ inherits from $H^{k-1}(D, \mathbb{C})$ a pure Hodge structure. This leads us to interpret the results of Leray in terms of mixed Hodge structures on the cohomology of the (noncompact) Kähler manifold $X \setminus D$: we introduce a weight filtration W on $\mathcal{E}_X^k\langle\log D\rangle$, which induces by formula (1.1) a filtration on $H^k(X \setminus D, \mathbb{C})$; we obtain that the nonzero quotients are

$$\frac{W_0 H^k(X \setminus D, \mathbb{C})}{W_{-1} H^k(X \setminus D, \mathbb{C})} \simeq \frac{H^k(X, \mathbb{C})}{\mathrm{im}\, \gamma_{D,X}^{k-2}}, \quad \frac{W_1 H^k(X \setminus D, \mathbb{C})}{W_0 H^k(X \setminus D, \mathbb{C})} \simeq \ker \gamma_{D,X}^{k-1}$$

The spectral sequence $E_2^{1,k}$ corresponding to the filtration W induced on the complex of global sections $\Gamma(X, \mathcal{E}_X^{\cdot}\langle\log D\rangle)$, degenerates at level 2 for trivial degree reasons, so that $E_2^{0,k}$ and $E_2^{1,k}$ coincide with the above quotients. The first term of the spectral sequence carries a natural pure Hodge structure, induced by the Hodge filtration on the complex of forms, and d_1 is a morphism of pure Hodge structures. It follows that also $E_2^{m,k}$, as the cohomology of $E_1^{m,k}$, carries a pure Hodge structure; hence $H^k(X \setminus D, \mathbb{C})$ carries a mixed Hodge structure.

1.2 The standard logarithmic De Rham complex

In this section, X is a complex analytic manifold of dimension n and D is a smooth divisor, i.e. a smooth hypersurface of X, so that at each point $x \in X$, one can find complex analytic coordinates $(z = z_1, \ldots, z_n)$ in a neighborhood U of x, such that the local equation of $D \cap U$ in U is $z = 0$.

1.2.1 Definition of $\mathcal{E}^{\cdot}_X\langle \log D\rangle$

If x is a point in $X \setminus D$, we define

$$\mathcal{E}^{\cdot}_{X,x}\langle \log D\rangle = \mathcal{E}^{\cdot}_{X,x}$$

i.e. the usual De Rham complex of differential forms on X at x. Let x be a point in D and U a neighborhood of x in X, such that the local equation of $D \cap U$ in U is $z = 0$. We define

$$\mathcal{E}^{\cdot}_{X,x}\langle \log D\rangle = \mathcal{E}^{\cdot}_{X,x}\left\{\frac{dz}{z}\right\} \tag{1.3}$$

i.e. the $\mathcal{E}^{\cdot}_{X,x}$-module generated by the logarithmic differential $\frac{dz}{z}$ of the equation of D through x. The differential d is defined as usual. That is, the complex $\big(\mathcal{E}^{\cdot}_{X,x}\langle \log D\rangle, d\big)$ induces a subcomplex of the De Rham complex of $X \setminus D$, i.e. there is an injection:

$$\big(\mathcal{E}^{\cdot}_{X,x}\langle \log D\rangle, d\big) \hookrightarrow \big(\rho_* \mathcal{E}^{\cdot}_{X \setminus D}, d\big) \tag{1.4}$$

where $\rho \colon X \setminus D \hookrightarrow X$ is the natural map.

Note that $\mathcal{E}^k_X\langle \log D\rangle$ *is a fine sheaf defined on all of* X. A section of $\mathcal{E}^k_X\langle \log D\rangle$ on an open set $U \subset X$ will be called a *logarithmic differential form* of degree k on U.

1.2.2 Filtration by the order of poles

We shall now define an increasing filtration W on $\mathcal{E}^{\cdot}_X\langle \log D\rangle$, called *the weight filtration*, in the following way

$$W_l \mathcal{E}^{\cdot}_X\langle \log D\rangle = \begin{cases} 0 & \text{for } l < 0 \\ \mathcal{E}^{\cdot}_X & \text{for } l = 0 \\ \mathcal{E}^{\cdot}_X\langle \log D\rangle & \text{for } l \geq 1 \end{cases} \tag{1.5}$$

It is clear that the filtration W is compatible with d:

$$d\big(W_l \mathcal{E}^k_X\langle \log D\rangle\big) \subset W_l \mathcal{E}^{k+1}_X\langle \log D\rangle \tag{1.6}$$

By (1.6) the above filtration induces a filtration $W_l H^k(X, \mathcal{E}^{\cdot}_X\langle\log D\rangle)$ in cohomology. The filtration (1.5) induces also a filtration on the complex of global sections $\Gamma(X, \mathcal{E}^{\cdot}_X\langle\log D\rangle)$ by

$$W_l\Gamma(X, \mathcal{E}^{\cdot}_X\langle\log D\rangle) = \Gamma(X, W_l\mathcal{E}^{\cdot}_X\langle\log D\rangle)$$

and a spectral sequence denoted by $E_r^{l,k}$ with first term

$$E_1^{l,k} = H^k\left(\frac{\Gamma(X, W_l\mathcal{E}^{\cdot}_X\langle\log D\rangle)}{\Gamma(X, W_{l-1}\mathcal{E}^{\cdot}_X\langle\log D\rangle)}\right) = H^k\left(X, \frac{W_l\mathcal{E}^{\cdot}_X\langle\log D\rangle}{W_{l-1}\mathcal{E}^{\cdot}_X\langle\log D\rangle}\right) \tag{1.7}$$

The second equality in the above formula holds because the $\frac{W_l\mathcal{E}^k_X\langle\log D\rangle}{W_{l-1}\mathcal{E}^k_X\langle\log D\rangle}$ are fine sheaves, so that we have isomorphisms

$$\frac{\Gamma(X, W_l\mathcal{E}^k_X\langle\log D\rangle)}{\Gamma(X, W_{l-1}\mathcal{E}^k_X\langle\log D\rangle)} = \Gamma\left(X, \frac{W_l\mathcal{E}^k_X\langle\log D\rangle}{W_{l-1}\mathcal{E}^k_X\langle\log D\rangle}\right)$$

The spectral sequence converges to the graded cohomology

$$\frac{W_l H^k(X, \mathcal{E}^{\cdot}_X\langle\log D\rangle)}{W_{l-1}H^k(X, \mathcal{E}^{\cdot}_X\langle\log D\rangle)} \tag{1.8}$$

By the definition (1.5) the quotients $\frac{W_l\mathcal{E}^{\cdot}_X\langle\log D\rangle}{W_{l-1}\mathcal{E}^{\cdot}_X\langle\log D\rangle}$ are

$$\frac{W_l\mathcal{E}^{\cdot}_X\langle\log D\rangle}{W_{l-1}\mathcal{E}^{\cdot}_X\langle\log D\rangle} = \begin{cases} 0 & \text{for } l \neq 0, 1 \\ \mathcal{E}^{\cdot}_X & \text{for } l = 0 \\ \dfrac{\mathcal{E}^{\cdot}_X\langle\log D\rangle}{\mathcal{E}^{\cdot}_X} & \text{for } l = 1 \end{cases} \tag{1.9}$$

The only nonzero $E_1^{l,k}$ are $E_1^{1,k} = H^k\left(X, \frac{\mathcal{E}^{\cdot}_X\langle\log D\rangle}{\mathcal{E}^{\cdot}_X}\right)$ and $E_1^{0,k} = H^k(X, \mathcal{E}^{\cdot}_X)$, so that the only relevant differential d_1 is

$$d_1\colon E_1^{1,k} = H^k\left(X, \frac{\mathcal{E}^{\cdot}_X\langle\log D\rangle}{\mathcal{E}^{\cdot}_X}\right) \to E_1^{0,k+1} = H^{k+1}(X, \mathcal{E}^{\cdot}_X) \tag{1.10}$$

Using (1.7), $E_1^{l,k}$ appears as the cohomology of the graded complex

$$\cdots\frac{\Gamma(X, W_l\mathcal{E}^k_X\langle\log D\rangle)}{\Gamma(X, W_{l-1}\mathcal{E}^k_X\langle\log D\rangle)} \xrightarrow{d_0} \frac{\Gamma(X, W_l\mathcal{E}^{k+1}_X\langle\log D\rangle)}{\Gamma(X, W_{l-1}\mathcal{E}^{k+1}_X\langle\log D\rangle)}\cdots$$

where d_0 is induced by d. We take an element $[[\pi]] \in E_1^{1,k}$. It corresponds to an element $\pi \in \Gamma(X, \mathcal{E}^k_X\langle\log D\rangle)$ such that $d\pi \in \Gamma(X, \mathcal{E}^{k+1}_X)$. Moreover the d_0 of the class of $d\pi$ is obviously 0, so $d\pi$ gives a cohomology class $[d\pi] \in H^{k+1}(X, \mathcal{E}^{\cdot}_X) = E_1^{0,k+1}$ depending only on $[[\pi]]$ and not on the other choices. Hence $d_1([[\pi]]) = [d\pi]$.

1.3 Residues (classical Leray theory)

Definition 1.1. We have introduced in part I, chapter 6 the (Leray) **residue**

$$\mathrm{Res}^k = \mathrm{Res}\colon \mathcal{E}_X^k\langle \log D\rangle \to \mathcal{E}_D^{k-1} \tag{1.11}$$

in the following way. Let ω be a section of $\mathcal{E}_X^k\langle \log D\rangle$ and let $(z = z_1, \ldots, z_n)$ be a coordinate system such that the local equation of $D \cap U$ in an open subset $U \subset X$ is $z = 0$, so that in U

$$\omega = \alpha \wedge \frac{dz}{z} + \beta \tag{1.12}$$

α (resp. β) being a regular differential form of degree $k-1$ (resp. of degree k) on U. We define

$$\mathrm{Res}^k \omega = \alpha|_D \tag{1.13}$$

It is easy to see that $\mathrm{Res}^k\omega$ is a well defined $(k-1)$-form on D, which does not depend on the choice of the local coordinates. In particular, for any open set $V \subset X$, the morphism Res^k induces a morphism of global sections on V

$$\mathrm{Res}^k\colon \Gamma\big(V, \mathcal{E}_X^k\langle \log D\rangle\big) \to \Gamma\big(V \cap D, \mathcal{E}_D^{k-1}\big) \tag{1.14}$$

Lemma 1.1. Res^k *commutes with the differentials d, ∂, and $\bar\partial$, and*

$$\mathrm{Res}^k\big(\mathcal{E}_X^k\big) = 0 \tag{1.15}$$

so that one has an induced map

$$\mathrm{Res}^k\colon \frac{\mathcal{E}_X^k\langle \log D\rangle}{\mathcal{E}_X^k} \to \mathcal{E}_D^{k-1} \tag{1.16}$$

which can be rewritten as

$$\mathrm{Res}^k\colon \frac{W_1\mathcal{E}_X^k\langle \log D\rangle}{W_0\mathcal{E}_X^k\langle \log D\rangle} \to \mathcal{E}_D^{k-1} \tag{1.17}$$

The proof of the above lemma is a simple local computation on the formula (1.12) (see part I, chapter 6, lemmas 6.1 and 6.2).

1.3.1 The residues in local cohomology

Lemma 1.2. Res^k *induces an isomorphism of the cohomology sheaves*

$$\underline{\mathrm{Res}}^k\colon \mathcal{H}^k\left(\frac{\mathcal{E}_X^{\cdot}\langle \log D\rangle}{\mathcal{E}_X^{\cdot}}\right) \to \mathcal{H}^{k-1}\mathcal{E}_D^{\cdot} \tag{1.18}$$

Proof. First we notice that $\mathcal{H}^{k-1}\mathcal{E}^{\cdot}_D = 0$ except for $k = 1$, where it is $\mathbb{C}_D$, because $\mathcal{E}^{\cdot}_D$ is the De Rham resolution of the constant sheaf on D. Moreover for $k < 1$, both sides of (1.18) are trivially 0. Finally, the statement of lemma 1.2 is purely local. So we can consider the situation where $X = U$ is a polydisk in $\mathbb{C}^n$ and D is defined by $z = 0$. We consider a k-form ω of $\mathcal{E}^k_X\langle\log D\rangle$ with poles of order ≤ 1; ω is given by formula (1.12), where we can suppose that $\beta = 0$, because we are working modulo $\mathcal{E}^k_X$. Let us suppose that ω defines a local cohomology class of the graded complex; this means that $d\omega$ is a $(k+1)$-form with no poles. Since

$$d\omega = d\alpha \wedge \frac{dz}{z} \tag{1.19}$$

it turns out that $d\alpha$ is in the ideal generated by z and dz. Clearly in α we can rule out all the components containing dz. Because we are working locally in a small polydisk, we can consider a Taylor expansion of α modulo the ideal generated by z and dz:

$$\alpha = \alpha_0 + \left(\bar{z}\alpha_1 + d\bar{z} \wedge \alpha_2\right) \equiv \alpha_0 + \alpha' \tag{1.20}$$

where α_0 contains only the coordinates $z_2, \cdots, z_n$ (here $z = z_1$), and their differentials, and α_1, α_2, do not contain z nor dz; but $d\alpha$ is in the ideal generated by z and dz so that from (1.20)

$$d\alpha_0 = 0, \qquad d\alpha' = 0 \tag{1.21}$$

a) If $k-1>0$, α_0 and α' have positive degree, so that, by Poincaré lemma in $n-1$ and n variables respectively we can assume that

$$\alpha_0 = d\theta, \qquad \alpha' = d\theta' \tag{1.22}$$

where θ is a differential containing only z_j for $j \neq 1$. Then we see that

$$\omega = d\left(\left(\theta + \theta'\right) \wedge \frac{dz}{z}\right)$$

which proves that the cohomology sheaf of the graded complex is 0 if $k-1>0$.

b) If $k-1=0$, equations (1.21) tell us that α is a constant and the cohomology sheaf of the graded complex reduces to $\mathbb{C}_D$, which is also $\mathcal{H}^0\mathcal{E}^{\cdot}_D$. □

As an immediate consequence of lemma 1.2, we obtain

Theorem 1.1. *For any open set V of X the morphisms $\underline{\mathrm{Res}}^k$ induce natural isomorphisms:*

$$\underline{\mathrm{Res}}^k : H^k\left(V, \frac{\mathcal{E}^{\cdot}_X\langle\log D\rangle}{\mathcal{E}^{\cdot}_X}\right) \to H^{k-1}(V \cap D, \mathcal{E}^{\cdot}_D) \tag{1.23}$$

Proof. On X, we consider the two complexes of fine sheaves $\frac{\mathcal{E}^{\cdot}_X\langle\log D\rangle}{\mathcal{E}^{\cdot}_X}$ and $\mathcal{E}^{\cdot}_D(-1)$ (extended by 0 outside D). Each of the complexes induces a spectral sequence of hypercohomology:

$$H^p\left(V, \mathcal{H}^q\left(\frac{\mathcal{E}^{\cdot}_X\langle\log D\rangle}{\mathcal{E}^{\cdot}_X}\right)\right) \Longrightarrow H^{p+q}\left(V, \frac{\mathcal{E}^{\cdot}_X\langle\log D\rangle}{\mathcal{E}^{\cdot}_X}\right)$$
$$H^p(V, \mathcal{H}^q\mathcal{E}^{\cdot}_D(-1)) \Longrightarrow H^{p+q}(V, \mathcal{E}^{\cdot}_D(-1))$$

By lemma 1.2 the cohomology sheaves $\mathcal{H}^q\left(\frac{\mathcal{E}^{\cdot}_X\langle\log D\rangle}{\mathcal{E}^{\cdot}_X}\right)$ and $\mathcal{H}^q\mathcal{E}^{\cdot}_D(-1)$ are isomorphic (by $\underline{\mathrm{Res}}^{\cdot}$). So in the limit we find that $H^k\left(V, \frac{\mathcal{E}^{\cdot}_X\langle\log D\rangle}{\mathcal{E}^{\cdot}_X}\right)$ and $H^k(V, \mathcal{E}^{\cdot}_D(-1))$ are isomorphic. But this last cohomology is $H^{k-1}(V \cap D, \mathcal{E}^{\cdot}_D)$ by De Rham theorem. □

Remark. By lemma 1.2 the morphisms (1.11), (1.16) (or (1.17)) are surjective; the morphism (1.16) is not injective.

For example the residue of the form $\omega = \bar{z}\frac{dz}{z}$ is zero, though $\omega \notin \mathcal{E}^1_X$.

1.3.2 The residues in global cohomology

On any open set $V \subset X$, we have the morphism Res^k of global sections given by (1.14), which commutes with d. This morphism induces a morphism of the filtered cohomology of the complexes, which we denote

$$\mathrm{RES}^k \colon H^k(V, \mathcal{E}^{\cdot}_X\langle\log D\rangle) \to H^{k-1}(V \cap D, \mathbb{C}) \tag{1.24}$$

Theorem 1.2. *The kernel of* RES^k *is* $W_0H^k(V, \mathcal{E}^{\cdot}_X\langle\log D\rangle)$.

Proof. Note that $W_0H^k(V, \mathcal{E}^{\cdot}_X\langle\log D\rangle)$ is the image in $H^k(V, \mathcal{E}^{\cdot}_X)$ under the map induced by the injection $\mathcal{E}^{\cdot}_X \to \mathcal{E}^{\cdot}_X\langle\log D\rangle$. Obviously, because of (1.15), the kernel of RES^k contains $W_0H^k(V, \mathcal{E}^{\cdot}_X\langle\log D\rangle)$. Conversely, let ω be a d-closed logarithmic k-form whose cohomology class $[\omega]$ goes to 0 through RES^k. Let

$$\overline{\omega} \in \Gamma\left(V, \frac{\mathcal{E}^k_X\langle\log D\rangle}{\mathcal{E}^k_X}\right) = \frac{\Gamma(V, \mathcal{E}^k_X\langle\log D\rangle)}{\Gamma(V, \mathcal{E}^k_X)}$$

be induced by ω. Our hypothesis says that ω has a residue 0 in cohomology of D, so that the isomorphism (1.23) of theorem 1.1 shows that

$$\overline{\omega} = d\overline{\gamma}, \qquad \overline{\gamma} \in \frac{\Gamma(V, \mathcal{E}^{k-1}_X\langle\log D\rangle)}{\Gamma(V, \mathcal{E}^{k-1}_X)}$$

This means that

$$\omega = d\gamma + \theta$$

with

$$\gamma \in \Gamma(V, \mathcal{E}_X^{k-1}\langle\log D\rangle), \qquad \theta \in \Gamma(V, \mathcal{E}_X^k)$$

hence the cohomology class $[\omega]$ is in $W_0 H^k(V, \mathcal{E}_X^{\cdot}\langle\log D\rangle)$. □

1.3.3 The cohomology of $X \setminus D$

Let W be any open set of X , in particular X itself. Then $(\mathcal{E}_{X\setminus D}^{\cdot}, d)$ can be used to calculate the cohomology of $W \setminus D$, by standard De Rham theorem.

Theorem 1.3. *One has a natural isomorphism induced by the morphism of inclusion $\rho\colon X \setminus D \to X$:*

$$H^k(W, \mathcal{E}_X^{\cdot}\langle\log D\rangle) \simeq H^k(W, \rho_* \mathcal{E}_{W\setminus D}^{\cdot}) \simeq H^k(W \setminus D, \mathbb{C}) \tag{1.25}$$

or in other words the cohomology of $W \setminus D$ can be calculated as the cohomology of the complex of sections $\left(\Gamma(W, \mathcal{E}_X^{\cdot}\langle\log D\rangle), d\right)$ of $\mathcal{E}_X^{\cdot}\langle\log D\rangle$.

First we note that the second isomorphism in the above formula (1.25) is a consequence of the theorem of De Rham for the standard De Rham complex.

Both complexes $\mathcal{E}_X^{\cdot}\langle\log D\rangle$ and $\rho_* \mathcal{E}_{X\setminus D}^{\cdot}$ induce spectral sequences on W:

$$H^p(W, \mathcal{H}^q \mathcal{E}_X^{\cdot}\langle\log D\rangle) \implies H^{p+q}(W, \mathcal{E}_X^{\cdot}\langle\log D\rangle) \tag{1.26}$$

$$H^p(W, \mathcal{H}^q \rho_* \mathcal{E}_{X\setminus D}^{\cdot}) \implies H^{p+q}(W, \rho_* \mathcal{E}_{X\setminus D}^{\cdot}) \tag{1.27}$$

It is sufficient to prove that the natural morphism of the cohomology sheaves

$$\mathcal{H}^q \mathcal{E}_X^{\cdot}\langle\log D\rangle \to \mathcal{H}^q \rho_* \mathcal{E}_{X\setminus D}^{\cdot}$$

is an isomorphism (so that the limits in the formulas (1.26) and (1.27) will be isomorphic). This is a local statement. The conclusion will follow from the following

Lemma 1.3. *Let U be an open set isomorphic to a polidysk in $\mathbb{C}^n$ such that the equation of $D \cap U$ is $z_1 = 0$. The morphism of complexes*

$$\Gamma(U, \mathcal{E}_X^{\cdot}\langle\log D\rangle) \to \Gamma(U \setminus D, \mathcal{E}_U^{\cdot})$$

induces an isomorphisms of their cohomologies.

Proof. Let Δ be the unit disk in $\mathbb{C}$; then $U \setminus D \simeq (\Delta \setminus (0)) \times \Delta^{n-1}$, so that $H^k(U \setminus D, \mathcal{E}_U^{\cdot})$ is 0 for $k \geq 2$ and is $\mathbb{C}$ for $k = 0, 1$. Hence it is enough to prove that also $H^k(U, \mathcal{E}_X^{\cdot}\langle\log D\rangle)$ is 0 for $k \geq 2$ and is $\mathbb{C}$ for $k = 0, 1$. The case $k = 0$ is trivial.

Let us remark that if $\xi \in \Gamma(U, \mathcal{E}_X^k)$ is a closed smooth form, by the classical De Rham theorem $\xi = d\mu$, $\mu \in \Gamma(U, \mathcal{E}_X^{k-1})$.

Let $\omega \in \Gamma(U, \mathcal{E}_X^k\langle\log D\rangle)$ with $d\omega = 0$. Let us suppose first $k \geq 2$. If $\omega \in W_0\Gamma(U, \mathcal{E}_X^k\langle\log D\rangle) = \Gamma(U, \mathcal{E}_X^k)$, then $\omega = d\eta$, $\eta \in \Gamma(U, \mathcal{E}_X^{k-1})$ by the above remark. If $\omega \in W_1\Gamma(U, \mathcal{E}_X^k\langle\log D\rangle)$, $\operatorname{Res}^k \omega$ is a closed $(k-1)$-form on $U \cap D$, which is exact because $H^{k-1}(U \cap D, \mathcal{E}_D^{\cdot}) = 0$; by lemma 1.2 there exists $\eta \in W_1\Gamma(U, \mathcal{E}_X^{k-1}\langle\log D\rangle)$ such that $\omega = d\eta + \xi$, $\xi \in \Gamma(U, \mathcal{E}_X^k)$. Now $d\omega = 0$ implies $d\xi = 0$, so that, as above, $\xi = d\mu$, $\mu \in \Gamma(U, \mathcal{E}_X^{k-1})$, and finally $\omega = d(\eta + \mu)$, that is: $H^k(U, \mathcal{E}_X^{\cdot}\langle\log D\rangle) = 0$ for $k \geq 2$.

For $k = 1$ similar methods show that ω is cohomologous, up to a multiplicative constant, to $\frac{dz_1}{z_1}$, so that $H^1(U, \mathcal{E}_X^{\cdot}\langle\log D\rangle) = \mathbb{C}$ (generated by the class of $\frac{dz_1}{z_1}$). □

1.4 Residues and mixed Hodge structures (the case of a smooth divisor)

1.4.1 Hodge filtrations and residues

If M is a complex manifold, we recall that any differential form π of degree k can be written using complex coordinates as

$$\pi = \sum_{|I|+|J|=k} \pi_{IJ}\, dz^I \wedge d\bar{z}^J \tag{1.28}$$

We say that π has type $\geq p$, if one has $|I| \geq p$ in the sum (1.28) and we define $F^p\mathcal{E}_M^k$ as the space of k-forms of type $\geq p$. This defines the decreasing Hodge filtration

$$\cdots F^p\mathcal{E}_M^k \supset F^{p+1}\mathcal{E}_M^k \supset \cdots$$

and d respects this filtration. The conjugate filtration is defined by saying that $\pi \in \bar{F}^q\mathcal{E}_M^k$ if $|J| \geq q$ in the sum (1.28) or in other words

$$\bar{F}^q\mathcal{E}_M^k = \overline{F^q\mathcal{E}_M^k}$$

In our case, we define on the De Rham complex of the manifold D, the usual Hodge filtration

$$F^p\mathcal{E}_D^{\cdot}$$

and the conjugate filtration

$$\bar{F}^q\mathcal{E}_D^k = \bar{F}^q\mathcal{E}_D^k \tag{1.29}$$

so that we can define a shifted filtration on the complex shifted by -1, namely

$$^{(-1)}F^p\mathcal{E}_D^k(-1) = F^{p-1}\mathcal{E}_D^{k-1} \tag{1.30}$$

as well as a shifted conjugate filtration

$$^{(-1)}\bar{F}^q\mathcal{E}_D^k(-1) = \bar{F}^{q-1}\mathcal{E}_D^{k-1} \tag{1.31}$$

If X is a Kähler compact manifold, D is also compact Kähler and the Hodge filtration of the De Rham complex induces a pure Hodge structure on the cohomology of D. Indeed, one has

$$H^k(D,\mathbb{C}) = \bigoplus_{a+b=k} F^a\bar{F}^b H^k(D,\mathbb{C}) \tag{1.32}$$

This means that the shifted filtrations of (1.30), (1.31) induce also a pure Hodge structure, namely (1.32) can be rewritten as

$$H^k(D,\mathbb{C}) = \bigoplus_{a+b=k} {}^{(-1)}F^{a+1}\ {}^{(-1)}\bar{F}^{b+1}H^k(D,\mathbb{C})$$

Renaming the indices, we obtain

$$H^{k-1}(D,\mathbb{C}) = \bigoplus_{a+b=k+1} {}^{(-1)}F^a\ {}^{(-1)}\bar{F}^b H^{k-1}(D,\mathbb{C}) \tag{1.33}$$

where we notice that the direct sum of (1.33) is taken over pairs (a,b) with $a+b=k+1$ (instead of $k-1$ as usual). That is, *$H^{k-1}(D,\mathbb{C})$ equipped with the (-1)-shifted filtrations, acquires a pure Hodge structure of weight $k+1$.*

We also define filtrations on $\mathcal{E}_X^{\cdot}\langle\log D\rangle$ by

$$\begin{aligned} {}^{(-1)}F^a\mathcal{E}_X^{\cdot}\langle\log D\rangle &= F^a\mathcal{E}_X^{\cdot}\langle\log D\rangle \\ {}^{(-1)}\bar{F}^b\mathcal{E}_X^{\cdot}\langle\log D\rangle &= \bar{F}^{b-1}\mathcal{E}_X^{\cdot}\langle\log D\rangle \end{aligned} \tag{1.34}$$

As usual in the second members of (1.34), F^a, $\bar{F}^b$ denote the standard Hodge filtrations.

Remark. The filtrations defined in (1.34) are not conjugate. Moreover the role of ${}^{(-1)}F^a$, ${}^{(-1)}\bar{F}^b$ is not symmetric. This is obviously due to the non symmetric role of $\frac{dz}{z}$ in the definition of logarithmic poles. Moreover the presence of $\frac{dz}{z}$ explains why we need shifted complexes and filtrations in order to obtain morphisms of filtered complexes, for example in the next lemma.

With the above notations, we deduce immediately

Lemma 1.4. *The residues induce morphisms of filtered spaces for the shifted filtrations*

$$\mathrm{Res}^{\cdot} : {}^{(-1)}F^p\mathcal{E}_X^{\cdot}\langle\log D\rangle \to {}^{(-1)}F^p\mathcal{E}_D^{\cdot}(-1)$$

$$\mathrm{Res}^{\cdot} : {}^{(-1)}F^p\left(\frac{\mathcal{E}_X^{\cdot}\langle\log D\rangle}{\mathcal{E}_X^{\cdot}}\right) \to {}^{(-1)}F^p\mathcal{E}_D^{\cdot}(-1)$$

and also for the $\bar{F}$ filtrations

$$\mathrm{Res}^{\cdot} : {}^{(-1)}\bar{F}^q\mathcal{E}_X^{\cdot}\langle\log D\rangle \to {}^{(-1)}\bar{F}^q\mathcal{E}_D^{\cdot}(-1)$$

$$\mathrm{Res}^{\cdot} : {}^{(-1)}\bar{F}^q\left(\frac{\mathcal{E}_X^{\cdot}\langle\log D\rangle}{\mathcal{E}_X^{\cdot}}\right) \to {}^{(-1)}\bar{F}^q\mathcal{E}_D^{\cdot}(-1)$$

1.4.2 Pure Hodge structure on $E_1^{l,k} = E_1^{l,k}(X)$

Recall from part I, chapter 6, that if M is a complex manifold of dimension n, and $j\colon S \to M$ is the embedding of a smooth complex hypersurface in M, the *Gysin map*

$$\gamma^k_{S,M}\colon H^k(S,\mathbb{C}) \to H^{k+2}(M,\mathbb{C})$$

can be obtained via Poincaré duality from the pullback

$$j^*\colon H^{2n-k-2}(M,\mathbb{C}) \to H^{2n-k-2}(S,\mathbb{C})$$

It follows that $\gamma^k_{S,M}$ is real, commutes to conjugation, and is a morphism of pure Hodge structures of degree 1.

In the sequel, X is a compact Kähler manifold.

Let us consider the spectral sequence $E_r^{l,k} = E_r^{l,k}(X)$ with first term (1.7), whose only nonzero elements are $E_1^{1,k} = H^k\left(X, \frac{\mathcal{E}^\cdot_X\langle\log D\rangle}{\mathcal{E}^\cdot_X}\right)$ and $E_1^{0,k} = H^k(X,\mathcal{E}^\cdot_X)$. Then the only relevant differential d_1 is

$$d_1\colon E_1^{1,k} = H^k\left(X, \frac{\mathcal{E}^\cdot_X\langle\log D\rangle}{\mathcal{E}^\cdot_X}\right) \to E_1^{0,k+1} = H^{k+1}(X,\mathcal{E}^\cdot_X) \tag{1.35}$$

Proposition 1.1. *Let X be a compact Kähler manifold.*

1) The term $E_1^{0,k} = H^k(X,\mathcal{E}^\cdot_X) = H^k(X,\mathbb{C})$ carries its natural pure Hodge structure of weight k defined by the Kähler compact manifold X. The term $E_1^{1,k} = H^k\left(X, \frac{\mathcal{E}^\cdot_X\langle\log D\rangle}{\mathcal{E}^\cdot_X}\right)$ carries a pure Hodge structure of weight $k+1$ induced by the shifted filtrations ${}^{(-1)}F$, ${}^{(-1)}\bar F$ defined by (1.34). Hence

$$E_1^{0,k} = \bigoplus_{p+q=k} F^p\bar F^q E_1^{0,k} \tag{1.36}$$

$$E_1^{1,k} = \bigoplus_{p+q=k+1} {}^{(-1)}F^p\, {}^{(-1)}\bar F^q E_1^{1,k} \tag{1.37}$$

In particular, the filtrations induced by ${}^{(-1)}F$, ${}^{(-1)}\bar F$ on $E_1^{l,k}$ are conjugate. The residue

$$\underline{\mathrm{Res}}^k\colon E_1^{1,k} \to H^{k-1}(D,\mathbb{C}) \tag{1.38}$$

induces an isomorphism of pure Hodge structures on $E_1^{1,k}$ and on $H^{k-1}(D,\mathbb{C})$ for the shifted Hodge filtrations ${}^{(-1)}F$, ${}^{(-1)}\bar F$.

2) The differential

$$d_1\colon E_1^{1,k} \to E_1^{0,k+1} \tag{1.39}$$

is a morphism of pure Hodge structures.

3) Let $\hat{d}_1$ be the differential induced by d_1 at the level of residues, namely the following diagram is commutative

$$\begin{array}{ccc} E_1^{1,k} & \xrightarrow{d_1} & E_1^{0,k+1} \\ {\scriptstyle \underline{\mathrm{Res}}^k}\downarrow & & \| \\ H^{k-1}(D,\mathbb{C}) & \xrightarrow{\hat{d}_1} & H^{k+1}(X,\mathbb{C}) \end{array} \tag{1.40}$$

Then $\hat{d}_1$ coincides with the Gysin map $\gamma_{D,X}^{k-1}$ of the embedding $D \to X$, and it commutes with the complex conjugation.

4) $\hat{d}_1$ is a morphism of pure Hodge structures for the shifted filtrations ${}^{(-1)}F$, ${}^{(-1)}\bar{F}$ on $H^{k-1}(D,\mathbb{C})$ and F, $\bar{F}$ on $H^{k+1}(X,\mathbb{C})$.

In order to prove proposition 1.1, we need a more precise realization of the inverse of the isomorphism $\underline{\mathrm{Res}}^k$ of (1.38) (or (1.23)), at a global level.

We consider the line bundle L on X associated to the divisor D and a holomorphic section σ of L whose zero set is exactly D. We introduce a hermitian metric $|\ |$ on L and define

$$\eta = -\frac{1}{2i\pi}\partial \log|\sigma|^2, \qquad d\eta = \Omega = -\frac{1}{2i\pi}\bar{\partial}\partial \log|\sigma|^2 \tag{1.41}$$

so that Ω is a C^∞ form on X which represents the Chern class $c_1(L)$.

Lemma 1.5. *Let α be a $(k-1)$-form on D.*

i. *If α has a well defined type, one can find a C^∞ extension $\tilde{\alpha}$ of α to X, having the same type as α. Moreover $\omega = \tilde{\alpha} \wedge \eta$ is in $\Gamma(X, \mathcal{E}_X^k\langle\log D\rangle)$ and ω has residue α on D.*

ii. *If α is a closed form on D, there exists a form $\tilde{\alpha}$ on X, closed in a neighborhood of D, such that if $\omega = \tilde{\alpha} \wedge \eta$, $d\omega$ is in $\Gamma(X, \mathcal{E}_X^{k+1})$ and ω defines an element $[[\omega]]_1 \in E_1^{1,k} = H^k\left(X, \frac{\mathcal{E}_X^{\cdot}\langle\log D\rangle}{\mathcal{E}_X^{\cdot}}\right)$. Moreover $d\omega$ defines the element $d_1[[\omega]]_1 = [[d\omega]]_1$ in $E_1^{0,k+1} = H^{k+1}(X, \mathcal{E}_X^{\cdot})$ and the residue of $d_1[[\omega]]_1$ on D is exactly the class $[\alpha] \in H^k(D, \mathcal{E}_D^{\cdot})$.*

Proof of Lemma 1.5. Let $(W_a)_{a\in A}$ be an open covering of X such that in each W_a intersecting D, one has a system of complex coordinates $z^{(a)}$:

$$z^{(a)} = \left(z_1^{(a)}, z''^{(a)}\right), \qquad z''^{(a)} = \left(z_2^{(a)}, \ldots, z_n^{(a)}\right)$$

such that the equation of $D \cap W_a$ in W_a is $z_1^{(a)} = 0$.

Let us consider $\alpha|_{D\cap W_a}$: it is a $(k-1)$-form which can be written in terms of the coordinates $z''^{(a)}$ only:

$$\alpha|_{D\cap W_a} = \sum_{|J|+|K|=k-1} \alpha_{JK}^{(a)}\left(z''^{(a)}\right)\left(dz''^{(a)}\right)^J \wedge \left(d\bar{z}''^{(a)}\right)^K \tag{1.42}$$

We can extend $\alpha|_{D\cap W_a}$ from $D \cap W_a$ to W_a simply using the second member of (1.42), as a form $\alpha^{(a)}$. If $(\beta_a)_{a\in A}$ is a partition of unity of X such that the support of β_a is in W_a, we can define an extension $\tilde{\alpha}$ of α by the formula

$$\tilde{\alpha} = \sum_{a\in A} \beta_a \alpha^{(a)} \tag{1.43}$$

If α has a well-defined type, $\alpha^{(a)}$ and $\tilde{\alpha}$ have the same type. It is obvious that the residue of $\tilde{\alpha} \wedge \eta$ on D is $\tilde{\alpha}|_D = \alpha$. This proves part (i) of lemma 1.5.

Let W be an open neighborhood of D in X which is a deformation retract of D. If α is closed, we can extend α to a closed form α' on W, and then (after possibly shrinking W) α' to a form $\tilde{\alpha}$ on X.

It follows

$$d\omega = d\tilde{\alpha} \wedge \eta \pm \tilde{\alpha} \wedge d\eta \tag{1.44}$$

and $d\tilde{\alpha} \wedge \eta$ is 0 on a neighborhood of D. Moreover, by the definition of η in (1.41), $d\eta = \Omega \in \Gamma(X, \mathcal{E}_X^{k+1})$ so that $d\omega \in \Gamma(X, \mathcal{E}_X^{k+1})$. Hence ω gives an element $[[\omega]]_1 \in E_1^{1,k} = H^k\left(X, \frac{\mathcal{E}_X^{\cdot}\langle\log D\rangle}{\mathcal{E}_X^{\cdot}}\right)$ such that

$$d_1[[\omega]]_1 = [[d\omega]]_1$$

This proves lemma 1.5. □

Proof of proposition 1.1.

Proof of 1). The filtrations ${}^{(-1)}F$, ${}^{(-1)}\bar{F}$ on the complex $\Gamma\left(X, \frac{\mathcal{E}_X^{\cdot}\langle\log D\rangle}{\mathcal{E}_X^{\cdot}}\right)$ induce corresponding filtrations on its cohomology , which is exactly $E_1^{1,k}$. By lemma 1.4, $\mathrm{Res}^{\cdot}$ is a morphism of filtered complexes for the shifted filtrations ${}^{(-1)}F$, ${}^{(-1)}\bar{F}$, and it induces the isomorphism $\underline{\mathrm{Res}}^{\cdot}$ at the level of cohomology, which is thus a morphism of filtered spaces

$$\begin{aligned} {}^{(-1)}F^pE_1^{1,k} &\to {}^{(-1)}F^pH^{k-1}(D,\mathbb{C}) \\ {}^{(-1)}\bar{F}^qE_1^{1,k} &\to {}^{(-1)}\bar{F}^qH^{k-1}(D,\mathbb{C}) \end{aligned}$$

By (1.33), $H^{k-1}(D,\mathbb{C})$ carries a pure Hodge structure of weight $k+1$. Since $\mathrm{Res}^{\cdot}$ respects the types, $\underline{\mathrm{Res}}^{\cdot}$ induces morphisms of subspaces

$$\underline{\mathrm{Res}}^k : {}^{(-1)}F^p\,{}^{(-1)}\bar{F}^qE_1^{1,k} \to {}^{(-1)}F^p\,{}^{(-1)}\bar{F}^qH^{k-1}(D,\mathbb{C}) \tag{1.45}$$

Since by theorem 1.1 $\underline{\mathrm{Res}}^k$ is an isomorphism of vector spaces between $E_1^{1,k}$ and $H^{k-1}(D,\mathbb{C})$, and $H^{k-1}(D,\mathbb{C})$ is a direct sum of the vector spaces in the second member of (1.45), it follows that (1.45) is an isomorphism. This also proves that the shifted filtrations ${}^{(-1)}F$ and ${}^{(-1)}\bar{F}$ induce a pure Hodge structure on $E_1^{1,k}$ isomorphic by $\underline{\mathrm{Res}}^k$ to the pure Hodge structure on $H^{k-1}(D,\mathbb{C})$ of the shifted Hodge filtrations (1.33).

Proof of 3). This is a consequence of lemma 1.5 and lemma 6.4 of part I, chapter 6. Since the Gysin map commutes with the complex conjugation, the same holds for $\hat{d}_1$.
Proof of 4). This is an immediate consequence of the fact that $\hat{d}_1$ is a Gysin map.
Proof of 2). It is an obvious consequence of the commutative diagram (1.40), the fact that $\hat{d}_1$ is a morphism of pure Hodge structures and that the residues are isomorphisms of Hodge structures. □

1.4.3 Mixed Hodge structure on $H^k(X \setminus D, \mathbb{C})$

The spectral sequence $E_r^{l,k}$ degenerates at E_2 and each term carries a pure Hodge structure:

Theorem 1.4. *Let X be a compact a Kähler manifold.*

1) *The differentials d_r of the spectral sequence of the filtration W_l are 0 for $r \geq 2$.*

2) *The nonzero second terms of the spectral sequence are $E_2^{1,k} = \ker d_1$ and $E_2^{0,k} = \frac{E_1^{0,k}}{\operatorname{im} d_1}$ where :*

$$d_1 \colon E_1^{1,k} = H^k\left(X, \frac{\mathcal{E}_X^{\cdot}\langle \log D\rangle}{\mathcal{E}_X^{\cdot}}\right) \to E_1^{0,k+1} = H^{k+1}(X, \mathcal{E}_X^{\cdot})$$

Both these terms carry a pure Hodge structure. More precisely

$$E_2^{1,k} = \bigoplus_{a+b=k+1} {}^{(-1)}F^a \, {}^{(-1)}\bar{F}^b E_2^{1,k} \tag{1.46}$$

$$E_2^{0,k} = \bigoplus_{a+b=k} F^a \bar{F}^b E_2^{0,k} \tag{1.47}$$

where ${}^{(-1)}F^a$, ${}^{(-1)}\bar{F}^b$, are the filtrations induced by $E_1^{1,k}$, $E_1^{0,k}$ respectively.

The proof is very easy. The spectral sequence degenerates at E_2 for degree reasons.

By proposition 1.1, the E_1 have pure Hodge structures defined by (1.36) and (1.37) and the differential d_1 is a morphism of pure Hodge structures, hence the cohomology of the complex $(E_1^{l,k}, d_1)$, which is $E_2^{l,k}$, carries a pure Hodge structure of the same weight $k+l$. This implies part (2) of the theorem.

From theorem 1.4 and the convergence of the spectral sequence to the graded cohomology, we deduce:

Theorem 1.5. *Let X be a compact Kähler manifold, $D \subset X$ a smooth divisor, $\gamma_{D,X}^{k-1} \colon H^{k-1}(D, \mathbb{C}) \to H^{k+1}(X, \mathbb{C})$ the Gysin map corresponding to the natural embedding $D \subset X$. The cohomology $H^k(X \setminus D, \mathbb{C})$ carries a* *mixed Hodge structure for the filtration W induced by* (1.5), *and the filtrations* ${}^{(-1)}F^a$, ${}^{(-1)}\bar{F}^b$, *induced by* (1.34). *The only non zero graded spaces* $\frac{W_l H^k(X \setminus D, \mathbb{C})}{W_{l-1} H^k(X \setminus D, \mathbb{C})}$ *occur for* $l = 0, 1$. *They are isomorphic to* $E_2^{0,k}$ *and* $E_2^{1,k}$ *thus they have a pure Hodge structure of weight* k *and* $k+1$ *respectively, as in* (1.46), (1.47).. *Precisely*

$$\frac{W_0 H^k(X \setminus D, \mathbb{C})}{W_{-1} H^k(X \setminus D, \mathbb{C})} \simeq \frac{H^k(X, \mathbb{C})}{\operatorname{im} \gamma_{D,X}^{k-2}}, \quad \frac{W_1 H^k(X \setminus D, \mathbb{C})}{W_0 H^k(X \setminus D, \mathbb{C})} \simeq \ker \gamma_{D,X}^{k-1}$$

Proof. Because $d_r = 0$ for $r \geq 2$, the graded cohomology is the second term of the spectral sequence:

$$\frac{W_l H^k(X \setminus D, \mathbb{C})}{W_{l-1} H^k(X \setminus D, \mathbb{C})} \simeq E_2^{l,k} \tag{1.48}$$

i.e. the graded spaces $\frac{W_l H^k(X \setminus D, \mathbb{C})}{W_{l-1} H^k(X \setminus D, \mathbb{C})}$ are isomorphic to $E_2^{l,k}(X)$, which carry a pure Hodge structure of weight $k+l$ for the filtration F induced by $E_1^{l,k}(X)$ ($E_2^{l,k}(X)$ is the cohomology of $(E_1^{l,k}(X), d_1)$). On the other hand, the filtration F on $H^k(X \setminus D, \mathbb{C})$ induces a filtration on the quotient $\frac{W_l H^k(X \setminus D, \mathbb{C})}{W_{l-1} H^k(X \setminus D, \mathbb{C})}$. We should show that the the two filtrations, under the isomorphism (1.47), coincide (and the same for $\bar{F}$). This will be made clear in the more general case of chapter 2. □

Remark. A shift of W by $-k$ is needed to normalize the weights in the quotients. In the shifted filtration $W_l' = W_{l-k}$ we obtain

$$\frac{W_l' H^k(X \setminus Q, \mathbb{C})}{W_{l-1}' H^k(X \setminus Q, \mathbb{C})} \simeq E_2^{l-k,k}(X)$$

that is, the quotient $\frac{W_l' H^k(X, \mathbb{C})}{W_{l-1}' H^k(X, \mathbb{C})}$ has weight l, as expected.

1.4.4 Functoriality

Let X, Y be compact manifolds, $D \subset X, F \subset Y$ smooth divisors. Let $f\colon (X, D) \to (Y, F)$ a morphism of pairs, i.e. a map $f\colon X \to Y$ such that $f(D) \subset F$. Then all the constructions and properties in the above paragraphs are functorial with respect to f. In particular

Theorem 1.6. *Let* $f\colon (X, D) \to (Y, F)$ *a morphism of pairs. Then*

$$f^* \mathcal{E}_Y^{\cdot} \langle \log F \rangle \subset \mathcal{E}_X^{\cdot} \langle \log D \rangle \tag{1.49}$$

$$f^* W_l \mathcal{E}_Y^{\cdot} \langle \log F \rangle \subset W_l \mathcal{E}_X^{\cdot} \langle \log D \rangle \tag{1.50}$$

$$f^* F^p \mathcal{E}_Y^{\cdot} \langle \log F \rangle \subset F^p \mathcal{E}_X^{\cdot} \langle \log D \rangle \tag{1.51}$$

$$f^* \bar{F}^q \mathcal{E}_Y^{\cdot} \langle \log F \rangle \subset \bar{F}^q \mathcal{E}_X^{\cdot} \langle \log D \rangle \tag{1.52}$$

Moreover the residue morphism commute with the pullback:

$$\mathrm{Res}^k f^*\omega = (f|_D)^*\mathrm{Res}^k\omega \quad \textit{for } \omega \in \mathcal{E}^{\cdot}_Y\langle \log F\rangle. \tag{1.53}$$

1.4.5 Other residues

In our situation (a manifold X and a smooth divisor $D \subset X$) the Leray residue Res^k defined in formulas (1.11), (1.13) is enough to handle the Hodge mixed structure in the cohomology $H^k(X \setminus D, \mathbb{C})$. In the next chapter we are going to face more complicated cases, and define a family of residues instead of a single one. Hence for future purposes we define even in our situation for every integer $l \geq 0$ a residue Res^k_l by

$$\mathrm{Res}^k_l = \begin{cases} \mathrm{Res}^k_0 = \text{identity: } W_0\mathcal{E}^k_X\langle \log D\rangle = \mathcal{E}^k_X \to \mathcal{E}^k_X & \text{for } l = 0 \\ \mathrm{Res}^k\colon W_1\mathcal{E}^k_X\langle \log D\rangle \to \mathcal{E}^{k-1}_D & \text{for } l = 1 \\ 0 & \text{for } l \geq 2 \end{cases} \tag{1.54}$$

The above definition can be summarized as follows. Let us define

$$D^{[0]} = X, D^{[1]} = D \text{ and } D^{[l]} = \emptyset \quad \text{for } l \geq 2 \tag{1.55}$$

then

$$\mathrm{Res}^k_l\colon W_l\mathcal{E}^k_X\langle \log D\rangle \to \mathcal{E}^{k-l}_{D^{[l]}} \tag{1.56}$$

and by lemma 1.1, formula (1.15), Res^k_l passes to the quotient

$$\mathrm{Res}^k_l\colon \frac{W_l\mathcal{E}^k_X\langle \log D\rangle}{W_{l-1}\mathcal{E}^k_X\langle \log D\rangle} \to \mathcal{E}^{k-l}_{D^{[l]}} \tag{1.57}$$

(we keep the same notation for the quotient mapping).

Then lemmas 1.1, 1.2 and theorem 1.1 can be restated as follows, as a preparation for the next chapter:

Theorem 1.7.

i. $\mathrm{Res}^k_l\colon W_l\mathcal{E}^k_X\langle \log D\rangle \to \mathcal{E}^{k-l}_{D^{[l]}}$ *commutes with the differentials* d, ∂, *and* $\bar\partial$.

ii. Res^k_l *induces an isomorphism of the cohomology sheaves*

$$\underline{\mathrm{Res}}^k_l\colon \mathcal{H}^k\left(\frac{W_l\mathcal{E}^{\cdot}_X\langle \log D\rangle}{W_{l-1}\mathcal{E}^{\cdot}_X\langle \log D\rangle}\right) \to \mathcal{H}^{k-l}\mathcal{E}^{\cdot}_{D^{[l]}}$$

iii. For any open set V *of* X *the morphism* $\underline{\mathrm{Res}}^k_l$ *induce natural isomorphisms*

$$\underline{\mathrm{Res}}^k_l\colon H^k\left(V, \frac{W_l\mathcal{E}^{\cdot}_X\langle \log D\rangle}{W_{l-1}\mathcal{E}^{\cdot}_X\langle \log D\rangle}\right) \to H^{k-l}(V \cap D^{[l]}, \mathcal{E}^{\cdot}_{D^{[l]}}).$$

Proof. For $l = 0$ the morphisms in (i), (ii),(iii) are the identity, for $l \geq 2$ they are zero. □

Chapter 2

Residues and mixed Hodge structures on noncompact manifolds

2.1 Introduction

In this chapter we define and study differential forms on a complex manifold which have logarithmic poles along a divisor with normal crossing.

Let X be a compact complex manifold of complex dimension $n = \dim_{\mathbb{C}} X$ and $D = D_1 \cup \cdots \cup D_N$ is a *divisor with normal crossing*; that means that each D_i is a smooth hypersurface of X, and at each point $x \in X$, there are at most n divisors D_j passing through x and which are transversal. In particular, given x, one can find complex analytic coordinates $(z_1, \ldots, z_n)$ in a neighborhood U of x, such that the local equation of $D \cap U$ in U is $z_1 \cdots z_s = 0$, s depending on x.

We are in fact interested in the study of the *open manifold* $W = X \setminus D$, so that the role of X is that of a smooth compactification of $X \setminus D$. Hence our setting includes the assumption for such compactification. Any affine, or quasi projective, algebraic manifold possesses a smooth compactification.

We define for any ordered multiindex $I = (i_1, \ldots, i_q)$ in $(1, \ldots, N)$

$$D_I = D_{i_1} \cap \cdots \cap D_{i_q}$$

and

$$D^{[q]} = \coprod_{|I|=q} D_I, \qquad D^{[0]} = X$$

where the sign $\coprod$ denotes the disjoint union. Then the $D^{[q]}$ are manifolds (not connected in general).

A *logarithmic differential k-form* on X (with poles of order $\leq l$) is a form ω on $X \setminus D$ which, in a sufficiently small neighborhood of any $x \in D$ can be written as

$$\omega = \sum_{|I| \leq l} \alpha_I \wedge \left(\frac{dz}{z}\right)^I \tag{2.1}$$

where $\left(\frac{dz}{z}\right)^I = \frac{dz_{i_1}}{z_{i_1}} \wedge \cdots \wedge \frac{dz_{i_l}}{z_{i_l}}$.

The differential $d\omega$ of a logarithmic form (with poles of order $\leq l$) is logarithmic (with poles of order $\leq l$). The above definition has a local nature: we can define a logarithmic form on $Y \setminus D$ for any open subset $Y \subset X$, hence the sheaf $\mathcal{E}^k_X\langle \log D\rangle$ of the logarithmic k-forms is well defined, and $\mathcal{E}^{\cdot}_X\langle \log D\rangle$ is a complex of fine sheaves on X.

If $\omega \in \Gamma\big(X, \mathcal{E}^k_X\langle \log D\rangle\big)$ has poles of order $\leq l$ so that (2.1) locally holds, we define the *l-residue* of ω, $\mathrm{Res}^k_l \omega \in \Gamma\big(D^{[l]}, \mathcal{E}^{k-1}_{D^{[l]}}\big)$:

$$\mathrm{Res}^k_l \omega|_{D_I} = \alpha_I|_{D_I} \quad \text{for } |I| = l \tag{2.2}$$

which is a global $(k-l)$–form on the compact manifold $D^{[l]}$. The residue commutes to the differentials d on X and $D^{[l]}$.

The logarithmic forms on any open set $Y \subset X$ are particular differential forms on $Y \setminus D$, hence we have an inclusion

$$\mathcal{E}^{\cdot}_X\langle \log D\rangle \subset \rho_* \mathcal{E}^{\cdot}_{X\setminus D}$$

where $\rho\colon X \setminus D \hookrightarrow X$ is the natural inclusion map. As in the case of Leray theory (chapter 1: D is smooth), the following statements hold: let U be an open neighborhood in X of a point $x \in D$, isomorphic to a polydisk in $\mathbb{C}^N$; then

- every closed differential form on $U \setminus D$ is cohomologous to a logarithmic form;
- every logarithmic differential form on $U \setminus D$ which is exact, is the differential of a logarithmic form.

The main consequence of the above result is that the cohomology of $X \setminus D$ is the cohomology of the complex of global sections $\Gamma\big(X, \mathcal{E}^{\cdot}_X\langle \log D\rangle\big)$:

$$H^k\big(X, \mathcal{E}^{\cdot}_X\langle \log D\rangle\big) \simeq H^k(X, \rho_* \mathcal{E}^{\cdot}_{X\setminus D}) \simeq H^k(X \setminus D, \mathbb{C})$$

We introduce a weight filtration W on $\mathcal{E}^k_X\langle \log D\rangle$, just defining $W_l\mathcal{E}^k_X\langle \log D\rangle$ as the subsheaf of $\mathcal{E}^k_X\langle \log D\rangle$ of the forms having poles of order $\leq l$, so that in fact the residue Res^k_l is defined on $W_l\mathcal{E}^k_X\langle \log D\rangle$. Since the residues commute to d, and

$$\mathrm{Res}^k_l\big(W_{l-1}\mathcal{E}^k_X\langle \log D\rangle\big) = 0$$

$\mathrm{Res}^{\cdot}_l$ can be considered as a morphism of complexes

$$\mathrm{Res}^{\cdot}_l \colon \frac{W_l\mathcal{E}^{\cdot}_X\langle \log D\rangle}{W_{l-1}\mathcal{E}^{\cdot}_X\langle \log D\rangle} \to \mathcal{E}^{\cdot}_{D^{[l]}}(-l) \tag{2.3}$$

which passes to the cohomology:

$$\underline{\mathrm{Res}^k_l} \colon H^k\left(X, \frac{W_l\mathcal{E}^{\cdot}_X\langle \log D\rangle}{W_{l-1}\mathcal{E}^{\cdot}_X\langle \log D\rangle}\right) \to H^{k-l}(D^{[l]}, \mathcal{E}^{\cdot}_{D^{[l]}}) \tag{2.4}$$

It turns out that $\underline{\mathrm{Res}}_l^k$ is an isomorphism (while (2.3) is surjective but not injective).

On the other hand, the cohomology spaces $H^k\left(X, \frac{W_l\mathcal{E}^\cdot_X\langle\log D\rangle}{W_{l-1}\mathcal{E}^\cdot_X\langle\log D\rangle}\right)$ in (2.4) are the terms $E_1^{l,k}(X)$ of the spectral sequence of the complex $\Gamma(X, \mathcal{E}^\cdot_X\langle\log D\rangle)$ with respect to the filtration W, which converges to $H^k(X \setminus D, \mathbb{C})$. The isomorphism (2.4) identifies $E_1^{l,k}(X)$ to the cohomology $H^{k-l}(D^{[l]}, \mathcal{E}^\cdot_{D^{[l]}}) = H^{k-l}(D^{[l]}, \mathbb{C})$ of the manifold $D^{[l]}$, which is compact and Kähler, so that the classical Hodge theory applies to it. Hence $E_1^{l,k}(X)$ inherits from $H^{k-l}(D^{[l]}, \mathbb{C})$ (suitably shifted), a pure Hodge structure of weight $k+l$, and d_1 is a morphism of pure Hodge structures. It follows that also $E_2^{l,k}$, as the cohomology of $(E_1^{l,k}, d_1)$, carries a pure Hodge structure of weight $k+l$.

The fundamental result is that the spectral sequence corresponding to the filtration W degenerates at level 2, so that $E_2^{l,k}(X)$ coincides with the graded quotient $\frac{W_l H^k(X\setminus D,\mathbb{C})}{W_{l-1}H^k(X\setminus D,\mathbb{C})}$. It follows that $H^k(X \setminus D, \mathbb{C})$ carries a mixed Hodge structure.

2.2 The standard logarithmic De Rham complex

In this section, X is a complex analytic manifold and $D = D_1 \cup \cdots \cup D_N$ is a *divisor with normal crossing*; that means that each D_i is a smooth hypersurface of X, and at each point $x \in X$, there are at most $n = \dim_{\mathbb{C}} X$ divisors D_j passing through x and which are transversal. In particular, given x, one can find complex analytic coordinates $(z_1, \ldots, z_n)$ in a neighborhood U of x, such that the local equation of $D \cap U$ in U is $z_1 \cdots z_s = 0$, s depending on x.

2.2.1 Definition of $\mathcal{E}^\cdot_X\langle\log D\rangle$

If x is a point in $X \setminus D$, we define

$$\mathcal{E}^\cdot_{X,x}\langle\log D\rangle = \mathcal{E}^\cdot_{X,x}$$

i.e. the usual De Rham complex of differential forms on X at x. Let x be a point in D and U a neighborhood of x in X, such that the local equation of $D \cap U$ in U is $z_1 \cdots z_s = 0$. We define

$$\mathcal{E}^\cdot_{X,x}\langle\log D\rangle = \mathcal{E}^\cdot_{X,x}\left\{\frac{dz_1}{z_1}, \ldots, \frac{dz_s}{z_s}\right\} \tag{2.5}$$

i.e. the $\mathcal{E}^\cdot_{X,x}$-module generated by the logarithmic differentials $\frac{dz_1}{z_1}, \ldots, \frac{dz_s}{z_s}$ of the equations of the components of D through x. The differential d is defined as usual. In particular, the complex $\big(\mathcal{E}^{\cdot}_{X,x}\langle\log D\rangle, d\big)$ induces a subcomplex of the De Rham complex of $X \setminus D$.

Note that $\mathcal{E}^k_X\langle\log D\rangle$ *is a fine sheaf defined on all of* X. A section of $\mathcal{E}^k_X\langle\log D\rangle$ on an open set $U \subset X$ will be called a *logarithmic differential form* of degree k on U.

One has an injection:

$$\big(\mathcal{E}^{\cdot}_X\langle\log D\rangle, d\big) \hookrightarrow \big(\rho_*\mathcal{E}^{\cdot}_{X\setminus D}, d\big) \tag{2.6}$$

where $\rho\colon X \setminus D \hookrightarrow X$ is the natural map.

We define the subsheaf $\mathcal{E}^{p,q}_X\langle\log D\rangle \subset \mathcal{E}^{p+q}_X\langle\log D\rangle$ of the logarithmic forms of type (p,q) by

$$\Gamma\big(U, \mathcal{E}^{p,q}_X\langle\log D\rangle\big) = \Gamma\big(U, \mathcal{E}^{p+q}_X\langle\log D\rangle\big) \cap \Gamma\big(U \setminus D, \mathcal{E}^{p,q}_X\big) \tag{2.7}$$

The operators ∂ and $\bar\partial$ act as follows:

$$\partial\colon \mathcal{E}^{p,q}_X\langle\log D\rangle \to \mathcal{E}^{p+1,q}_X\langle\log D\rangle \tag{2.8}$$

$$\bar\partial\colon \mathcal{E}^{p,q}_X\langle\log D\rangle \to \mathcal{E}^{p,q+1}_X\langle\log D\rangle \tag{2.9}$$

giving rise to complexes $\big(\mathcal{E}^{\cdot,q}_X\langle\log D\rangle, \partial\big)$ and $\big(\mathcal{E}^{p,\cdot}_X\langle\log D\rangle, \bar\partial\big)$.

We define also the subcomplex $\Omega^{\cdot}_X\langle\log D\rangle \subset \mathcal{E}^{\cdot}_X\langle\log D\rangle$ of meromorphic differential forms with logarithmic poles on D:

$$\begin{aligned} \Omega^k_{X,x}\langle\log D\rangle &= \Omega^k_{X,x} \quad (x \in X \setminus D) \\ \Omega^k_{X,x}\langle\log D\rangle &= \Omega^k_{X,x}\left\{\tfrac{dz_1}{z_1}, \ldots, \tfrac{dz_s}{z_s}\right\} \quad (x \in D) \end{aligned} \tag{2.10}$$

$\big(\Omega^{\cdot}_X\langle\log D\rangle, d\big)$ is a complex of sheaves of $\mathcal{O}_X$- modules. Let us remark that the sheaves $\Omega^{\cdot}_X\langle\log D\rangle$ are not fine.

The following immediate result will be important in the sequel.

Proposition 2.1. *One has a natural isomorphisms*

$$\Omega^p_X\langle\log D\rangle \otimes_{\mathcal{O}_X} \mathcal{E}^{0,q}_X \simeq \mathcal{E}^{p,q}_X\langle\log D\rangle \tag{2.11}$$

compatible with $\bar\partial$, *in the sense that*

$$\bar\partial(\alpha \otimes \beta) = \alpha \otimes \bar\partial\beta \tag{2.12}$$

2.2.2 Filtration by the order of poles

Let U be a small neighborhood of a point $x \in D$ with coordinates $(z_1, \ldots, z_n)$ such that the local equation of $D \cap U$ in U is $z_1 \cdots z_s = 0$. Such coordinate system is called *a system of adapted coordinates for* (X, D) *in* U.

We shall denote

$$\left(\frac{dz}{z}\right)^I = \left(\frac{dz_{i_1}}{z_{i_1}} \wedge \cdots \wedge \frac{dz_{i_l}}{z_{i_l}}\right) \tag{2.13}$$

where $I = (i_1, \ldots, i_l)$ is an ordered multiindex contained in $(1, \ldots, s)$. We know that $\mathcal{E}^{\cdot}_{X,x}\langle \log D\rangle$ is generated by the $\left(\frac{dz}{z}\right)^I$ over the module $\mathcal{E}^{\cdot}_{X,x}$ for $|I| \le s$ (s depends on x).

We shall now define an increasing filtration on $\mathcal{E}^{\cdot}_X\langle \log D\rangle$, called *the weight filtration*, in the following way

$$W_l\mathcal{E}^{\cdot}_X\langle \log D\rangle = 0 \quad (l < 0) \tag{2.14}$$

$$W_0\mathcal{E}^{\cdot}_X\langle \log D\rangle = \mathcal{E}^{\cdot}_X \tag{2.15}$$

while for $l > 0$

$$\left(W_l\mathcal{E}^{\cdot}_X\langle \log D\rangle\right)_x = \mathcal{E}^{\cdot}_{X,x}\left[\left(\frac{dz}{z}\right)^I\right], \quad |I| \le l \quad (l > 0) \tag{2.16}$$

is the $\mathcal{E}^{\cdot}_{X,x}$-submodule generated by the $\left(\frac{dz}{z}\right)^I$ for $|I| \le l$. One has $l \le \chi_{X,D}$, where $\chi_{X,D}$ is defined as follows. If we denote by $s(x)$ the number s so that $D \subset U(x)$ is given by $z_1 \cdots z_s = 0$, we define

$$\chi_{X,D} = \max_{x\in X}\{s(x)\} \tag{2.17}$$

hence

$$\chi_{X,D} \le \dim X$$

It is clear that

$$d(W_l\mathcal{E}^{\cdot}_X\langle \log D\rangle) \subset W_l\mathcal{E}^{\cdot}_X\langle \log D\rangle \tag{2.18}$$

Thus the above filtration induces a filtration $W_lH^k(V, \mathcal{E}^{\cdot}_X\langle \log D\rangle)$ in cohomology.

Remark. $W_l\mathcal{E}^{\cdot}_X\langle \log D\rangle = 0$ for $l < 0$, and $W_l\mathcal{E}^{\cdot}_X\langle \log D\rangle = \mathcal{E}^{\cdot}_X\langle \log D\rangle$ for $l \ge \chi_{X,D}$.

We remark also that $W_l\mathcal{E}^k_X\langle \log D\rangle$ is a fine sheaf, so that for any open subset $V \subset X$ the following equality holds:

$$\Gamma\left(V, \frac{W_l\mathcal{E}^k_X\langle \log D\rangle}{W_{l-1}\mathcal{E}^k_X\langle \log D\rangle}\right) = \frac{\Gamma(V, W_l\mathcal{E}^k_X\langle \log D\rangle)}{\Gamma(V, W_{l-1}\mathcal{E}^k_X\langle \log D\rangle)} \tag{2.19}$$

2.2.3 The filtration W on $\Omega^k_X\langle \log D\rangle$

The filtration W on $\mathcal{E}^k_X\langle \log D\rangle$ induces filtrations on the subsheaves $\Omega^k_X\langle \log D\rangle$ and $\mathcal{E}^{p,k-p}_X\langle \log D\rangle$. Then

Theorem 2.1. *There are natural isomorphisms:*

$$W_l\Omega^p_X\langle \log D\rangle \otimes_{\mathcal{O}_X} \mathcal{E}^{0,q}_X \simeq W_l\mathcal{E}^{p,q}_X\langle \log D\rangle \tag{2.20}$$

$$\frac{W_l\Omega^p_X\langle \log D\rangle}{W_{l-1}\Omega^p_X\langle \log D\rangle} \otimes_{\mathcal{O}_X} \mathcal{E}^{0,q}_X \simeq \frac{W_l\mathcal{E}^{p,q}_X\langle \log D\rangle}{W_{l-1}\mathcal{E}^{p,q}_X\langle \log D\rangle} \tag{2.21}$$

Proof. The isomorphism (2.20) is clear by proposition 2.1.
Next we consider the exact sequence

$$0 \longrightarrow W_{l-1}\Omega_X^p\langle \log D\rangle \longrightarrow W_l\Omega_X^p\langle \log D\rangle \longrightarrow \frac{W_l\Omega_X^p\langle \log D\rangle}{W_{l-1}\Omega_X^p\langle \log D\rangle} \longrightarrow 0 \quad (2.22)$$

We use the following result of Malgrange [Ma]: the sheaf $\mathcal{E}_X^{0,0} = \mathcal{E}_X^0$ of differentiable functions on X is flat over the sheaf $\mathcal{O}_X$ (that is, for every $x \in X$ the $\mathcal{O}_{X,x}$-module $\mathcal{E}_{X,x}^{0,0}$ is flat). The sheaf $\mathcal{E}_X^{0,q}$ is locally free over the sheaf $\mathcal{E}_X^{0,0}$; It follows that $\mathcal{E}_X^{0,q}$ is also flat over the sheaf $\mathcal{O}_X$. Thus the exact sequence (2.22) remains exact after tensorization with $\otimes_{\mathcal{O}_X}\mathcal{E}_X^{0,q}$:

$$0 \longrightarrow W_{l-1}\Omega_X^p\langle \log D\rangle \otimes_{\mathcal{O}_X} \mathcal{E}_X^{0,q} \longrightarrow W_l\Omega_X^p\langle \log D\rangle \otimes_{\mathcal{O}_X} \mathcal{E}_X^{0,q} \longrightarrow$$
$$\longrightarrow \frac{W_l\Omega_X^p\langle \log D\rangle}{W_{l-1}\Omega_X^p\langle \log D\rangle} \otimes_{\mathcal{O}_X} \mathcal{E}_X^{0,q} \longrightarrow 0$$

or, taking into account (2.20):

$$0 \longrightarrow W_{l-1}\mathcal{E}_X^{p,q}\langle \log D\rangle \longrightarrow W_l\mathcal{E}_X^{p,q}\langle \log D\rangle \longrightarrow$$
$$\longrightarrow \frac{W_l\Omega_X^p\langle \log D\rangle}{W_{l-1}\Omega_X^p\langle \log D\rangle} \otimes_{\mathcal{O}_X} \mathcal{E}_X^{0,q} \longrightarrow 0 \quad (2.23)$$

Comparing (2.23) with the quotient sequence

$$0 \longrightarrow W_{l-1}\mathcal{E}_X^{p,q}\langle \log D\rangle \longrightarrow W_l\mathcal{E}_X^{p,q}\langle \log D\rangle \longrightarrow \frac{W_l\mathcal{E}_X^{p,q}\langle \log D\rangle}{W_{l-1}\mathcal{E}_X^{p,q}\langle \log D\rangle} \longrightarrow 0$$

we obtain (2.21). □

2.2.4 The De Rham complex of a divisor

We consider again a manifold X, $D = D_1 \cup \cdots \cup D_N$ a divisor with normal crossing and we define for any ordered multiindex $I = (i_1, \ldots, i_q)$ in $(1, \ldots, N)$

$$D_I = D_{i_1} \cap \cdots \cap D_{i_q} \quad (2.24)$$

and

$$D^{[q]} = \coprod_{|I|=q} D_I, \quad D^{[0]} = X \quad (2.25)$$

where the sign $\coprod$ denotes the disjoint union. Then the $D^{[q]}$ are manifolds (not connected in general).

We define a double complex

$$\mathcal{E}^{p,q}_{X,D} = \mathcal{E}^{p}_{D[q]}, \quad \mathcal{E}^{p,0}_{X,D} = \mathcal{E}^{p}_{X} \tag{2.26}$$

with two differentials:

$$d \colon \mathcal{E}^{p,q}_{X,D} \to \mathcal{E}^{p+1,q}_{X,D} \tag{2.27}$$

which is the De Rham differential on $\mathcal{E}^{p}_{D[q]}$ or $\mathcal{E}^{p}_{X}$, and

$$\delta \colon \mathcal{E}^{p,q}_{X,D} \to \mathcal{E}^{p,q+1}_{X,D} \tag{2.28}$$

such that if $\phi \in \mathcal{E}^{p,q}_{X,D}$, so that $\phi = (\phi_I)_{|I|=q}$ and ϕ_I is a $p-$form on D_I, then

$$\delta(\phi)_{j_1,\ldots,j_{q+1}} = \sum_{l=1}^{q} (-1)^l \phi_{j_1,\ldots,\hat{j}_l,\ldots,j_{q+1}} | D_{(j_1,\ldots,j_{q+1})} \tag{2.29}$$

and if $q = 0$, ϕ is a p-form on X and $\delta(\phi)_i = \phi | D_i$.

Clearly:

Lemma 2.1. *δ commutes with d.*

We define then the diagonal complex

$$\Lambda^k_D = \bigoplus_{p+q=k+1,\ q>0} \mathcal{E}^{p,q}_{X,D} \tag{2.30}$$

with the differential $d_D = d + (-1)^k \delta$.

2.3 Residues (smooth case)

Definition 2.1. We introduce for $l \geq 0$ the **residue (or l-residue)** map

$$\mathrm{Res}^k_l \colon W_l \mathcal{E}^k_X \langle \log D \rangle \to \mathcal{E}^{k-1}_{D[l]} \tag{2.31}$$

in the following way. Let ω be a section of $W_l \mathcal{E}^k_X \langle \log D \rangle$ and let $(z_1, \ldots, z_n)$ be an adapted coordinate system for (X, D) so that in this coordinate system

$$\omega = \sum_{|I| \leq l} \alpha_I \wedge \left(\frac{dz}{z} \right)^I \tag{2.32}$$

(remember (2.16)). We define the l-residue of ω, $\mathrm{Res}^k_l \omega \in \mathcal{E}^{k-1}_{D[l]}$:

$$\mathrm{Res}^k_l \omega |_{D_I} = \alpha_I |_{D_I} \quad \text{for } |I| = l \quad (\text{if } l > 0) \tag{2.33}$$

For $l = 0$, $W_0 \mathcal{E}^k_X \langle \log D \rangle = \mathcal{E}^k_X$ and $\mathrm{Res}^k_0 \omega = \omega \in \mathcal{E}^k_X$.

In other words,

$$\left[\mathrm{Res}_l^k\left(\alpha_I \wedge \left(\frac{dz}{z}\right)^I\right)\right]|_{D_J} = \begin{cases} \alpha_I|_{D_I} & \text{for } J = I \\ 0 & \text{for } J \neq I \end{cases} \tag{2.34}$$

It is easy to see that $\mathrm{Res}_l^k \omega$ is a well defined $(k-l)$-form on $D^{[l]}$, which does not depend on the choice of the local adapted coordinates. In particular, for any open set $V \subset X$, the morphism Res_l^k induces a morphism of global sections on V

$$\mathrm{Res}_l^k : \Gamma(V, W_l \mathcal{E}_X^k \langle \log D \rangle) \to \Gamma(V \cap D^{[l]}, \mathcal{E}_{D^{[l]}}^{k-1}) \tag{2.35}$$

From the definition we find immediately:

Lemma 2.2. Res_l^k *commutes with the differentials* d, ∂ *and* $\bar\partial$, *and*

$$\mathrm{Res}_l^k(W_{l-1}\mathcal{E}_X^k\langle \log D\rangle) = 0 \tag{2.36}$$

so that one has an induced map

$$\mathrm{Res}_l^k : \frac{W_l \mathcal{E}_X^k \langle \log D \rangle}{W_{l-1}\mathcal{E}_X^k\langle \log D\rangle} \to \mathcal{E}_{D^{[l]}}^{k-1} \tag{2.37}$$

which is a morphism of complexes

$$\mathrm{Res}_l^{\cdot} : \frac{W_l \mathcal{E}_X^{\cdot} \langle \log D \rangle}{W_{l-1}\mathcal{E}_X^{\cdot}\langle \log D\rangle} \to \mathcal{E}_{D^{[l]}}^{\cdot}(-l) \tag{2.38}$$

Remark. As it will be proved later (lemma 2.8, i)), the residue morphisms (2.31) and (2.37) are surjective. Note that (2.37) is not injective. For example, the first residue of the form $\omega = \bar z_1 \frac{dz_1}{z_1}$ is zero, though $\omega \notin W_0$.

If ω is a logarithmic form of type (p,q), $\omega \in W_l \mathcal{E}_X^{p,q}\langle \log D\rangle$, its residue $\mathrm{Res}_l^{p+q}\omega$ is a form of type $(p-l, q)$ on $D^{[l]}$; if moreover ω is a holomorphic logarithmic form $\in W_l\Omega_X^p\langle \log D\rangle$, its residue is a holomorphic $p-l$ form on $D^{[l]}$. Hence, by lemma 2.2 the $\mathrm{Res}_l^{\cdot}$ induce the following morphisms of complexes:

$$\mathrm{Res}_l^{\cdot+q} : \left(\frac{W_l \mathcal{E}_X^{\cdot,q}\langle \log D\rangle}{W_{l-1}\mathcal{E}_X^{\cdot,q}\langle \log D\rangle}, \partial\right) \to (\mathcal{E}_{D^{[l]}}^{\cdot,q}(-l), \partial) \tag{2.39}$$

$$\mathrm{Res}_l^{p+\cdot} : \left(\frac{W_l \mathcal{E}_X^{p,\cdot}\langle \log D\rangle}{W_{l-1}\mathcal{E}_X^{p,\cdot}\langle \log D\rangle}, \bar\partial\right) \to (\mathcal{E}_{D^{[l]}}^{p-l,\cdot}, \bar\partial) \tag{2.40}$$

$$\mathrm{Res}_l^{\cdot} : \left(\frac{W_l \Omega_X^{\cdot}\langle \log D\rangle}{W_{l-1}\Omega_X^{\cdot}\langle \log D\rangle}, \partial\right) \to (\Omega_{D^{[l]}}^{\cdot}(-l), \partial) \tag{2.41}$$

Proposition 2.2. *The residue* $\mathrm{Res}_l^{\cdot}$ *induces an isomorphism of the* ∂*-complexes* (2.41).

It is clear that (2.41) is surjective. In order to prove the injectivity, we must show that, given an adapted system of coordinates $(z_1, \ldots, z_n)$ for D in X, a holomorphic form $\omega = \sum_{|I|\leq l} \alpha_I \wedge \left(\frac{dz}{z}\right)^I$ with zero l-residue has poles of order at most $l-1$. The assumption on ω means $\alpha_I|_{D_I} = 0$ for all I with $|I| = l$. Since α_I is a holomorphic form, and $z_{i_1} \cdots z_{i_l} = 0$ is the local equation of D_I in X, we obtain that

$$\alpha_I = z_{i_1} \cdots z_{i_l} \beta_I$$

where β_I is also holomorphic. Then $\omega = \sum_{|I|\leq l-1} \alpha_I \wedge \left(\frac{dz}{z}\right)^I + \sum_{|I|=l} \beta_I \wedge dz^I$ has poles of order at most $l-1$.

2.3.1 The residues in local cohomology

Lemma 2.3. Res_l^k *induces an isomorphism of the cohomology sheaves*

$$\underline{\mathrm{Res}}_l^k : \mathcal{H}^k\left(\frac{W_l\mathcal{E}_X^{\cdot}\langle\log D\rangle}{W_{l-1}\mathcal{E}_X^{\cdot}\langle\log D\rangle}\right) \to \mathcal{H}^{k-l}\mathcal{E}_{D^{[l]}}^{\cdot} \tag{2.42}$$

Proof. First we notice that $\mathcal{H}^{k-l}\mathcal{E}_{D^{[l]}}^{\cdot} = 0$ except for $k = l$, where it is $\mathbb{C}_{D^{[l]}}$, because $\mathcal{E}_{D^{[l]}}^{\cdot}$ is the De Rham resolution of the constant sheaf on $D^{[l]}$. Moreover for $k < l$, both sides of (2.42) are trivially 0. Finally, the statement of lemma 2.3 is purely local. So we can consider the situation where $X = U$ is a polydisk in $\mathbb{C}^n$ and D is defined by $z_1 \cdots z_s = 0$. We consider a k-form ω of $\mathcal{E}_X^k\langle\log D\rangle$ with poles of order $\leq l$ and $l \leq s$; ω is given by the formula (2.32) where the multiindices $I \subset (1, \ldots, s)$ and α_I has degree $k-|I|$. Because we are working $\mod W_{l-1}$, we can assume that the sum in (2.32) is on the multiindices I with $|I| = l$. If ω defines a local cohomology class of the graded complex, $d\omega$ is a $(k+1)$-form with poles of order $\leq l-1$. But now

$$d\omega = \sum_{|I|\leq l} d\alpha_I \wedge \left(\frac{dz}{z}\right)^I \tag{2.43}$$

which implies that $d\alpha_I$ is in the ideal generated by the z_i and dz_i for $i \in I$. Clearly, in α_I we can forget all the components containing at least one dz_i for $i \in I$. Because we are working in a small polydisk, we can consider a Taylor expansion of α_I modulo the ideal generated by the z_i and dz_i for $i \in I$:

$$\alpha_I = \alpha_{0,I} + \sum_{i\in I}(\bar{z}_i\beta_I + d\bar{z}_i \wedge \gamma_I) \equiv \alpha_{0,I} + \alpha'_I \quad (\mod z_i, dz_i) \tag{2.44}$$

where $\alpha_{0,I}$ contains only the coordinates z_j, $j \notin I$ and their differentials, and β_I, γ_I, do not contain the z_i and dz_i for $i \in I$; but $d\alpha_I$ is in the ideal generated by the z_i and dz_i for $i \in I$ so that from (2.44):

$$d(\alpha_{0,I}) = 0, \qquad d\alpha'_I = 0 \tag{2.45}$$

a) If $k - l > 0$, $\alpha_{0,I}$ and α'_I have positive degree, so that by Poincaré lemma for $k - |I|$ and k variables respectively we get

$$\alpha_{0,I} = d\theta_I, \qquad \alpha'_I = d\theta'_I \tag{2.46}$$

where θ_I is a differential form containing only the z_j for $j \notin I$. Then we see that

$$\omega = d\left(\sum_{|I|=l} (\theta_I + \theta'_I) \wedge \left(\frac{dz}{z}\right)^I\right)$$

which proves that the cohomology sheaf of the graded complex is 0 if $k - l > 0$.

b) If $k - l = 0$, the equations in (2.45) tell us that α_I is locally constant on D_I and the cohomology sheaf of the graded complex reduces to $\mathbb{C}_{D^{[l]}} = \mathcal{H}^0\mathcal{E}^{\cdot}_{D^{[l]}}$.

We notice that for $l = 0$, one has

$$W_0\mathcal{E}^{\cdot}_X\langle\log D\rangle = \mathcal{E}^{\cdot}_X$$
$$W_{-1}\mathcal{E}^{\cdot}_X\langle\log D\rangle = 0$$
$$\mathcal{E}^{\cdot}_{D^{[0]}} = \mathcal{E}^{\cdot}_X$$

and the mapping in (2.42) is the identity. □

As a consequence of lemma 2.3, we obtain

Theorem 2.2. *For any open set V of X the morphism $\underline{\mathrm{Res}}_l^k$ induces a natural isomorphism*

$$\underline{\mathrm{Res}}_l^k : H^k\left(V, \frac{W_l\mathcal{E}^{\cdot}_X\langle\log D\rangle}{W_{l-1}\mathcal{E}^{\cdot}_X\langle\log D\rangle}\right) \to H^{k-l}(V \cap D^{[l]}, \mathcal{E}^{\cdot}_{D^{[l]}}) \tag{2.47}$$

Proof. On X, we consider the two complexes of fine sheaves $\frac{W_l\mathcal{E}^{\cdot}_X\langle\log D\rangle}{W_{l-1}\mathcal{E}^{\cdot}_X\langle\log D\rangle}$ and $\mathcal{E}^{\cdot}_{D^{[l]}}(-l)$ (extended by 0 outside $D^{[l]}$). Each complex induces a spectral sequence:

$$H^p\left(V, \mathcal{H}^q\left(\frac{W_l\mathcal{E}^{\cdot}_X\langle\log D\rangle}{W_{l-1}\mathcal{E}^{\cdot}_X\langle\log D\rangle}\right)\right) \Longrightarrow H^{p+q}\left(V, \frac{W_l\mathcal{E}^{\cdot}_X\langle\log D\rangle}{W_{l-1}\mathcal{E}^{\cdot}_X\langle\log D\rangle}\right)$$
$$H^p(V, \mathcal{H}^q\mathcal{E}^{\cdot}_{D^{[l]}}(-l)) \Longrightarrow H^{p+q}(V, \mathcal{E}^{\cdot}_{D^{[l]}}(-l))$$

By lemma 2.3 the cohomology sheaves $\mathcal{H}^q\left(\frac{W_l\mathcal{E}^{\cdot}_X\langle\log D\rangle}{W_{l-1}\mathcal{E}^{\cdot}_X\langle\log D\rangle}\right)$ and $\mathcal{H}^q\mathcal{E}^{\cdot}_{D^{[l]}}(-l)$ are isomorphic (by $\underline{\mathrm{Res}}_l^{\cdot}$). So in the limit of spectral sequences we find that the cohomologies $H^k\left(V, \frac{W_l\mathcal{E}^{\cdot}_X\langle\log D\rangle}{W_{l-1}\mathcal{E}^{\cdot}_X\langle\log D\rangle}\right)$ and $H^k(V, \mathcal{E}^{\cdot}_{D^{[l]}}(-l))$ are isomorphic. But this last cohomology is $H^{k-l}(V \cap D^{[l]}, \mathcal{E}^{\cdot}_{D^{[l]}})$. □

2.3.2 The residues in global cohomology

On any open set $V \subset X$, we have the morphisms Res_l^k of global sections given by (2.35), which commute with d. These morphisms induce morphisms of the filtered cohomology of the complexes, which we denote

$$\mathrm{RES}_l^k \colon W_l H^k(V, \mathcal{E}_X^{\cdot}\langle \log D\rangle) \to H^{k-l}(V \cap D^{[l]}, \mathbb{C}) \tag{2.48}$$

Lemma 2.4. *The kernel of* RES_l^k *is* $W_{l-1}H^k(V, \mathcal{E}_X^{\cdot}\langle \log D\rangle)$.

Proof. Because of (2.36), the kernel of RES_l^k contains $W_{l-1}H^k(V, \mathcal{E}_X^{\cdot}\langle \log D\rangle)$. Conversely, let ω be a d-closed logarirhmic k-form with poles of order $\leq l$ whose cohomology class $[\omega]$ goes to 0 through RES_l^k. So ω induces an element in the quotient

$$\overline{\omega} \in \Gamma\left(V, \frac{W_l\mathcal{E}_X^k\langle \log D\rangle}{W_{l-1}\mathcal{E}_X^k\langle \log D\rangle}\right) = \frac{\Gamma(V, W_l\mathcal{E}_X^k\langle \log D\rangle)}{\Gamma(V, W_{l-1}\mathcal{E}_X^k\langle \log D\rangle)}$$

(the above equality holds by (2.19)). Our hypothesis says that ω has a residue 0 in cohomology of $D^{[l]}$, so that the isomorphism (2.47) of theorem 2.2 shows that

$$\overline{\omega} = d\overline{\gamma}, \quad \overline{\gamma} \in \frac{\Gamma(V, W_l\mathcal{E}_X^{k-1}\langle \log D\rangle)}{\Gamma(V, W_{l-1}\mathcal{E}_X^{k-1}\langle \log D\rangle)}$$

This means that

$$\omega = d\gamma + \theta$$

with

$$\gamma \in \Gamma(V, W_l\mathcal{E}_X^{k-1}\langle \log D\rangle), \quad \theta \in \Gamma(V, W_{l-1}\mathcal{E}_X^{k}\langle \log D\rangle)$$

so that the cohomology class $[\omega]$ is in $W_{l-1}H^k(V, \mathcal{E}_X^{\cdot}\langle \log D\rangle)$. □

As an immediate consequence, using the above lemma and a descending recursion, we have

Theorem 2.3. *Let* $[\omega]$ *be a class in* $W_mH^k(V, \mathcal{E}_X^{\cdot}\langle \log D\rangle)$, *and* $l < m$. *Then* $[\omega] \in W_lH^k(V, \mathcal{E}_X^{\cdot}\langle \log D\rangle)$ *if and only if*

$$\mathrm{RES}_m^k[\omega] = \mathrm{RES}_{m-1}^k[\omega] = \cdots = \mathrm{RES}_{l+1}^k[\omega] = 0$$

2.3.3 The cohomology of $X \setminus D$

Let W be any open set of X, in particular X itself. Then $(\mathcal{E}_X^{\cdot}\langle \log D\rangle, d)$ can be used to calculate the cohomology of $W \setminus D$, by the following theorem.

Theorem 2.4. *One has a natural isomorphism induced by the morphism of inclusion* $\rho\colon X \setminus D \to X$:

$$H^k(W, \mathcal{E}_X^{\cdot}\langle \log D\rangle) \simeq H^k(W, \rho_*\mathcal{E}_{X\setminus D}^{\cdot}) \simeq H^k(W \setminus D, \mathbb{C}) \tag{2.49}$$

or in other words the cohomology of $W \setminus D$ *can be calculated as the cohomology of the complex of sections* $(\Gamma(W, \mathcal{E}_X^{\cdot}\langle \log D\rangle), d)$ *of* $\mathcal{E}_X^{\cdot}\langle \log D\rangle$.

First we note that the second isomorphism in the above formula (2.49) is a consequence of the theorem of De Rham for the standard De Rham complex. Each of the complexes $\mathcal{E}_X^{\cdot}\langle\log D\rangle$ and $\rho_*\mathcal{E}_{X\setminus D}^{\cdot}$ induces a spectral sequence of hypercohomology on W:

$$H^p(W, \mathcal{H}^q\mathcal{E}_X^{\cdot}\langle\log D\rangle) \Longrightarrow H^{p+q}(W, \mathcal{E}_X^{\cdot}\langle\log D\rangle) \tag{2.50}$$

$$H^p(W, \mathcal{H}^q\rho_*\mathcal{E}_{X\setminus D}^{\cdot}) \Longrightarrow H^{p+q}(W, \rho_*\mathcal{E}_{X\setminus D}^{\cdot}) \tag{2.51}$$

It is sufficient to prove that the natural morphism of the cohomology sheaves

$$\mathcal{H}^q\mathcal{E}_X^{\cdot}\langle\log D\rangle \to \mathcal{H}^q\rho_*\mathcal{E}_{X\setminus D}^{\cdot}$$

is an isomorphism (so that the limits in the formulas (2.50) and (2.51) will be isomorphic). This is a local statement. Then the conclusion follows from the following

Lemma 2.5. *Let U be an open set isomorphic to a polidysk in $\mathbb{C}^n$ such that the equation of $D \cap U$ is $z_1 \cdots z_s = 0$. The morphism of complexes*

$$\Gamma(U, \mathcal{E}_X^{\cdot}\langle\log D\rangle) \to \Gamma(U \setminus D, \mathcal{E}_U^{\cdot})$$

induces isomorphisms on their cohomologies.

Proof. Let $\mathbb{C}\left\{\frac{dz_1}{z_1}, \ldots, \frac{dz_s}{z_s}\right\} = \mathbb{C}\left\{\left\{\frac{dz_j}{z_j}\right\}\right\}$ be the free differential algebra over $\mathbb{C}$ generated by $\frac{dz_1}{z_1}, \ldots, \frac{dz_s}{z_s}$ and having differential $d = 0$. The algebra $\mathbb{C}\left\{\left\{\frac{dz_j}{z_j}\right\}\right\}$ coincides with its cohomology, and it carries an obvious filtration by the order of poles $W_l\mathbb{C}\left\{\left\{\frac{dz_j}{z_j}\right\}\right\}$. There is a commutative diagram of complexes

$$\begin{array}{ccc}
\Gamma(U, \mathcal{E}_X^{\cdot}\langle\log D\rangle) & \longrightarrow & \Gamma(U \setminus D, \mathcal{E}_U^{\cdot}) \\
{\scriptstyle\mu}\nwarrow & & \nearrow{\scriptstyle\nu} \\
& \mathbb{C}\left\{\left\{\dfrac{dz_j}{z_j}\right\}\right\} &
\end{array}$$

and ν_* is an isomorphism in cohomology by the usual De Rham theorem. Thus it will suffice to show that μ_* is an isomorphism in cohomology.

Passing to the graded spaces we obtain for every l a commutative diagram

$$\begin{array}{ccc}
\dfrac{W_l\Gamma(U, \mathcal{E}_X^{\cdot}\langle\log D\rangle)}{W_{l-1}\Gamma(U, \mathcal{E}_X^{\cdot}\langle\log D\rangle)} & \xrightarrow{\ \mathrm{Res}_l^{\cdot}\ } & \Gamma(U \cap D^{[l]}, \mathcal{E}_{D^{[l]}}^{\cdot}(-l)) \\
{\scriptstyle\mu_l}\nwarrow & & \nearrow{\scriptstyle res_l^{\cdot}} \\
& \dfrac{W_l\mathbb{C}\left\{\left\{\frac{dz_j}{z_j}\right\}\right\}}{W_{l-1}\mathbb{C}\left\{\left\{\frac{dz_j}{z_j}\right\}\right\}} &
\end{array}$$

where $res_l^{\cdot}$ is the classical Cauchy residue, which induces an isomorphism in cohomology. Since $\mathrm{Res}_l^{\cdot}$ too is an isomorphism in cohomology (theorem 2.2), we deduce that also μ_l induces an isomorphism on cohomology. By an easy induction argument on l it follows that

$$\mu\colon W_l\mathbb{C}\left\{\left\{\frac{dz_j}{z_j}\right\}\right\} \to W_l\Gamma(U, \mathcal{E}_X^{\cdot}\langle\log D\rangle)$$

is an isomorphism on cohomology. For $l = s$ we obtain the lemma. □

2.4 Residues and mixed Hodge structures (the smooth case)

2.4.1 Hodge filtrations and residues

The morphisms of residues defined as in (2.31) and (2.37) become morphisms of complexes

$$\mathrm{Res}_l^{\cdot}\colon W_l\mathcal{E}_X^{\cdot}\langle\log D\rangle \to \mathcal{E}_{D^{[l]}}^{\cdot}(-l), \quad \mathrm{Res}_l^{\cdot}\colon \frac{W_l\mathcal{E}_X^{\cdot}\langle\log D\rangle}{W_{l-1}\mathcal{E}_X^{\cdot}\langle\log D\rangle} \to \mathcal{E}_{D^{[l]}}^{\cdot}(-l) \quad (2.52)$$

and preserve the degrees of the complexes.

Let $(\Lambda^{\cdot}, d)$ be a filtered complex with a certain filtration Φ. We define *the shifted filtration* ${}^{(n)}\Phi$ *on the shifted complex* $C^{\cdot}(n)$ by

$$^{(n)}\Phi^p C^k(n) = \Phi^{p+n}C^{k+n} \tag{2.53}$$

In our case, we define on the De Rham complexes of the manifolds $D^{[r]}$, the usual Hodge filtration

$$F^p\mathcal{E}_{D^{[r]}}^{\cdot}$$

which induces a shifted filtration on the shifted complex :

$$^{(-r)}F^p\mathcal{E}_{D^{[r]}}^k(-r) = F^{p-r}\mathcal{E}_{D^{[r]}}^{k-r} \tag{2.54}$$

and we define the conjugate filtration

$$^{(-r)}\bar{F}^q\mathcal{E}_{D^{[r]}}^k(-r) = \bar{F}^{q-r}\mathcal{E}_{D^{[r]}}^{k-r} \tag{2.55}$$

If X is a compact Kähler manifold, each $D^{[r]}$ is also compact Kähler and the Hodge filtration of the De Rham complex induces a pure Hodge structure on the cohomology of $D^{[r]}$. Indeed, one has

$$H^k(D^{[r]}, \mathbb{C}) = \bigoplus_{a+b=k} F^a\bar{F}^bH^k(D^{[r]}, \mathbb{C}) \tag{2.56}$$

This means that the shifted filtrations of (2.54), (2.55) induce also a pure Hodge structure, namely (2.56) can be rewritten as

$$H^k(D^{[r]},\mathbb{C}) = \bigoplus_{a+b=k} {}^{(-r)}F^{a+r}\ {}^{(-r)}\bar{F}^{b+r} H^k(D^{[r]},\mathbb{C})$$

Renaming the indices, we obtain

$$H^{k-r}(D^{[r]},\mathbb{C}) = \bigoplus_{a+b=k+r} {}^{(-r)}F^{a}\ {}^{(-r)}\bar{F}^{b} H^{k-r}(D^{[r]},\mathbb{C}) \tag{2.57}$$

where we notice that the direct sum of (2.57) is taken over pairs (a,b) with $a+b=k+r$ (instead of $k-r$ as usual). That is, $H^{k-r}(D^{[r]},\mathbb{C})$ equipped with the $(-r)$-shifted filtrations, acquires a pure Hodge structure of weight $k+r$.

We also define filtrations on $\mathcal{E}_X^{\cdot}\langle\log D\rangle$ by

$$ {}^{(-r)}F^a\mathcal{E}_X^{\cdot}\langle\log D\rangle = F^a\mathcal{E}_X^{\cdot}\langle\log D\rangle \tag{2.58}$$

$$ {}^{(-r)}\bar{F}^b\mathcal{E}_X^{\cdot}\langle\log D\rangle = \bar{F}^{b-r}\mathcal{E}_X^{\cdot}\langle\log D\rangle \tag{2.59}$$

As usual in the second members of (2.58), (2.59) $F^a, \bar{F}^b$ denote the standard Hodge filtrations.

Remark. The filtrations defined in (2.58), (2.59) are not conjugate. Moreover the role of ${}^{(-r)}F^a$, ${}^{(-r)}\bar{F}^b$ is not symmetric; in particular ${}^{(-r)}F^a$ does not depend on r, so it is a true filtration on the complex $\mathcal{E}_X^{\cdot}\langle\log D\rangle$, while ${}^{(-r)}\bar{F}^b$ depends on r and is adapted to the subcomplex $W_r\mathcal{E}_X^{\cdot}\langle\log D\rangle$ and to the quotient complex $\frac{W_r\mathcal{E}_X^{\cdot}\langle\log D\rangle}{W_{r-1}\mathcal{E}_X^{\cdot}\langle\log D\rangle}$.

With the above notations, we deduce immediately

Lemma 2.6. *For any l, the residues induce morphisms of filtered spaces for the shifted filtrations*

$$\mathrm{Res}_l^{\cdot} : {}^{(-l)}F^pW_l\mathcal{E}_X^{\cdot}\langle\log D\rangle \to {}^{(-l)}F^p\mathcal{E}_{D^{[l]}}^{\cdot}(-l)$$

$$\mathrm{Res}_l^{\cdot} : {}^{(-l)}F^p\left(\frac{W_l\mathcal{E}_X^{\cdot}\langle\log D\rangle}{W_{l-1}\mathcal{E}_X^{\cdot}\langle\log D\rangle}\right) \to {}^{(-l)}F^p\mathcal{E}_{D^{[l]}}^{\cdot}(-l)$$

and also for the $\bar{F}$ filtrations

$$\mathrm{Res}_l^{\cdot} : {}^{(-l)}\bar{F}^qW_l\mathcal{E}_X^{\cdot}\langle\log D\rangle \to {}^{(-l)}\bar{F}^q\mathcal{E}_{D^{[l]}}^{\cdot}(-l)$$

$$\mathrm{Res}_l^{\cdot} : {}^{(-l)}\bar{F}^q\left(\frac{W_l\mathcal{E}_X^{\cdot}\langle\log D\rangle}{W_{l-1}\mathcal{E}_X^{\cdot}\langle\log D\rangle}\right) \to {}^{(-l)}\bar{F}^q\mathcal{E}_{D^{[l]}}^{\cdot}(-l)$$

The following easy result will be useful in the sequel.

Lemma 2.7. *There is a natural isomorphism*

$$\frac{F^p\mathcal{E}_X^k\langle\log D\rangle}{F^{p+1}\mathcal{E}_X^k\langle\log D\rangle} = \mathcal{E}_X^{p,k-p}\langle\log D\rangle$$

2.4.2 Pure Hodge structure on $E_1^{l,k}(X)$

Recall that if M is a complex manifold of dimension n, and $j\colon S \to M$ is the embedding of a smooth complex hypersurface in M, the *Gysin map*

$$\gamma_{S,M}\colon H^k(S,\mathbb{C}) \to H^{k+2}(M,\mathbb{C})$$

can be obtained via Poincaré duality from the pullback

$$j^*\colon H^{2n-k-2}(M,\mathbb{C}) \to H^{2n-k-2}(S,\mathbb{C})$$

It follows that $\gamma_{S,M}$ is real, commutes to conjugation, and it is a morphism of pure Hodge structures of degree 1.

In the sequel, X is a compact Kähler manifold.

Let us consider the spectral sequence $E_r^{l,k}(X)$ of the complex $\Gamma(X, \mathcal{E}_X^{\cdot}\langle\log D\rangle)$ with respect to the filtration W, whose first term is

$$E_1^{l,k}(X) = H^k\left(\frac{\Gamma(X, W_l\mathcal{E}_X^{\cdot}\langle\log D\rangle)}{\Gamma(X, W_{l-1}\mathcal{E}_X^{\cdot}\langle\log D\rangle)}\right) = H^k\left(X, \frac{W_l\mathcal{E}_X^{\cdot}\langle\log D\rangle}{W_{l-1}\mathcal{E}_X^{\cdot}\langle\log D\rangle}\right)$$

The second equality in the above formula holds because of the isomorphisms (2.19).

The spectral sequence converges to the graded cohomology

$$\frac{W_l H^k(X, \mathcal{E}_X^{\cdot}\langle\log D\rangle)}{W_{l-1}H^k(X, \mathcal{E}_X^{\cdot}\langle\log D\rangle)} \tag{2.60}$$

One can calculate the differential d_1

$$d_1\colon E_1^{l,k}(X) \to E_1^{l-1,k+1}(X) \tag{2.61}$$

First, we consider the differential d_0 of the graded complex

$$d_0\colon \frac{\Gamma(X, W_l\mathcal{E}_X^{k}\langle\log D\rangle)}{\Gamma(X, W_{l-1}\mathcal{E}_X^{k}\langle\log D\rangle)} \to \frac{\Gamma(X, W_l\mathcal{E}_X^{k+1}\langle\log D\rangle)}{\Gamma(X, W_{l-1}\mathcal{E}_X^{k+1}\langle\log D\rangle)}$$

$E_1^{l,k}(X)$ is exactly the cohomology of this complex. We take an element $[[\pi]] \in E_1^{l,k}(X)$. It corresponds to an element $\pi \in \Gamma(X, W_l\mathcal{E}_X^{k}\langle\log D\rangle)$ such that $d\pi \in W_{l-1}\mathcal{E}_X^{k+1}\langle\log D\rangle$. Moreover the d_0 of the class of $d\pi$ is obviously 0, so it gives a cohomology class which is an element of $E_1^{l-1,k+1}(X)$, which depends only on $[[\pi]]$ and not on any other choices.

Proposition 2.3. *Let X be a Kähler compact manifold.*

1) The first term $E_1^{l,k}(X)$ of the spectral sequence has a pure Hodge structure of weight $k+l$ induced by the shifted filtrations ${}^{(-l)}F$, ${}^{(-l)}\bar{F}$

$$E_1^{l,k}(X) = \bigoplus_{p+q=k+l} {}^{(-l)}F^p\, {}^{(-l)}\bar{F}^q E_1^{l,k}(X) \tag{2.62}$$

In particular, the filtrations induced by ${}^{(-l)}F$, ${}^{(-l)}\bar{F}$ *on* $E_1^{l,k}(X)$ *are conjugate.*
Moreover the residue

$$\underline{\mathrm{Res}}_l^{\cdot}\colon E_1^{l,\cdot}(X) \to H^{\cdot-l}(D^{[l]},\mathbb{C}) \tag{2.63}$$

induces an isomorphism of pure Hodge structures of weight $k+l$ *on* $E_1^{l,k}(X)$ *and on* $H^{k-l}(D^{[l]},\mathbb{C})$ *for the shifted Hodge filtrations* ${}^{(-l)}F$, ${}^{(-l)}\bar{F}$ *on* $D^{[l]}$.

2) *The differential*

$$d_1\colon E_1^{l,k}(X) \to E_1^{l-1,k+1}(X) \tag{2.64}$$

is a morphism of pure Hodge structures.

3) *Let us define* $\hat{d}_1$ *as the differential induced by* d_1 *at the level of residues, namely such that the following diagram is commutative*

$$\begin{array}{ccc} E_1^{l,k}(X) & \xrightarrow{\quad d_1 \quad} & E_1^{l-1,k+1}(X) \\ {\scriptstyle \underline{\mathrm{Res}}_l^k}\downarrow & & \downarrow{\scriptstyle \underline{\mathrm{Res}}_{l-1}^{k+1}} \\ H^{k-l}(D^{[l]},\mathbb{C}) & \xrightarrow{\quad \hat{d}_1 \quad} & H^{k+2-l}(D^{[l-1]},\mathbb{C}) \end{array} \tag{2.65}$$

Let $\{[\alpha_I]; |I| = l\}$ *be cohomology classes on* $\{D_I; |I| = l\}$, α_I *being a closed form of degree* $k-l$ *on* D_I. *Then* $\hat{d}_1(\{[\alpha_I]\}$ *is given as follows. For any* J *such that* $|J| = l-1$, $D_J \neq \emptyset$, *we have*

$$\hat{d}_1(\{[\alpha_I]\}_J = \sum_{I\supset J,|I|=l} \epsilon_{I,J}\gamma_{I,J}[\alpha_I] \tag{2.66}$$

where

$$\gamma_{I,J}\colon H^{k-l}(D_I,\mathbb{C}) \to H^{k+2-l}(D_J,\mathbb{C}) \tag{2.67}$$

is the Gysin map for D_I *considered as a hypersurface in* D_J *and* $\epsilon_{I,J}$ *is the signature of the permutation reordering* (j,J) *as* I *where* $I = J \cup \{j\}$, *so that* $z_j = 0$ *is the equation of* D_I *in* D_J; $\hat{d}_1$ *commutes with the complex conjugation.*

4) $\hat{d}_1$ *is a morphism of pure Hodge structures of weight* $k+l$ *for the shifted Hodge filtrations* ${}^{(-l)}F$, ${}^{(-l)}\bar{F}$ *on the cohomology* $H^{k-l}(D^{[l]},\mathbb{C})$ *and* ${}^{(-l+1)}F$, ${}^{(-l+1)}\bar{F}$ *on the cohomology* $H^{k+1-(l-1)}(D^{[l-1]},\mathbb{C})$.

5) *If we consider* $H^{2n-k-l}(D^{[l]},\mathbb{C})$ *as the Poincaré dual of* $H^{k-l}(D^{[l]},\mathbb{C})$, *the adjoint of* $\hat{d}_1$

$$\hat{d}_1^*\colon H^{2n-k-l}(D^{[l-1]},\mathbb{C}) \to H^{2n-k-l}(D^{[l]},\mathbb{C})$$

is exactly the morphism which is induced in cohomology by the morphism

$$\delta\colon \Lambda_D^k \to \Lambda_D^{k+1}$$

defined by (2.29) *for the diagonal complex* (2.30) *of the divisor* D.

In order to prove proposition 2.3, we need a more precise realization of the inverse of the isomorphism $\underline{\mathrm{Res}}_l^k$ of (2.63) (or (2.47)), at a global level.

We consider the line bundle L_i on X associated to the divisor D_i and a section σ_i of L_i whose zero set is exactly D_i. We introduce a hermitian metric $|\ |$ on L_i and define

$$\eta_i = -\frac{1}{2i\pi}\partial \log |\sigma_i|^2, \quad \Omega_i = -\frac{1}{2i\pi}\bar{\partial}\partial \log |\sigma_i|^2 = d\eta_i \tag{2.68}$$

so that Ω_i is a C^∞ form which represents the Chern class $c_1(L_i)$. Finally, if I is a multiindex such that $D_I \neq \emptyset$ we define

$$\eta^I = \bigwedge_{i\in I} \eta_i$$

Note that because $d\eta_i$ is a smooth form, we have

$$d\eta^I \in W_{l-1}\mathcal{E}_X^{\cdot}\langle \log D\rangle, l = |I| \tag{2.69}$$

Lemma 2.8. *Let I be a multiindex with $|I| = l$, $D_I \neq \emptyset$ and let α be a $(k-l)$-form on D_I.*

i) If α has a well defined type, one can find a C^∞ extension $\tilde{\alpha}$ of α to X, having the same type as α. Moreover $\omega = \tilde{\alpha}\wedge\eta^I$ is in $\Gamma(X, W_l\mathcal{E}_X^k\langle\log D\rangle)$ and ω has residue α on D_I and 0 on D_J for $J \neq I$.

ii) If α is a closed form on D_I, there exists a form $\tilde{\alpha}$ on X, closed in a neighborhood of D_I, such that if $\omega = \tilde{\alpha}\wedge\eta^I$, $d\omega$ is in $\Gamma(X, W_{l-1}\mathcal{E}_X^{k+1}\langle\log D\rangle)$, so that ω defines an element $[[\omega]]_1 \in E_1^{l,k}(X)$. Moreover $d\omega$ defines the element $d_1[[\omega]]_1 = [[d\omega]]_1$ in $E_1^{l-1,k+1}(X)$. If J is a multiindex with $|J| = l-1$, $J \subset I$, the residue of $d_1[[\omega]]_1$ on D_J is exactly $\epsilon_{I,J}\gamma_{I,J}([\alpha])$ where $\epsilon_{I,J}$ is the signature of the permutation which reorders (j,J) as I where $I = J\cup\{j\}$, so that $z_j = 0$ is the equation of D_I in D_J.

Proof of Lemma 2.8. Let $(W_a)_{a\in A}$ be an open covering of X such that in each W_a intersecting D_I one has an adapted system of complex coordinates $z^{(a)}$. We denote

$$z^{(a)} = \left(z'^{(a)}, z''^{(a)}\right), \quad z'^{(a)} = \left(z_1^{(a)}, \dots, z_l^{(a)}\right), \quad z''^{(a)} = \left(z_{l+1}^{(a)}, \dots, z_n^{(a)}\right)$$

and the equations for $D_I \cap W_a$ are

$$z_1^{(a)} = \cdots = z_l^{(a)} = 0$$

Let us consider $\alpha|_{D_I\cap W_a}$: it is a $(k-l)$-form which can be written in terms of the coordinates $z''^{(a)}$ only:

$$\alpha|_{D_I\cap W_a} = \sum_{|J|+|K|=k-l} \alpha_{JK}^{(a)}(z''^{(a)})\left(dz''^{(a)}\right)^J \wedge \left(d\bar{z}''^{(a)}\right)^K \tag{2.70}$$

We can extend $\alpha|_{D_I \cap W_a}$ from $D_I \cap W_a$ to W_a simply using the second member of (2.70), as a form $\alpha^{(a)}$. If $(\beta_a)_{a\in A}$ is a partition of unity of X such that the support of β_a is in W_a, we can define an extension $\tilde{\alpha}$ of α by the formula

$$\tilde{\alpha} = \sum_{a\in A} \beta_a \alpha^{(a)} \tag{2.71}$$

If α has a well-defined type, $\alpha^{(a)}$ and $\tilde{\alpha}$ have the same type. It is obvious that the residue of $\tilde{\alpha} \wedge \eta^I$ on D_I is $\tilde{\alpha}|_{D_I} = \alpha$ and that the residue of $\tilde{\alpha} \wedge \eta^I$ an another D_J with $J \neq I$ is 0. This proves part (i) of lemma 2.8.

Let W be an open neighborhood of D_I in X which is a deformation retract of D_I. If α is closed, we can extend α to a closed form α' on W, and then (after shrinking W) α' to a form $\tilde{\alpha}$ on X.

It follows

$$d\omega = d\tilde{\alpha} \wedge \eta^I \pm \tilde{\alpha} \wedge d\eta^I \tag{2.72}$$

and $d\tilde{\alpha}\wedge\eta^I$ is 0 on a neighborhood of D_I. By (2.69) we obtain that $d\omega$ belongs to $\Gamma(X, W_{l-1}\mathcal{E}_X^{k+1}\langle \log D\rangle)$. By the definition of $E_1^{l,k}(X)$, ω gives an element $[[\omega]]_1 \in E_1^{l,k}(X)$ such that

$$d_1[[\omega]]_1 = [[d\omega]]_1$$

Moreover if $|J| = l-1$, $J \subset I$, one can define the residue on D_J of $d\omega$ by taking in (2.72) the terms in $\left(\frac{dz}{z}\right)^J$. Using the fact that $d\eta_i = \Omega_i$ is C^∞, this residue is the coefficient of η^J in

$$\epsilon_{I,J}(d\tilde{\alpha} \wedge \eta_j \wedge \eta^J \pm \tilde{\alpha} \wedge \Omega_j \wedge \eta^J) \tag{2.73}$$

where $z_j = 0$ is the local equation of D_I in the manifold D_J and thus $I = \{j, J\}$. In other words the residue is

$$\epsilon_{I,J} d\left((\tilde{\alpha} \wedge \eta_j)|_{D_J}\right) = \epsilon_{I,J}\gamma_{I,J}(\alpha)$$

This proves lemma 2.8. □

Proof of proposition 2.3.

Proof of 1). The filtrations ${}^{(-l)}F$, ${}^{(-l)}\bar{F}$ on the complex $\Gamma\left(X, \frac{W_l\mathcal{E}_X^{\cdot}\langle \log D\rangle}{W_{l-1}\mathcal{E}_X^{\cdot}\langle \log D\rangle}\right)$ induce corresponding filtrations on its cohomology, which is exactly $E_1^{l,k}(X)$. By lemma 2.6, $\mathrm{Res}_l^{\cdot}$ is a morphism of filtered complexes for the shifted filtrations ${}^{(-l)}F$, ${}^{(-l)}\bar{F}$, and induces the isomorphism $\underline{\mathrm{Res}}_l^{\cdot}$ at the level of cohomology, which is thus a morphism of filtered spaces

$$ {}^{(-l)}F^pE_1^{l,k}(X) \to {}^{(-l)}F^pH^{k-l}(D^{[l]}, \mathbb{C})$$
$$ {}^{(-l)}\bar{F}^qE_1^{l,k}(X) \to {}^{(-l)}\bar{F}^qH^{k-l}(D^{[l]}, \mathbb{C})$$

By (2.57), $H^{k-l}(D^{[l]},\mathbb{C})$ carries a pure Hodge structure. Since $\mathrm{Res}_l^{\cdot}$ respects the types, $\underline{\mathrm{Res}}_l^{\cdot}$ induces morphisms of subspaces

$$\underline{\mathrm{Res}}_l^k : {}^{(-l)}F^p\,{}^{(-l)}\bar{F}^q E_1^{l,k}(X) \to {}^{(-l)}F^p\,{}^{(-l)}\bar{F}^q H^{k-l}(D^{[l]},\mathbb{C}) \qquad (2.74)$$

Since, by theorem 2.2, $\underline{\mathrm{Res}}_l^k$ is an isomorphism of vector spaces between $E_1^{l,k}(X)$ and $H^{k-l}(D^{[l]},\mathbb{C})$, and $H^{k-l}(D^{[l]},\mathbb{C})$ is a direct sum of the vector spaces in the second member of (2.74), it follows that (2.74) is an isomorphism. This also proves that the shifted filtrations ${}^{(-l)}F$ and ${}^{(-l)}\bar{F}$ induce a pure Hodge structure on $E_1^{l,k}(X)$ isomorphic by $\underline{\mathrm{Res}}_l^k$ to the pure Hodge structure on $H^{k-l}(D^{[l]},\mathbb{C})$ of the shifted Hodge filtrations (2.57).
Proof of 3). By lemma 2.7 and lemma 6.4 of part I, chapter 6, $\hat{d}_1$ is a sum of Gysin maps with coefficients ± 1. The Gysin maps commute with the complex conjugation, hence the same holds for $\hat{d}_1$.
Proof of 4). This is an immediate consequence of the fact that $\hat{d}_1$ is a sum of Gysin maps with coefficients ± 1.
Proof 2). It is an obvious consequence of the commutative diagram (2.65), the fact that $\hat{d}_1$ is a morphism of pure Hodge structures and that the residues are isomorphisms of Hodge structures.
Proof of 5). We have seen that $\hat{d}_1$ is an alternate sum of Gysin maps. So, the adjoint of $\hat{d}_1$ given by (2.66) is exactly the morphism δ of (2.29)

$$\delta\left(\{[\omega_J]\}\right)_I = \sum_{J\subset I} \epsilon_{I,J}[\omega_J]|[D_I$$

where $[\omega_J]|_{D_I}$ is the pullback in cohomology

$$H^k(D_J,\mathbb{C}) \to H^k(D_I,\mathbb{C})$$

and $|I| = l$, $|J| = l-1$ □

2.4.3 Degeneration of the spectral sequence

We recall that we are supposing that X is a compact Kähler manifold. Then, all the manifolds D_I and $D^{[l]}$ are compact Kähler.

In this section, we prove that the spectral sequence $E_r^{l,k}(X)$ degenerates at the term E_2 and that each term $E_2^{l,k}(X)$ carries a pure Hodge structure.

Theorem 2.5. *Let X be a compact Kähler manifold.*

1) *The differentials d_r of the spectral sequence of the filtration W_l are 0 for $r \geq 2$.*

2) *The second term of the spectral sequence $E_2^{l,k}(X)$, considered as the cohomology of the complex $(E_1^{l,k}(X), d_1)$, carries a pure Hodge structure of weight $k+l$*

$$E_2^{l,k}(X) = \bigoplus_{a+b=k+l} {}^{(-l)}F^a\,{}^{(-l)}\bar{F}^b E_2^{l,k}(X) \qquad (2.75)$$

where ${}^{(-l)}F^a$, ${}^{(-l)}\bar{F}^b$ *are induced by the corresponding filtrations on* $E_1^{l,k}(X)$.

Proof.

Proof. First we prove 2). By proposition 2.3, 1), 2) each $E_1^{l,k}(X)$ carries a pure Hodge structure of weight $k+l$, and d_1 is a morphism of pure Hodge structures, hence the cohomology of the complex $(E_1^{l,k}(X), d_1)$, (which is $E_2^{l,k}(X)$), carries a pure Hodge structure of weight $k+l$ for the induced filtrations. □

In order to prove 1) we need two lemmas.

Lemma 2.9.

1) *Using the residue isomorphisms, the diagram* (2.65) *induces a commutative diagram*

$$\begin{array}{ccc} E_2^{l,k}(X) & \xrightarrow{d_2} & E_2^{l-2,k+1}(X) \\ \Big\downarrow{\scriptstyle \underline{\mathrm{Res}}_l^k} & & \Big\downarrow{\scriptstyle \underline{\mathrm{Res}}_{l-2}^{k+1}} \\ \hat{E}_2^{l,k}(X) & \xrightarrow{\hat{d}_2} & \hat{E}_2^{l-2,k+1}(X) \end{array} \tag{2.76}$$

where

$$\hat{E}_2^{l,k}(X) = \frac{\ker\left\{ \hat{d}_1 \colon H^{k-l}(D^{[l]}, \mathbb{C}) \to H^{k+2-l}(D^{[l-1]}, \mathbb{C}) \right\}}{\hat{d}_1 H^{k-2-l}(D^{[l+1]}, \mathbb{C})} \tag{2.77}$$

and $\hat{d}_2$ *is induced by* d_2 *via the residue isomorphisms.*

2) $\hat{E}_2^{l,k}(X)$ *has a pure Hodge structure of weight* $k+l$, *as the cohomology space defined by* (2.77):

$$\hat{E}_2^{l,k}(X) = \bigoplus_{p+q=k+l} {}^{(-l)}F^p\, {}^{(-l)}\bar{F}^q \hat{E}_2^{l,k}(X) \tag{2.78}$$

Proof of lemma 2.9.

1) From the diagram (2.65) and the fact that $\underline{\mathrm{Res}}_l^k$ are isomorphisms, one sees immediately that they induce isomorphisms between the cohomologies $E_2^{l,k}(X)$ of the complexes $(E_1^{l,k}(X), d_1)$ and the cohomologies $\hat{E}_2^{l,k}(X)$ of the complexes $(\hat{E}_1^{l,k}(X), \hat{d}_1)$, where $\hat{E}_1^{l,k}(X) = H^{k-l}(D^{[l]}, \mathbb{C})$.

2) Because $\hat{d}_1$ is a morphism of pure Hodge structures (see proposition 2.3), the cohomologies $\hat{E}_2^{l,k}(X)$ of the complexes $(\hat{E}_1^{l,k}(X), \hat{d}_1)$ carry a pure Hodge structure as in (2.78). □

Lemma 2.10. *Let* $\hat{E}_1^{l,k}(X)^* = H^{k-l}(D^{[l]}, \mathbb{C})^* \simeq H^{2n-k-l}(D^{[l]}, \mathbb{C})$ *be the Poincaré dual and let* $\hat{d}_1^*$ *be the corresponding adjoint of* $\hat{d}_1$:

$$\hat{d}_1^* \colon H^{2n-k-l}(D^{[l+1]}, \mathbb{C}) \to H^{2n-k-l}(D^{[l]}, \mathbb{C})$$

which is $\hat{d}_1^* = \delta$ *(see proposition 2.3, 5)). Then*

1) the dual $\hat{E}_2^{l,k}(X)^*$ *can be identified with the homology of the cocomplex* $(\hat{E}_1^{l,k}(X)^*, \hat{d}_1^*)$ *and thus*

$$\hat{E}_2^{l,k}(X)^* = \frac{\ker\{\delta\colon H^{2n-k-l}(D^{[l]},\mathbb{C}) \to H^{2n-k-l}(D^{[l+1]},\mathbb{C})\}}{\delta H^{2n-k-l}(D^{[l-1]},\mathbb{C})} \tag{2.79}$$

and it carries a pure Hodge structure of weight $2n-k-l$.

2) the adjoint $\hat{d}_2^*$ *of* $\hat{d}_2$

$$\hat{d}_2^*\colon \hat{E}_2^{l,k}(X)^* \to \hat{E}_2^{l+2,k-1}(X)^* \tag{2.80}$$

is identically 0.

Proof.

1) We consider the complex

$$\cdots \longrightarrow \hat{E}_1^{l+1,k-1}(X) \xrightarrow{\hat{d}_1} \hat{E}_1^{l,k}(X) \xrightarrow{\hat{d}_1} \hat{E}_1^{l-1,k+1}(X) \longrightarrow \cdots$$

The dual of the cohomology of this complex is the homology of the dual cocomplex

$$\cdots H^{2n-k-l}(D^{[l+1]},\mathbb{C}) \longleftarrow H^{2n-k-l}(D^{[l]},\mathbb{C}) \longleftarrow H^{2n-k-l}(D^{[l-1]},\mathbb{C}) \cdots$$

which is exactly (2.79). Obviously $H^{2n-k-l}(D^{[l]},\mathbb{C})$ carries the standard Hodge structure induced by the standard Hodge filtrations F^p, $\bar{F}^q$, of $D^{[l]}$, and δ is a morphism of filtered spaces and thus

$$\delta\colon F^p\bar{F}^q H^{2n-k-l}(D^{[l]},\mathbb{C}) \to F^p\bar{F}^q H^{2n-k-l}(D^{[l-1]},\mathbb{C})$$

so that $\hat{E}_2^{l,k}(X)^*$, as the cohomology given by (2.79), carries an induced pure Hodge structure of weight $2n-k-l$.

2) Let us consider the dual $\hat{d}_2^*$ of $\hat{d}_2$. Using the diagram (2.65), and coming back to the level of residues, we see that $\hat{d}_2^*$ can be constructed as follows. Let $[[\omega]]_2 \in \hat{E}_2^{l,k}(X)^*$. Using (2.79) we see that we start with a collection $\{\omega_I\}$ of closed $(2n-k-l)$-forms on D_I for $|I| = l$ such that $\delta(\{\omega_I\})$ (in the sense of (2.29)) is 0 in cohomology on the D_J for $|J| = l+1$. This means that for all J with $|J| = l+1$

$$\delta(\{\omega_I\})_J = d\phi_J \tag{2.81}$$

where ϕ_J is a $(2n-k-l-1)$-form on D_J. Then, $\delta(\{\phi_J\})$ is a collection of $(2n-k-l-1)$-forms on the D_K with $|K| = l+2$. By (2.29) d commutes with δ, hence

$$d\left((\delta(\{\phi_J\})\right)_K = (\delta(d\{\phi_J\}))_K = (\delta\{\delta(\omega_I)\})_K = 0$$

because $\delta^2 = 0$, so that $\delta(\{\phi_J\})$ induces an element of $H^{2n-k-l-1}(D^{[l+2]}, \mathbb{C})$ and thus an element of $\hat{E}_2^{l+2,k-1}(X)^*$ which is

$$\hat{d}_2^*([[\omega]]_2) = [[\delta(\{\phi_J\})]]_2 \tag{2.82}$$

We are going to prove that this element is 0. Let $[[\omega]]_2 \in F^p\bar{F}^q\hat{E}_2^{l,k}(X)^*$ with $p+q = 2n-k-l$, so that all the ω_I are (p,q)-forms. By theorem 5.11 of part I, chapter 5, we can find a solution ϕ_J of type $(p, q-1)$ of the equation (2.81), and thus $\delta(\{\phi_J\})$ has type $(p, q-1)$ and

$$\hat{d}_2^*([[\omega]]_2) \in F^p\bar{F}^{q-1}\hat{E}_2^{l+2,k-1}(X)^* \tag{2.83}$$

But by the same theorem 5.11 we can also find a solution ϕ'_J of type $(p-1, q)$ of the equation (2.81) so that

$$\hat{d}_2^*([[\omega]]_2) \in F^{p-1}\bar{F}^{q}\hat{E}_2^{l+2,k-1}(X)^* \tag{2.84}$$

Since the dual space $\hat{E}_2^{l+2,k-1}(X)^*$ carries a pure Hodge structure of weight $2n-k-l-1$ and $p+q-1 = 2n-k-l-1$ we have

$$F^p\bar{F}^{q-1}\hat{E}_2^{l+2,k-1}(X)^* \cap F^{p-1}\bar{F}^{q}\hat{E}_2^{l+2,k-1}(X)^* = 0 \tag{2.85}$$

hence $\hat{d}_2^*([[\omega]]_2) = 0$. □

End of proof of theorem 2.5. We assume that $d_2 = \cdots = d_{r-1} = 0$. This implies that $E_r^{l,k}(X) = E_2^{l,k}(X)$. We construct

$$\hat{d}_r^* \colon \hat{E}_r^{l,k}(X)^* \to \hat{E}_r^{l+r,k-1}(X)^*$$

and prove that it is 0. Let us point out that $\hat{E}_r^{l,k}(X)^* = \hat{E}_2^{l,k}(X)^*$ so that it has a pure Hodge structure of weight $2n-k-l$. The construction of $\hat{d}_r^*$ is a generalization of that $\hat{d}_2^*$ of in lemma 2.10.

An element $[[\omega]]_r \in \hat{E}_r^{l,k}(X)^*$ is a collection of closed $(2n-k-l)$-forms $\{\omega_I\}, |I| = l$ on D_I, which define successively elements $[[\omega]]_1 \in \hat{E}_1^{l,k}(X)^*$ with $\hat{d}_1^*[[\omega]]_1 = 0$, then $[[\omega]]_s \in \hat{E}_s^{l,k}(X)^*$ with $\hat{d}_s^*[[\omega]]_s = 0$ for $2 \le s \le r-1$. This means that we have

$$\delta(\{\omega_I\})_{J_1} = d\phi_{J_1}^{(1)} \text{ on } D_{J_1} \quad \text{with } |J_1| = l+1 \tag{2.86}$$

$$\delta(\{\phi_{J_1}^{(1)}\})_{J_2} = d\phi_{J_2}^{(2)} \text{ on } D_{J_2} \quad \text{with } |J_2| = l+2 \tag{2.87}$$

$$\cdots$$

$$\delta(\{\phi_{J_{r-2}}^{(r-2)}\})_{J_{r-1}} = d\phi_{J_{r-1}}^{(r-1)} \quad \textit{on} \quad D_{J_{r-1}} \quad \textit{with} \quad |J_{r-1}| = l+r-1 \tag{2.88}$$

then we consider $\delta(\{\phi_{J_{r-1}}^{(r-1)}\})$ as an element of $H^{2n-k-l-r+1}(D^{[l+r]}, \mathbb{C})$ so that

$$\hat{d}_r^*([[\omega]]_r) = [[\delta(\{\phi_{J_{r-1}}^{(r-1)}\})]]_r \in \hat{E}_r^{l+r,k-1}(X)^*$$

If we start with $\{\omega_I\}$ of type $(p,q), p+q = 2n-k-l$ we can solve the equations (2.86), (2.87),..., (2.88) with forms of type $(p,q-1)$, $(p,q-2)$, $\cdots$, $(p,q-r+1)$ and obtain $\hat{d}_r^*([[\omega]]_r)$ of type $(p,q-r+1)$; but we can also solve the above equations with forms of type $(p-1,q)$, $(p-2,q)$, $\cdots$, $(p-r+1,q)$ and obtain $\hat{d}_r^*([[\omega]]_r)$ of type $(p-r+1,q)$. Because $r>1$, this implies that $\hat{d}_r^*([[\omega]]_r)=0$, exactly as in the equation (2.86). □

From theorem 2.5, and the convergence of the spectral sequence to the graded cohomology, we deduce:

Theorem 2.6. *Let X be a compact Kähler manifold, and $D \subset X$ a divisor with normal crossings. The cohomology $H^k(X\backslash D, \mathbb{C})$, provided with the weight filtration W shifted by $-k$, carries a mixed Hodge structure. More precisely, the graded spaces $\frac{W_l H^k(X\backslash D,\mathbb{C})}{W_{l-1}H^k(X\backslash D,\mathbb{C})}$ are isomorphic to $E_2^{l,k}(X)$ and thus have a pure Hodge structure as in* (2.75).

(In the shifted filtration $W'_m = W_{m-k}$ the quotient $\frac{W'_l H^k(X\backslash D,\mathbb{C})}{W'_{l-1}H^k(X\backslash D,\mathbb{C})} \simeq E_2^{l-k,k}(X)$ has weight l, as needed in the definition of mixed Hodge structure).

Proof. Because $d_r = 0$ for $r \geq 2$, the graded cohomology is the second term of the spectral sequence:

$$\frac{W_l H^k(X \setminus D, \mathbb{C})}{W_{l-1}H^k(X \setminus D, \mathbb{C})} \simeq E_2^{l,k} \tag{2.89}$$

i.e. the graded space $\frac{W_l H^k(X\backslash D,\mathbb{C})}{W_{l-1}H^k(X\backslash D,\mathbb{C})}$ is isomorphic to $E_2^{l,k}(X)$, which carries a pure Hodge structure of weight $k+l$ for the filtration F induced by $E_1^{l,k}(X)$, using that $E_2^{l,k}(X)$ is the cohomology of $(E_1^{l,k}(X), d_1)$. On the other hand the filtration F on $H^k(X \setminus D, \mathbb{C})$ induces a filtration on the quotient $\frac{W_l H^k(X\backslash D,\mathbb{C})}{W_{l-1}H^k(X\backslash D,\mathbb{C})}$. It remains to show that the the two filtrations, under the isomorphism (2.89), coincide (and the same for $\bar{F}$). This will be made clear later (see theorem 2.11 below). □

2.5 The strictness of d_0 and d with respect to the Hodge filtration

2.5.1 The conjugate complex

It is clear that the complex $\mathcal{E}_X^{\cdot}\langle \log D\rangle$ is not closed under conjugation. We can define *the conjugate complex* $\mathcal{E}_X^{\cdot}\langle\overline{\log D}\rangle$ by

$$\Gamma(U, \mathcal{E}_X^k\langle\overline{\log D}\rangle) = \left\{ \omega \in \Gamma(U \setminus D, \mathcal{E}_X^k) \mid \overline{\omega} \in \Gamma(U, \mathcal{E}_X^k\langle \log D\rangle) \right\} \tag{2.90}$$

so that a form $\omega \in \mathcal{E}_X^k\langle\overline{\log D}\rangle$ writes locally as

$$\omega = \sum_I \alpha_I \wedge \left(\frac{d\bar{z}}{\bar{z}}\right)^I \tag{2.91}$$

where $(z_1, \ldots, z_n)$ is an adapted coordinate system for (X, D).
Note that

$$\overline{d\omega} = d\overline{\omega} \tag{2.92}$$

On the conjugate complex we can also define:

- The conjugate weight filtration $\overline{W}_l$:

$$\omega \in \overline{W}_l \mathcal{E}_X^k\langle\overline{\log D}\rangle \iff \overline{\omega} \in W_l \mathcal{E}_X^k\langle\log D\rangle$$

- The conjugate residues $\overline{\mathrm{Res}_l^k}$:

$$\overline{\mathrm{Res}_l^k} : \overline{W}_l \mathcal{E}_X^k\langle\overline{\log D}\rangle \to \mathcal{E}_{D[l]}^{p-l}$$

If ω is locally written as in (2.91), with $|I| \leq l$, then by definition

$$\overline{\mathrm{Res}_l^k} \omega|_{D_I} = \alpha_I|_{D_I} \quad \text{for } |I| = l \tag{2.93}$$

Using (2.91), (2.92), (2.93), it is easy to express the results in lemma 2.2, lemma 2.3, theorem 2.2 replacing $W_l \mathcal{E}_X^k\langle\log D\rangle$ with $\overline{W}_l \mathcal{E}_X^k\langle\overline{\log D}\rangle$. In particular let us note the following immediate consequence.

Proposition 2.4. *The spectral sequence with respect to the filtration $\overline{W}_l$ of the conjugate complex $\Gamma(X, \mathcal{E}_X^{\cdot}\langle\overline{\log D}\rangle)$ coincides, for $r \geq 1$, with the spectral sequence $E_r^{l,k}(X)$ with respect to the filtration W_l of the complex $\Gamma(X, \mathcal{E}_X^{\cdot}\langle\log D\rangle)$. The map $\omega \to \overline{\omega}$ from $\Gamma(X, \mathcal{E}_X^{\cdot}\langle\log D\rangle)$ to $\Gamma(X, \mathcal{E}_X^{\cdot}\langle\overline{\log D}\rangle)$ induces a conjugation on each term $E_r^{l,k}(X)$, $r \geq 1$; moreover the differentials d_r commute to conjugations.*

The Hodge filtrations on $\mathcal{E}_X^{\cdot}\langle\overline{\log D}\rangle$ are defined, following (2.58), (2.59):

$$\begin{aligned} {}^{(-r)}F^b \mathcal{E}_X^{\cdot}\langle\overline{\log D}\rangle &= F^{b-r} \mathcal{E}_X^{\cdot}\langle\overline{\log D}\rangle \\ {}^{(-r)}\bar{F}^a \mathcal{E}_X^{\cdot}\langle\overline{\log D}\rangle &= \bar{F}^a \mathcal{E}_X^{\cdot}\langle\overline{\log D}\rangle \end{aligned} \tag{2.94}$$

As usual in the second members of (2.94) $F^b, \bar{F}^a$ denote the standard Hodge filtrations. Then it is clear that

Proposition 2.5. *The conjugation $\omega \to \overline{\omega}$ transforms ${}^{(-r)}F^a \mathcal{E}_X^{\cdot}\langle\log D\rangle$ to ${}^{(-r)}\bar{F}^a \mathcal{E}_X^{\cdot}\langle\overline{\log D}\rangle$ and ${}^{(-r)}\bar{F}^b \mathcal{E}_X^{\cdot}\langle\log D\rangle$ to ${}^{(-r)}F^b \mathcal{E}_X^{\cdot}\langle\overline{\log D}\rangle$*

2.5.2 Strictness of d_0

In this section we want to prove that *the differential in the W-graded complex of global sections*

$$d_0 \colon \frac{W_l\Gamma(X, \mathcal{E}_X^k\langle\log D\rangle)}{W_{l-1}\Gamma(X, \mathcal{E}_X^k\langle\log D\rangle)} \to \frac{W_l\Gamma(X, \mathcal{E}_X^{k+1}\langle\log D\rangle)}{W_{l-1}\Gamma(X, \mathcal{E}_X^{k+1}\langle\log D\rangle)}$$

is strict for the filtration induced by F, defined by formula (2.58).
Let us denote

$$\mathrm{Gr}_l^W \mathcal{E}_X^{\cdot}\langle\log D\rangle = \frac{W_l\mathcal{E}_X^{\cdot}\langle\log D\rangle}{W_{l-1}\mathcal{E}_X^{\cdot}\langle\log D\rangle} \tag{2.95}$$

By formula (2.19) and the equality $W_l\Gamma(X, \mathcal{E}_X^{\cdot}\langle\log D\rangle) = \Gamma(X, W_l\mathcal{E}_X^{\cdot}\langle\log D\rangle)$ we obtain

$$\Gamma(X, \mathrm{Gr}_l^W \mathcal{E}_X^k\langle\log D\rangle) = \frac{W_l\Gamma(X, \mathcal{E}_X^{\cdot}\langle\log D\rangle)}{W_{l-1}\Gamma(X, \mathcal{E}_X^k\langle\log D\rangle)} \tag{2.96}$$

so that

$$d_0 \colon \Gamma(X, \mathrm{Gr}_l^W \mathcal{E}_X^k\langle\log D\rangle) \to \Gamma(X, \mathrm{Gr}_l^W \mathcal{E}_X^{k+1}\langle\log D\rangle) \tag{2.97}$$

By theorem 1.1 of part I, chapter 1 we must show that *the spectral sequence of the complex $\Gamma(X, \mathrm{Gr}_l^W \mathcal{E}_X^{\cdot}\langle\log D\rangle)$ with respect to the filtration induced by F degenerates at level* 1. Let us denote $L^k = \mathcal{E}_X^k\langle\log D\rangle$ and

$$E_r^{l,k}(\mathrm{Gr}_l^W L^{\cdot}, F) \tag{2.98}$$

the spectral sequence. Then $E_0^{l,k}(\mathrm{Gr}_l^W L^{\cdot}, F)$ is the graded space, with respect to F, of $\Gamma(X, \mathrm{Gr}_l^W L^{\cdot})$; precisely, taking into account that the involved sheaves are fine, and that F must be turned to an increasing filtration:

$$E_0^{m,k}(\mathrm{Gr}_l^W L^{\cdot}, F) = \Gamma\left(X, \frac{F^{-m}\mathrm{Gr}_l^W \mathcal{E}_X^k\langle\log D\rangle}{F^{-m+1}\mathrm{Gr}_l^W \mathcal{E}_X^k\langle\log D\rangle}\right) = \Gamma(X, \mathrm{Gr}_m^F \mathrm{Gr}_l^W L^k) \tag{2.99}$$

(clearly the only interesting cases occur for $m \leq 0$).
The following very easy result, is known as **Zassenhaus lemma**:

Proposition 2.6. *Let $L^{\cdot}$ be a complex of $\mathbb{C}$-vector spaces provided with two filtrations F and W. Then there is a natural isomorphism of the double graded complexes*

$$\mathrm{Gr}_m^F \mathrm{Gr}_l^W L^{\cdot} = \mathrm{Gr}_l^W Gr_m^F L^{\cdot}$$

An easy computation shows:

Lemma 2.11. *One has by lemma 2.7*

$$\mathrm{Gr}_m^F \mathcal{E}_X^k\langle\log D\rangle = \frac{F^{-m}\mathcal{E}_X^k\langle\log D\rangle}{F^{-m+1}\mathcal{E}_X^k\langle\log D\rangle} = \mathcal{E}_X^{-m,k+m}\langle\log D\rangle \tag{2.100}$$

and the differential $\mathrm{Gr}^F_m\mathcal{E}^k_X\langle\log D\rangle \to \mathrm{Gr}^F_m\mathcal{E}^{k+1}_X\langle\log D\rangle$ *coincides with*

$$\bar\partial\colon \mathcal{E}_X^{-m,k+m}\langle\log D\rangle \to \mathcal{E}_X^{-m,k+m+1}\langle\log D\rangle$$

The differential

$$\mathrm{Gr}^W_l\mathrm{Gr}^F_m L^k \to \mathrm{Gr}^W_l\mathrm{Gr}^F_m L^{k+1}$$

is induced by $\bar\partial$*:*

$$\bar\partial\colon \frac{W_l\mathcal{E}_X^{-m,k+m}\langle\log D\rangle}{W_{l-1}\mathcal{E}_X^{-m,k+m}\langle\log D\rangle} \to \frac{W_l\mathcal{E}_X^{-m,k+m+1}\langle\log D\rangle}{W_{l-1}\mathcal{E}_X^{-m,k+m+1}\langle\log D\rangle}$$

Theorem 2.7. *The complex* $(\mathrm{Gr}^W_l\mathrm{Gr}^F_m\mathcal{E}^{\cdot}_X\langle\log D\rangle,\bar\partial)$ *is a resolution of the sheaf* $\mathrm{Gr}^W_l\Omega_X^{-m}\langle\log D\rangle$.

Proof. Let us consider the Dolbeault resolution of $\mathcal{O}_X$ (theorem 3.4 of part I, chapter 3):

$$0 \longrightarrow \mathcal{O}_X \longrightarrow \mathcal{E}_X^{0,0} \xrightarrow{\bar\partial} \mathcal{E}_X^{0,1} \xrightarrow{\bar\partial} \cdots \xrightarrow{\bar\partial} \mathcal{E}_X^{0,n} \longrightarrow 0$$

By Malgrange [Ma] each sheaf $\mathcal{E}_X^{0,q}$ is flat over $\mathcal{O}_X$, so that by proposition 1.6 of part I, chapter 1 the above exact sequence remains exact after tensorization with $\mathrm{Gr}^W_l\Omega_X^{-m}\langle\log D\rangle\otimes_{\mathcal{O}_X}$:

$$0 \longrightarrow \mathrm{Gr}^W_l\Omega_X^{-m}\langle\log D\rangle \cdots \xrightarrow{\bar\partial} \mathrm{Gr}^W_l\Omega_X^{-m}\langle\log D\rangle\otimes_{\mathcal{O}_X}\mathcal{E}_X^{0,q} \xrightarrow{\bar\partial} \cdots 0 \quad (2.101)$$

By theorem 2.1, formula (2.21), and (2.100), we have

$$\mathrm{Gr}^W_l\Omega_X^{-m}\langle\log D\rangle\otimes_{\mathcal{O}_X}\mathcal{E}_X^{0,q} \simeq \mathrm{Gr}^W_l\mathrm{Gr}^F_m\mathcal{E}_X^{q-m}\langle\log D\rangle$$

so that (2.101) gives the exact sequence

$$\begin{aligned} &0 \longrightarrow \mathrm{Gr}^W_l\Omega_X^{-m}\langle\log D\rangle \longrightarrow \mathrm{Gr}^W_l\mathrm{Gr}^F_m\mathcal{E}_X^{-m}\langle\log D\rangle \xrightarrow{\bar\partial} \cdots \\ &\cdots \mathrm{Gr}^W_l\mathrm{Gr}^F_m\mathcal{E}_X^{q-m}\langle\log D\rangle \xrightarrow{\bar\partial} \cdots \xrightarrow{\bar\partial} \mathrm{Gr}^W_l\mathrm{Gr}^F_m\mathcal{E}_X^{n-m}\langle\log D\rangle \longrightarrow 0 \end{aligned} \quad (2.102)$$

which is the expected resolution. □

Let us come back to the spectral sequence (2.98). By (2.99), and lemma 2.11, the term $E_1^{m,k}(\mathrm{Gr}^W_l L^{\cdot},F)$, which is the cohomology of the complex $E_0^{m,k}(\mathrm{Gr}^W_l L^{\cdot},F)$, coincides with the cohomology of the complex of the global sections $(\Gamma(X,\mathrm{Gr}^W_l\mathrm{Gr}^F_m\mathcal{E}^{\cdot}_X\langle\log D\rangle),\bar\partial)$. Hence by theorem 2.7 it is the cohomology on X of the sheaf $\mathrm{Gr}^W_l\Omega_X^{-m}\langle\log D\rangle$. In other terms:

$$E_1^{m,k}(\mathrm{Gr}^W_l L^{\cdot},F) = H^k(X,\mathrm{Gr}^W_l\Omega_X^{-m}\langle\log D\rangle) \quad (2.103)$$

moreover d_1 is induced by ∂:

$$\partial\colon H^k(X, \mathrm{Gr}_l^W \Omega_X^{-m}\langle \log D\rangle) \to H^{k+1}(X, \mathrm{Gr}_l^W \Omega_X^{-m+1}\langle \log D\rangle) \qquad (2.104)$$

Proposition 2.2 gives the isomorphism of complexes (2.41), that is

$$\mathrm{Res}_l^{\cdot}\colon (\mathrm{Gr}_l^W \Omega_X^{\cdot}\langle \log D\rangle, \partial) \simeq (\Omega_{D^{[l]}}^{\cdot}(-l), \partial)$$

Finally we find that *the first term* $E_1^{m,k}(\mathrm{Gr}_l^W L^{\cdot}, F)$ *of the spectral sequence is isomorphic (up to a shift), through the residues, to the first term*

$$E_1^{m+l,k} = H^k(D^{[l]}, \Omega_{D^{[l]}}^{-m-l})$$

of the classical Hodge spectral sequence for the compact Kähler manifold $D^{[l]}$. *Moreover the isomorphism extends to the corresponding* d_1, *hence to the terms* $E_2, \ldots, E_r$.

By theorem 5.10 of part I, chapter 5 the Hodge spectral sequence on $D^{[l]}$ degenerates at E_1, so the same will be true for $E_r^{m,k}(\mathrm{Gr}_l^W L^{\cdot}, F)$.

We can conclude:

Theorem 2.8. *The differential*

$$d_0\colon \frac{W_l\Gamma(X, \mathcal{E}_X^k\langle \log D\rangle)}{W_{l-1}\Gamma(X, \mathcal{E}_X^k\langle \log D\rangle)} \to \frac{W_l\Gamma(X, \mathcal{E}_X^{k+1}\langle \log D\rangle)}{W_{l-1}\Gamma(X, \mathcal{E}_X^{k+1}\langle \log D\rangle)} \qquad (2.105)$$

is strict for the filtration induced by F, defined by the formula (2.58).

Remark. The proof of theorem 2.8 does not depend on theorem 2.5 (the degeneration of the spectral sequence with respect to the filtration W). In fact, as we shall see in the most general case (chapter 4), theorem 2.5 can be proved also using the theorem 2.8.

2.5.3 The recursive and the direct filtrations on $E_2^{l,k}(X)$

We recall that we are supposing that X is a Kähler manifold.

On each term $E_r^{l,k}(X)$ of the spectral sequence of the complex $\Gamma(X, \mathcal{E}_X^{\cdot}\langle \log D\rangle)$ with respect to the filtration W_l there are two kind of Hodge filtrations:

- The *direct filtrations* ${}^{(-l)}F_1$, ${}^{(-l)}\bar{F}_1$, induced by the filtrations ${}^{(-l)}F$ and ${}^{(-l)}\bar{F}$ of the complex $\mathcal{E}_X^{\cdot}\langle \log D\rangle$ (see (2.58), (2.59)); precisely here $E_r^{l,k}(X)$ must be considered as a quotient of the subspace of approximate cocycles $Z_r^{l,k} \subset \Gamma(X, W_l\mathcal{E}_X^k\langle \log D\rangle)$.
- The *recursive filtrations* ${}^{(-l)}F_2$, ${}^{(-l)}\bar{F}_2$ induced recursively on $E_r^{l,k}(X)$ considered as the cohomology of the complex $(E_{r-1}^{l,k}(X), d_{r-1})$.

It is clear that on $E_1^{l,k}(X)$ *the filtration* ${}^{(-l)}F_1$ *(resp.* ${}^{(-l)}\bar{F}_1$*) is identical to the filtration* ${}^{(-l)}F_2$ *(resp.* ${}^{(-l)}\bar{F}_2$*). Moreover it is easy to prove that* $F_1 \subset F_2$, $\bar{F}_1 \subset \bar{F}_2$.

The filtrations which appear in the Hodge structure of $E_1^{l,k}(X)$ (2.62) and $E_2^{l,k}(X)$ (2.75) are ${}^{(-l)}F_2$ and ${}^{(-l)}\bar{F}_2$.

Next we want to prove that in fact $F_1 = F_2$, $\bar{F}_1 = \bar{F}_2$.

Lemma 2.12. *The filtrations* ${}^{(-l)}F_1$ *and* ${}^{(-l)}\bar{F}_1$*, as well as* ${}^{(-l)}F_2$ *and* ${}^{(-l)}\bar{F}_2$*, are conjugate.*

The statement for ${}^{(-l)}F_1$, ${}^{(-l)}\bar{F}_1$ follows from propositions 2.4 and 2.5. For ${}^{(-l)}F_2$ and ${}^{(-l}\bar{F}_2$ it is enough to prove it in the case of $E_1^{l,k}(X)$, where it is obvious because $F_1 = F_2$, $\bar{F}_1 = \bar{F}_2$ in $E_1^{l,k}(X)$.

Theorem 2.9. *On* $E_r^{l,k}(X)$ *the filtration* ${}^{(-l)}F_1$ *(resp.* ${}^{(-l)}\bar{F}_1$*) is identical to the filtration* ${}^{(-l)}F_2$ *(resp.* ${}^{(-l)}\bar{F}_2$*).*

By lemma 2.12 it is sufficient to prove $F_1 = F_2$; since we already know that $F_1 \subset F_2$, we must show $F_2 \subset F_1$. We shall denote by ${}^{(-l)}F$ the filtrations ${}^{(-l)}F_1$, ${}^{(-l)}F_2$ when they coincide. For simplicity we write $L^k = \Gamma(X, \mathcal{E}_X^k\langle\log D\rangle)$.

We proceed by induction on r: for $r = 1$ we know that $F_1 = F_2$ on $E_1^{l,k}(X)$. Thus we suppose $F_1 = F_2 = F$ on $E_r^{l,k}(X)$.

Let $\alpha \in F_2^a E_{r+1}^{l,k}$; there is a representative in $F^a E_r^{l,k} \cap \ker d_r$, i.e. an element $x \in F^a Z_r^{l,k} \subset F^a W_l L^k$ with $d_r[x]_r = 0$; we have $dx \in W_{l-r} L^{k+1} \cap F^a W_l L^{k+1}$ so that

$$dx \in F^a W_{l-r} L^{k+1} \tag{2.106}$$

$d_r[x]_r = 0$ gives

$$dx = dx_1 + z_1, \quad x_1 \in Z_{r-1}^{l-1,k}, \quad z_1 \in Z_{r-1}^{l-r-1,k+1} \tag{2.107}$$

Then successively we get by theorem 2.5 , 1): $d_{r-1}[x_1]_{r-1} = 0$ so that

$$dx_1 = dx_2 + z_2, \quad x_2 \in Z_{r-2}^{l-2,k}, \quad z_2 \in Z_{r-2}^{l-r-1,k+1} \tag{2.108}$$

$$\cdots$$

$$dx_{r-2} = dx_{r-1} + z_{r-1}, \quad x_{r-1} \in Z_1^{l-r+1,k}, \quad z_{r-1} \in Z_1^{l-r-1,k+1} \tag{2.109}$$

From the above equalities we find that

$$dx_{r-1} = dx \mod Z_0^{l-r-1,k+1} = W_{l-r-1} L^{k+1} \tag{2.110}$$

Since $x_{r-1} \in Z_1^{l-r+1,k}$, we can compute $d_1[x_{r-1}]_1$. It follows from (2.110) and (2.106) that $d_1[x_{r-1}]_1 \in F^a E_1^{l-r,k}$. But d_1 is a morphism of pure Hodge structures, hence it is strict. Therefore there exists $x'_{r-1} \in F^a Z_1^{l-r+1,k}$ with $d_1[x'_{r-1}]_1 = d_1[x_{r-1}]_1$, that is

$$dx_{r-1} = dx'_{r-1} + dy + z', \quad y \in Z_0^{l-r,k}, \quad z' \in Z_0^{l-r-1,k+1} \tag{2.111}$$

Note that

$$dx'_{r-1} \in F^a W_{l-r+1} L^{k+1} \cap W_{l-r} L^{k+1} \subset F^a W_{l-r} L^{k+1} \tag{2.112}$$

We obtain an element $[y]_0 \in E_0^{l-r,k}$, and from (2.110), (2.111), (2.112) we get $d_0[y]_0 \in F^a E_0^{l-r,k}$. Since d_0 is strict by theorem 2.8, we find $y' \in F^a W_{l-r} L^k$ with $d_0[y]_0 = d_0[y']_0$ or

$$dy = dy' + z'', \quad z'' \in W_{l-r-1} L^{k+1} \tag{2.113}$$

From (2.106) $\cdots$ (2.113) we obtain

$$x - x'_{r-1} - y' \in F^a W_l L^k$$

and

$$d(x - x'_{r-1} - y') \in W_{l-r-1} L^{k+1}$$

so that $x - x'_{r-1} - y' \in F^a Z_{r+1}^{l,k}$ and its class in $E_{r+1}^{l,k}$ is α. This proves that $F_1 = F_2$ on $E_{r+1}^{l,k}$.

By theorem 2.4 the cohomology space $H^k(X \setminus D, \mathbb{C})$ is isomorphic to the cohomology $H^k(X, \mathcal{E}_X^{\cdot}\langle \log D \rangle$. Let us denote by F_c the Hodge filtration on $H^k(X \setminus D, \mathbb{C})$ given by (2.58).

Precisely

$$F_c^p H^k(X \setminus D, \mathbb{C}) = F^p H^k(X, \mathcal{E}_X^{\cdot} < logD >) \tag{2.114}$$

Theorem 2.10. *Let us suppose that X is a Kähler manifold. Let F_d be the direct (or the recursive) filtration on $E_2^{l,k}$, and F_c be the filtration induced on $E_2^{l,k}$, under the isomorphism (2.89), by the filtration F_c given by (2.114). Then $F_d = F_c$.*

Proof. As in the proof of the previous theorem, let $L^k = \Gamma(X, \mathcal{E}_X^k \langle \log D \rangle)$. The isomorphism (2.89) means the following. Let $x \in Z_2^{l,k}$; then there exists $x' \in W_l L^k$ with $dx' = 0$ and

$$x' = x + d\tilde{x} + z, \quad \tilde{x} \in Z_1^{l-1,k-1}, \quad z \in Z_1^{l-1,k}$$

or

$$x' \equiv x + z$$

where the symbol $\equiv$ means cohomologous.

Let $\alpha \in F_c^p E_2^{l,k}$; there exists $x \in F^p W_l L^k$, with $dx = 0$, which induces α in $E_2^{l,k} \simeq \frac{W_l H^k(X \setminus D, \mathbb{C})}{W_{-1} H^k(X \setminus D, \mathbb{C})}$. It is clear that $x \in Z_2^{l,k}$, so that $x \in F^p W_l L^k \cap Z_2^{l,k} = F^p Z_2^{l,k}$. Hence $\alpha \in F_d^p E_2^{l,k}$. This proves $F_c \subset F_d$.
Conversely, if $\alpha \in F_d^p E_2^{l,k}$ there exists $x \in F^p Z_2^{l,k}$ inducing α. By the isomorphism (2.89) there exists $x' \in W_l L^k$ with $dx' = 0$ and

$$x' \equiv x + z, \quad z \in Z_1^{l-1,k} \tag{2.115}$$

let $[z]_1 \in E_1^{l-1,k}$ be the class of z. By (2.115) we have

$$dz = -dx \in F^p W_l L^{k+1}$$

and since $dz \in W_{l-2} L^{k+1}$ we obtain

$$dz \in W_{l-2} L^{k+1} \cap F^p W_l L^{k+1} \subset F^p W_{l-2} L^{k+1}$$

Thus $d_1[z]_1 \in F^p E_1^{l-2,k+1}$. Since d_1 is a strict morphism for F, we find $x_1 \in F^p Z_1^{l-1,k}$ with $d_1[z]_1 = d_1[x]_1$, or $d_1[z - x_1]_1 = 0$, that is, $z - x_1$ gives a class in $E_2^{l-1,k}$. By the isomorphism (2.89) for $l-1$, there is $x'' \in W_{l-1} L^k$, $dx'' = 0$, with $x'' \equiv z - x_1 - z_1$, $z_1 \in Z_1^{l-2,k}$, or

$$z \equiv x'' + x_1 + z_1$$

and by (2.115)

$$x' \equiv x + x'' + x_1 + z_1$$

Let us write $H^k = H^k(X \setminus D, \mathbb{C})$. We remark that the cohomology class of x'' belongs to $W_{l-1} H^k$; thus we have

$$x' \equiv (x + x_1) + z_1 \qquad \text{mod } W_{l-1} H^k$$

We note that $x + x_1 \in F^p W_l L^k$. We proceed as above, finding that $d_1[z_1]_1 \in F^p E_1^{l-3,k+1}$ and we obtain

$$x' \equiv (x + x_1 + x_2) + z_2 \qquad \text{mod } W_{l-1} H^k, \quad x_2 \in F^p Z_1^{l-2,k}, z_2 \in Z_1^{l-3,k}$$

Going on, because $W_s L^{\cdot} = 0$ for $s << 0$ we finally write

$$x' \equiv x + x_1 + x_2 + \cdots + x_s \qquad \text{mod } W_{l-1} H^k$$

with $x_j \in F^p Z_1^{l-j,k}$ so that $x' \in F^p W_l L^k$ and x', through $W_l H^k$, induces α. This proves $F_d \subset F_c$. □

Therefore we can state a more precise version of the theorem 2.6:

Theorem 2.11. *Let us suppose that X is a compact Kähler manifold. We provide the cohomology $H^k(X \setminus D, \mathbb{C})$ with the weight filtration W shifted by $-k$, the Hodge filtration F induced by the complex of global sections of $\mathcal{E}_X^{\cdot}\langle \log D \rangle$, and the filtration $\bar{F}$ conjugate to F. Then $H^k(X \setminus D, \mathbb{C})$ carries a mixed Hodge structure. More precisely, the graded spaces $\frac{W_l H^k(X \setminus D, \mathbb{C})}{W_{l-1} H^k(X \setminus D, \mathbb{C})}$ are isomorphic to $E_2^{l,k}(X)$ and thus have a pure Hodge structure of weight $k+l$. The filtration induced by F on $E_2^{l,k}(X)$ coincides with the direct and the recursive filtration.*

Remark. The shift of W by $-k$ is needed to normalize the weights in the quotients. In the shifted filtration $W'_l = W_{l-k}$ we obtain

$$\frac{W'_l H^k(X \setminus Q, \mathbb{C})}{W'_{l-1} H^k(X \setminus Q, \mathbb{C})} \simeq E_2^{l-k,k}(X)$$

that is, the quotient $\frac{W'_l H^k(X, \mathbb{C})}{W'_{l-1} H^k(X, \mathbb{C})}$ has weight l, as expected.

2.5.4 Strictness of d

We quote here the following theorems, which will be proved in full generality in chapter 4:

Theorem 2.12. *Let X be a compact Kähler manifold, $D \subset X$ a divisor with normal crossings. The differential $d\colon \Gamma(X, \mathcal{E}^k_X\langle \log D\rangle) \to \Gamma(X, \mathcal{E}^{k+1}_X\langle \log D\rangle)$ is strict for the Hodge filtration F defined by* (2.58).

By theorem 1.1 of part I, chapter 1, the above theorem is equivalent to

Theorem 2.13. *Let $L^\cdot = \Gamma(X, \mathcal{E}^k_X\langle \log D\rangle)$. The spectral sequence $E^{m,k}_r(L^\cdot, F)$ associated to the filtration F on $L^\cdot$ degenerates at E_1.*

Theorem 2.14. *Let X be a compact Kähler manifold, $D \subset X$ a divisor with normal crossings. Let $\omega \in \Gamma(X, \mathcal{E}^k_X\langle \log D\rangle)$ such that $d\omega \in W_m\Gamma(X, \mathcal{E}^k_X\langle \log D\rangle)$; there exists $\theta \in W_{m+1}\Gamma(X, \mathcal{E}^k_X\langle \log D\rangle)$, with $d\theta = d\omega$.*

As a consequence of theorem 2.13 we can prove that global meromorphic logarithmic forms are closed:

Theorem 2.15. *Let X be a compact Kähler manifold, $D \subset X$ a divisor with normal crossings. A form $\omega \in \Gamma(X, \mathcal{E}^k_X\langle \log D\rangle)$ which is holomorphic on $X \setminus D$ is closed.*

Proof. Let $L^\cdot = \Gamma(X, \mathcal{E}^k_X\langle \log D\rangle)$. The form ω belongs to F^kL^k; since $F^{k+1}L^k = 0$, it is also an element of $E_0^{-k,k}(L^\cdot, F) = F^kL^k$, and $d_0\omega = \bar\partial\omega = 0$ because ω is holomorphic on $X \setminus D$. As a consequence, ω detects a class $[\omega]_1 \in E_1^{-k,k}(L^\cdot, F)$. By theorem 2.13 there is an isomorphism

$$\frac{F^kH^k(X \setminus D, \mathbb{C})}{F^{k+1}H^k(X \setminus D, \mathbb{C})} = F^kH^k(X \setminus D, \mathbb{C}) \simeq E_1^{-k,k}(L^\cdot, F)$$

This means that there exists a closed form $\theta \in F^k\Gamma(X, \mathcal{E}^k_X\langle \log D\rangle)$ such that $\omega - \theta$ is exact. It follows $d\omega = d\theta = 0$. □

Chapter 3

Mixed Hodge structures on noncompact spaces: the basic example

3.1 Introduction

In this chapter, X is a quasi-smooth complex space. This means that there exists a resolution of singularities of X, i.e. a commutative diagram:

$$\begin{array}{ccc} \tilde{E} & \xrightarrow{i} & \tilde{X} \\ {\scriptstyle q}\downarrow & & \downarrow{\scriptstyle \pi} \\ E & \xrightarrow{j} & X \end{array} \tag{3.1}$$

where $E \subset X$ is a nowhere dense closed subspace, containing the singularities of X, $j \colon E \to X$ is the natural inclusion, $\tilde{X}$ is a smooth manifold and π is a proper modification inducing an isomorphism $\tilde{X} \setminus \tilde{E} \simeq X \setminus E$, with the additional assumption that E **and** $\tilde{E}$ **are smooth manifolds.**

Let $Q \subset X$ be a codimension 1 subspace, with the following property: the subspaces

$$\tilde{Q} = \pi^{-1}(Q) \subset \tilde{X}, \; M = E \cap Q \subset E, \; \tilde{M} = \tilde{E} \cap \tilde{Q} \subset \tilde{E}$$

are smooth divisors in $\tilde{X}$, E and $\tilde{E}$ respectively.

In pratice the above property is almost never fulfilled. Nevertheless it seems useful to deal first with the above case, which can be followed step by step, in order to better understand the general case (chapter 4). This will be the basic example in introducing complexes of differential forms on X, which are logarithmic along Q; in some sense, we can say that it is the singular case of the Leray situation treated in chapter 1.

We easily understand how to construct a complex $\Lambda^{\cdot}_X\langle \log Q\rangle$ of logarithmic differential forms on X; a k-form is a triple of forms: a logarithmic k-form on on $\tilde{X} \setminus \tilde{Q}$ (as defined in chapter 1), a logarithmic k-form on $E \setminus M$ and a $(k-1)$-form on $\tilde{E} \setminus \tilde{M}$. On the manifolds $\tilde{X}$, E and $\tilde{E}$ everything is well known by chapter 1. Then the diagram (3.1) allows us to give definitions for

X. The weight filtration on the complex of logarithmic forms on X with poles along Q is defined also as a direct sum, combining the order of poles together with the rank of the space in the hypercovering of the complex; $\tilde{X}$ and E have rank 0, while $\tilde{E}$ has rank 1.

Let us consider the the spectral sequence $E_r^{m,k}$ attached to the weight filtration (the main ingredient for the mixed Hodge theory). Then (under the right assumption on $X \setminus Q$, in particular if $X \setminus Q$ is quasi-projective):

1) $E_0^{m,k}$, $d_0 \colon E_0^{m,k} \to E_0^{m,k+1}$, $E_1^{m,k}$ are direct sums of the corresponding objects on $\tilde{X}$, E and $\tilde{E}$;

2) the spectral sequence degenerates at the level 2: $d_r = 0$, hence $E_r^{m,k} = E_2^{m,k}$, for $r \geq 2$;

3) the $E_2^{m,k}$ carry a pure Hodge structure, and they are isomorphic to the graded quotients $\frac{W_m H^k(X\setminus Q,\mathbb{C})}{W_{m-1}H^k(X\setminus Q,\mathbb{C})}$ of the cohomology $H^k(X \setminus Q, \mathbb{C})$ with respect to the weight filtrations.

So we must add, to the separate knowledge of the properties of $\tilde{X}$, E and $\tilde{E}$, only the computation of $d_1 \colon E_1^{m,k} \to E_1^{m-1,k}$. We find an explicit formula for d_1 (see formula (3.59)).

3.2 The standard logarithmic De Rham complex

Let X, Q be as in the introduction. Then, the sheaves (of logarithmic forms on a manifold along a smooth divisor) $\mathcal{E}^{\cdot}_{\tilde{X}}\langle\log \tilde{Q}\rangle$, $\mathcal{E}^{\cdot}_{E}\langle\log M\rangle$ and $\mathcal{E}^{\cdot}_{\tilde{E}}\langle\log \tilde{M}\rangle$ are well defined by chapter 1. Moreover, since $i^{-1}(\tilde{Q}) = \tilde{M}$ and $q^{-1}(M) = \tilde{M}$, the pullback of forms

$$i^* : \mathcal{E}^{\cdot}_{\tilde{X}}\langle\log \tilde{Q}\rangle \to \mathcal{E}^{\cdot}_{\tilde{E}}\langle\log \tilde{M}\rangle$$
$$q^* : \mathcal{E}^{\cdot}_{E}\langle\log M\rangle \to \mathcal{E}^{\cdot}_{\tilde{E}}\langle\log \tilde{M}\rangle$$

are defined, and commute to differentials.

We define the complex $\Lambda^{\cdot}_X\langle\log Q\rangle$ of *logarithmic differential forms on X along Q* by

$$\Lambda^{\cdot}_X\langle\log Q\rangle = \pi_*\mathcal{E}^{\cdot}_{\tilde{X}}\langle\log \tilde{Q}\rangle \oplus j_*\mathcal{E}^{\cdot}_{E}\langle\log M\rangle \oplus (j \circ q)_*\mathcal{E}^{\cdot}_{\tilde{E}}\langle\log \tilde{M}\rangle(-1)$$

More precisely, let U be an open set of X, $\tilde{U} = \pi^{-1}(U) \subset \tilde{X}$; then

$$\Lambda^k_X\langle\log Q\rangle(U) = \mathcal{E}^k_{\tilde{X}}\langle\log \tilde{Q}\rangle(\tilde{U}) \oplus \mathcal{E}^k_{E}\langle\log M\rangle(U \cap E)) \oplus \mathcal{E}^{k-1}_{\tilde{E}}\langle\log \tilde{M}\rangle(\tilde{U} \cap \tilde{E})$$

that is, a k-logarithmic differential form on U is a triple of logarithmic forms (ω, σ, θ) on $\tilde{U} \subset \tilde{X}$, $U \cap E \subset E$, $\tilde{U} \cap \tilde{E} \subset \tilde{E}$ respectively along the corresponding divisors, with the usual rule that ω and σ have the same degree k, while θ has degree $k-1$.

The differential is defined by

$$d\colon \Lambda^k_X\langle\log Q\rangle = \mathcal{E}^k_{\tilde{X}}\langle\log\tilde{Q}\rangle \oplus \mathcal{E}^k_E\langle\log M\rangle \oplus \mathcal{E}^{k-1}_{\tilde{E}}\langle\log\tilde{M}\rangle \to$$
$$\to \Lambda^{k+1}_X\langle\log Q\rangle = \mathcal{E}^{k+1}_{\tilde{X}}\langle\log\tilde{Q}\rangle \oplus \mathcal{E}^{k+1}_E\langle\log M\rangle \oplus \mathcal{E}^k_{\tilde{E}}\langle\log\tilde{M}\rangle$$
$$d(\omega,\sigma,\theta) = (d\omega, d\sigma, d\theta + (-1)^k(i^*\omega - q^*\sigma) \tag{3.2}$$

where $i^*\colon \mathcal{E}^k_{\tilde{X}}\langle\log\tilde{Q}\rangle \to \mathcal{E}^k_{\tilde{E}}\langle\log\tilde{M}\rangle$ and $q^*\colon \mathcal{E}^k_E\langle\log M\rangle \to \mathcal{E}^k_{\tilde{E}}\langle\log\tilde{M}\rangle$ are the pullback of forms.

A trivial computation shows that $d^2 = 0$ so that $\Lambda^\cdot_X\langle\log Q\rangle$ is a complex.

Note that $\Lambda^k_X\langle\log Q\rangle$ is a fine sheaf defined on all of X. A section of $\Lambda^k_X\langle\log Q\rangle$ on an open set $U \subset X$ will be called a *logarithmic differential form* of degree k on U.

Notation. From now on we will write for simplicity

$$\Lambda^\cdot_X\langle\log Q\rangle = \mathcal{E}^\cdot_{\tilde{X}}\langle\log\tilde{Q}\rangle \oplus \mathcal{E}^\cdot_E\langle\log M\rangle \oplus \mathcal{E}^\cdot_{\tilde{E}}\langle\log\tilde{M}\rangle(-1) \tag{3.3}$$

instead of $\Lambda^\cdot_X\langle\log Q\rangle = \pi_*\mathcal{E}^\cdot_{\tilde{X}}\langle\log\tilde{Q}\rangle \oplus j_*\mathcal{E}^\cdot_E\langle\log M\rangle \oplus (j\circ q)_*\mathcal{E}^\cdot_{\tilde{E}}\langle\log\tilde{M}\rangle(-1)$.

3.2.1 The cohomology of $X \setminus Q$

Let us consider the space $X\setminus Q$. It is a quasi-smooth space, whose associated diagram, induced from diagram (3.1), is

$$\begin{array}{ccc} \tilde{E}\setminus\tilde{M} & \xrightarrow{\ i\ } & \tilde{X}\setminus\tilde{Q} \\ {\scriptstyle q}\downarrow & & \downarrow{\scriptstyle \pi} \\ E\setminus M & \xrightarrow{\ j\ } & X\setminus Q \end{array} \tag{3.4}$$

so that by part II, chapter 2, we can introduce the sheaf of differential forms

$$\Lambda^\cdot_{X\setminus Q} = \mathcal{E}^\cdot_{\tilde{X}\setminus\tilde{Q}} \oplus \mathcal{E}^\cdot_{E\setminus M} \oplus \mathcal{E}^\cdot_{\tilde{E}\setminus\tilde{Q}}(-1)$$

which is a resolution of $\mathbb{C}_{X\setminus Q}$ by theorem 2.1 of part II, chapter 2.

One has an obvious injective morphism of complexes:

$$(\Lambda^\cdot_X\langle\log Q\rangle, d) \hookrightarrow (\rho_*\Lambda^\cdot_{X\setminus Q}, d) \tag{3.5}$$

where $\rho\colon X\setminus Q \hookrightarrow X$ is the natural map.

Let W be any open set of X, in particular X itself.

Theorem 3.1. *The morphism of inclusion $\rho\colon X\setminus Q \hookrightarrow X$ induces an isomorphism*

$$H^k(W, \Lambda^\cdot_X\langle\log Q\rangle) \simeq H^k(W, j_*\Lambda^\cdot_{X\setminus Q}) \simeq H^k(W\setminus Q, \mathbb{C}) \tag{3.6}$$

or in other words the cohomology of $W\setminus Q$ can be calculated as the cohomology of the complex of sections $(\Gamma(W, \Lambda^\cdot_X\langle\log Q\rangle), d)$ of $\Lambda^\cdot_X\langle\log Q\rangle$.

Proof. Each of the complexes $\Lambda^{\cdot}_X\langle\log Q\rangle$ and $\rho_*\Lambda^{\cdot}_{X\setminus Q}$ induce a spectral sequence of hypercohomology on W:

$$H^p(W, \mathcal{H}^q\Lambda^{\cdot}_X\langle\log Q\rangle) \Longrightarrow H^{p+q}(W, \Lambda^{\cdot}_X\langle\log Q\rangle) \tag{3.7}$$
$$H^p(W, \mathcal{H}^q\rho_*\Lambda^{\cdot}_{X\setminus Q}) \Longrightarrow H^{p+q}(W, \rho_*\Lambda^{\cdot}_{X\setminus Q}) \tag{3.8}$$

It is sufficient to prove that the natural morphism of the cohomology sheaves

$$\mathcal{H}^q\Lambda^{\cdot}_X\langle\log Q\rangle \to \mathcal{H}^q\rho_*\Lambda^{\cdot}_{X\setminus Q} \tag{3.9}$$

is an isomorphism (so that the limits in the formulas (3.7) and (3.8) will be isomorphic). This is a local statement.
Proof of the surjectivity in (3.9). Let $x \in Q$ and $\mu = (\omega, \sigma, \theta) \in \Lambda^q_X(U \setminus Q)$ be a d-closed section of $\rho_*\Lambda^q_{X\setminus Q}$ in an open neighborhood U of x. In particular $d\sigma = 0$; by lemma 1.3 of chapter 1 the morphism of the cohomology sheaves

$$\mathcal{H}^q\mathcal{E}^{\cdot}_E\langle\log M\rangle \to \mathcal{H}^q(\rho_{|E})_*\mathcal{E}^{\cdot}_{E\setminus M}$$

is an isomorphism, so there exist sections $\sigma' \in \mathcal{E}^q_E\langle\log M\rangle(U \cap E)$ and $\sigma'' \in \mathcal{E}^{q-1}_E((U\cap E)\setminus M)$ with $\sigma - d\sigma'' = \sigma'$ (after possibly shrinking U); replacing μ by $\mu - d(0, \sigma'', 0)$ we can suppose that $\sigma \in \mathcal{E}^q_E\langle\log M\rangle(U\cap E)$. A similar argument, starting with $d\omega = 0$, shows that we can also suppose $\omega \in \mathcal{E}^q_{\tilde{X}}\langle\log\tilde{Q}\rangle(\tilde{U})$. Since $(-1)^q(i^*\omega - q^*\sigma) = -d\theta$, using the injectivity in theorem 1.3, formula (1.25), of chapter 1, on $\tilde{E}$, we find $\theta' \in \mathcal{E}^{q-1}_{\tilde{E}}\langle\log\tilde{M}\rangle(q^{-1}(U\cap E))$ such that $(-1)^q(i^*\omega - q^*\sigma) = -d\theta'$. Again by the same theorem (surjectivity for $\tilde{E}$), $d(\theta - \theta') = 0$ implies that we can write $\theta - \theta' = d\tilde{\theta} + \theta''$, where $\theta'' \in \mathcal{E}^{q-1}_{\tilde{E}}\langle\log\tilde{M}\rangle(q^{-1}(U\cap E))$ and $\tilde{\theta} \in \mathcal{E}^{q-2}_{\tilde{E}}(q^{-1}(U\cap E)\setminus\tilde{Q}))$. Finally

$$(\omega, \sigma, \theta) - d(0, 0, \tilde{\theta}) = (\omega, \sigma, \theta' + \theta'')$$

which ends the proof of the surjectivity in (3.9).
Proof of the injectivity in (3.9). Let U be an open neighborhood of $x \in Q$ and $\mu = (\omega, \sigma, \theta) \in \Lambda^q_X\langle\log Q\rangle(U)$ such that there exists $(\omega', \sigma', \theta') \in \Lambda^{q-1}_X(U \setminus Q)$ with $(\omega, \sigma, \theta) = d(\omega', \sigma', \theta')$. Then in particular $\sigma = d\sigma'$ so that (by the injectivity in formula (1.25) of chapter 1 on E) there is a $\sigma'' \in \mathcal{E}^{q-1}_E\langle\log M\rangle(U\cap E)$ with $\sigma = d\sigma''$ (after possibly shrinking U); we can replace μ with $\mu - d(0, \sigma'', 0)$, so that we can suppose that $\sigma = 0$. A similar argument on $\tilde{X}$ (starting with $d\omega = 0$) leads us to suppose also $\omega = 0$. Then we have $\theta = d\theta'$ and the surjectivity on $\tilde{E}$ gives a $\theta'' \in \mathcal{E}^{q-2}_{\tilde{E}}\langle\log\tilde{M}\rangle(q^{-1}(U\cap E))$ with $\theta = d\theta''$. The conclusion follows. □

3.2.2 Filtration W by the order of poles

Let us recall from chapter 1 that if Y is a smooth manifold and D is a smooth divisor on Y we have defined a filtration W on $\mathcal{E}_Y^{\cdot}\langle\log D\rangle$

$$W_l\mathcal{E}_Y^{\cdot}\langle\log D\rangle = \begin{cases} 0 & \text{if } l<0 \\ \mathcal{E}_Y^{\cdot} & \text{if } l=0 \\ \mathcal{E}_Y^{\cdot}\langle\log D\rangle & \text{if } l\geq 1 \end{cases} \tag{3.10}$$

Let us consider the complex (3.3).

We define the increasing filtration called *the weight filtration*:

$$\begin{aligned} &W_l\Lambda_X^{\cdot}\langle\log Q\rangle = \\ &= W_l\mathcal{E}_{\tilde{X}}^{\cdot}\langle\log\tilde{Q}\rangle \oplus W_l\mathcal{E}_E^{\cdot}\langle\log M\rangle \oplus {}^{(1)}W_l\mathcal{E}_{\tilde{E}}^{\cdot}\langle\log\tilde{M}\rangle(-1) \end{aligned} \tag{3.11}$$

where ${}^{(1)}W_l$ is the filtration shifted by + 1, namely

$${}^{(1)}W_l\mathcal{E}_{\tilde{E}}^{k}\langle\log\tilde{M}\rangle(-1) = W_{l+1}\mathcal{E}_{\tilde{E}}^{k-1}\langle\log\tilde{M}\rangle \tag{3.12}$$

In (3.11) $W_l\mathcal{E}_{\tilde{X}}^{\cdot}\langle\log\tilde{Q}\rangle$, $W_l\mathcal{E}_E^{\cdot}\langle\log M\rangle$ and $W_l\mathcal{E}_{\tilde{E}}^{\cdot}\langle\log\tilde{M}\rangle$ are the weight filtrations as in (3.10). The above filtration has the property that $(\Lambda_X^k\langle\log Q\rangle, d)$ is a filtered complex:

$$d(W_l\Lambda_X^k\langle\log Q\rangle) \subset W_l\Lambda_X^{k+1}\langle\log Q\rangle$$

because of the formula (3.2) for d, and the fact that the pullback i^* and q^* preserve the filtrations W on the corresponding complexes (see chapter 1).

From definition (3.11) we deduce:

Lemma 3.1. *We have*

$$\begin{aligned} &W_l\Lambda_X^k\langle\log Q\rangle = \\ &= \begin{cases} 0 & \text{if } l<-1 \\ (0)\oplus(0)\oplus\mathcal{E}_{\tilde{E}}^{k-1} & \text{if } l=-1 \\ \mathcal{E}_{\tilde{X}}^k\oplus\mathcal{E}_E^k\oplus\mathcal{E}_{\tilde{E}}^{k-1}\langle\log\tilde{M}\rangle & \text{if } l=0 \\ \mathcal{E}_{\tilde{X}}^k\langle\log\tilde{Q}\rangle\oplus\mathcal{E}_E^k\langle\log M\rangle\oplus\mathcal{E}_{\tilde{E}}^{k-1}\langle\log\tilde{M}\rangle & \text{if } l\geq 1 \end{cases} \end{aligned} \tag{3.13}$$

Hence the graded complex with respect to the filtration W_l is given by

$$\frac{W_l\Lambda_X^k\langle\log Q\rangle}{W_{l-1}\Lambda_X^k\langle\log Q\rangle} = \begin{cases} (0) & \text{if } l<-1 \\ (0)\oplus(0)\oplus\mathcal{E}_{\tilde{E}}^{k-1} & \text{if } l=-1 \\ \mathcal{E}_{\tilde{X}}^k\oplus\mathcal{E}_E^k\oplus\dfrac{\mathcal{E}_{\tilde{E}}^{k-1}\langle\log\tilde{M}\rangle}{\mathcal{E}_{\tilde{E}}^{k-1}} & \text{if } l=0 \\ \dfrac{\mathcal{E}_{\tilde{X}}^k\langle\log\tilde{Q}\rangle}{\mathcal{E}_{\tilde{X}}^k}\oplus\dfrac{\mathcal{E}_E^k\langle\log M\rangle}{\mathcal{E}_E^k}\oplus(0) & \text{if } l=1 \\ (0) & \text{if } l\geq 2 \end{cases} \tag{3.14}$$

We remark that $W_l\Lambda_X^k\langle\log Q\rangle$ is a fine sheaf, so that for any open subset $V \subset X$ the following equality holds:

$$\Gamma\left(V, \frac{W_l\Lambda_X^k\langle\log Q\rangle}{W_{l-1}\Lambda_X^k\langle\log Q\rangle}\right) = \frac{\Gamma(V, W_l\Lambda_X^k\langle\log Q\rangle)}{\Gamma(V, W_{l-1}\Lambda_X^k\langle\log Q\rangle)} \tag{3.15}$$

The filtration W induces a filtration in cohomology: $W_lH^k(X, \Lambda_X^{\cdot}\langle\log Q\rangle)$ and, using the natural isomorphism of theorem 3.1, formula (3.6), it induces also a filtration $W_lH^k(X\backslash Q, \mathbb{C})$. As usual, there is a spectral sequence denoted by $E_r^{l,k}$ with first term

$$E_1^{l,k} = H^k\left(X, \frac{W_l\Lambda_X^{\cdot}\langle\log Q\rangle}{W_{l-1}\Lambda_X^{\cdot}\langle\log Q\rangle}\right) \tag{3.16}$$

which is the cohomology of the graded complex, and the spectral sequence converges to the graded cohomology

$$\frac{W_lH^k(X, \Lambda_X^{\cdot}\langle\log Q\rangle)}{W_{l-1}H^k(X, \Lambda_X^{\cdot}\langle\log Q\rangle)} \tag{3.17}$$

One can calculate the differential d_1

$$d_1\colon E_1^{l,k} \to E_1^{l-1,k+1} \tag{3.18}$$

First, we consider the differential d_0 of the graded complex

$$d_0\colon \frac{\Gamma(X, W_l\Lambda_X^k\langle\log Q\rangle)}{\Gamma(X, W_{l-1}\Lambda_X^k\langle\log Q\rangle)} \to \frac{\Gamma(X, W_l\Lambda_X^{k+1}\langle\log Q\rangle)}{\Gamma(X, W_{l-1}\Lambda_X^{k+1}\langle\log Q\rangle)} \tag{3.19}$$

Using (3.15) and (3.16), $E_1^{l,k}$ appears as the cohomology of this complex. Let $[[\pi]] \in E_1^{l,k}$. It corresponds to an element $\pi \in \Gamma(X, W_l\Lambda_X^k\langle\log Q\rangle)$ such that $d\pi \in W_{l-1}\Lambda_X^{k+1}\langle\log Q\rangle$. Moreover the d_0 of the class of $d\pi$ is obviously 0, so it gives an element of $E_1^{l-1,k+1}$ which depends only on $[[\pi]]$ and not on the other choices.

Taking into account the formula (3.14), there are only two relevants d_1, namely for $l = 0, 1$.

3.3 Residues (quasi-smooth case)

In this section we define residues for $\Lambda_X^{\cdot}\langle\log Q\rangle$. The residues are defined on subspaces $W_l\Lambda_X^{\cdot}\langle\log Q\rangle$ and live on the smooth manifolds $\tilde{Q}$, M and $\tilde{M}$. Let us recall from chapter 1 that if Y is a smooth manifold and D is a smooth divisor on Y we have defined

$$\mathrm{Res}^k\colon \mathcal{E}_Y^k\langle\log D\rangle \to \mathcal{E}_D^{k-1}$$

and Res^k has the property that for $\omega \in \mathcal{E}_X^k$, $\mathrm{Res}^k(\omega) = 0$. Hence by the definition of the filtration W (formula (3.10)) Res^k can be written as a map

$$\mathrm{Res}^k\colon W_1\mathcal{E}_Y^k\langle\log D\rangle \to \mathcal{E}_D^{k-1}$$

and induces a morphism

$$\mathrm{Res}^k\colon \frac{W_1\mathcal{E}_Y^k\langle\log D\rangle}{W_0\mathcal{E}_Y^k\langle\log D\rangle} \to \mathcal{E}_D^{k-1}$$

We have defined Res_l^k for any integer $l \geq 0$ by

$$\mathrm{Res}_l^k = \begin{cases} \mathrm{Res}_0^k = \text{identity}\colon W_0\mathcal{E}_Y^k\langle\log D\rangle = \mathcal{E}_Y^k \to \mathcal{E}_Y^k \quad \text{for } l = 0 \\ \mathrm{Res}^k\colon W_1\mathcal{E}_Y^k\langle\log D\rangle \to \mathcal{E}_D^{k-1} \quad \text{for } l = 1 \\ 0 \quad \text{for } l \geq 2 \end{cases} \tag{3.20}$$

The above definition can be summarized as follows. Let us define

$$D^{[0]} = Y, D^{[1]} = D \text{ and } D^{[l]} = \emptyset \quad \text{for } l \geq 2 \tag{3.21}$$

then

$$\mathrm{Res}_l^k\colon W_l\mathcal{E}_Y^k\langle\log D\rangle \to \mathcal{E}_{D^{[l]}}^{k-l} \tag{3.22}$$

Let us go back to a quasi-smooth complex space X, a subspace Q as in section 3.2 and the corresponding sheaf $\Lambda_X^{\cdot}\langle\log Q\rangle$.

Definition 3.1. We introduce for any integer $l \geq 0$ the residue map

$$\mathrm{Res}_l^k\colon W_l\Lambda_X^{\cdot}\langle\log Q\rangle \to \mathcal{E}_{\tilde{Q}^{[l]}}^{k-l} \oplus \mathcal{E}_{M^{[l]}}^{k-l} \oplus \mathcal{E}_{\tilde{M}^{[l+1]}}^{k-1-(l+1)} \tag{3.23}$$

in the following way. Let (ω, σ, θ) be a section of $W_l\Lambda_X^{\cdot}\langle\log Q\rangle = W_l\mathcal{E}_{\tilde{X}}^{\cdot}\langle\log \tilde{Q}\rangle \oplus W_l\mathcal{E}_E^{\cdot}\langle\log M\rangle \oplus W_{l+1}\mathcal{E}_{\tilde{E}}^{\cdot}\langle\log \tilde{M}\rangle(-1)$. Then we define

$$\mathrm{Res}_l^k(\omega, \sigma, \theta) = (\mathrm{Res}_l^k(\omega), \mathrm{Res}_l^k(\sigma), \mathrm{Res}_{l+1}^{k-1}\theta)) \tag{3.24}$$

More precisely from the definition (3.21) we obtain (forgetting the trivial zero components)

$$\begin{cases} \mathrm{Res}_0^k(\omega, \sigma, \theta) = (\omega, \sigma, \mathrm{Res}^{k-1}(\theta)) \in \mathcal{E}_{\tilde{X}}^k \oplus \mathcal{E}_E^k \oplus \mathcal{E}_{\tilde{M}}^{k-2} \\ \mathrm{Res}_1^k(\omega, \sigma, \theta) = (\mathrm{Res}^k(\omega), \mathrm{Res}^k(\sigma)) \in \mathcal{E}_{\tilde{Q}}^{k-1} \oplus \mathcal{E}_M^{k-1} \\ \mathrm{Res}_l^k\omega = 0 \quad l \geq 2 \end{cases} \tag{3.25}$$

Lemma 3.2. *The restriction of* Res_l^k *to* $W_{l-1}\Lambda_X^k\langle\log Q\rangle$ *is identically zero, hence* Res_l^k *induces a morphism (still denoted* Res_l^k*)*

$$\mathrm{Res}_l^k\colon \frac{W_l\mathcal{E}_X^k\langle\log Q\rangle}{W_{l-1}\mathcal{E}_X^k\langle\log Q\rangle} \to \mathcal{E}_{\tilde{Q}^{[l]}}^{k-l} \oplus \mathcal{E}_{M^{[l]}}^{k-l} \oplus \mathcal{E}_{\tilde{M}^{[l+1]}}^{k-1-(l+1)} \tag{3.26}$$

This follows from the definition (3.24) and the analogous result in the smooth case (theorem 1.7 of chapter 1).

3.4 The residue complex

In the above definition of residue mappings it is not yet clear where Res_l^k takes its values. To make it precise, taking into account the second member of (3.23) we define for a fixed l **the residue complex** $\mathrm{Res}_l^{\cdot}\Lambda_X\langle\log Q\rangle$ of $\Lambda_X\langle\log Q\rangle$:

$$\mathrm{Res}_l^k\Lambda_X\langle\log Q\rangle = \mathcal{E}_{\tilde{Q}^{[l]}}^{k-l} \oplus \mathcal{E}_{M^{[l]}}^{k-l} \oplus \mathcal{E}_{\tilde{M}^{[l+1]}}^{k-1-(l+1)} \tag{3.27}$$

whose differential $d\colon \mathrm{Res}_l^k\Lambda_X\langle\log Q\rangle \to \mathrm{Res}_l^{k+1}\Lambda_X\langle\log Q\rangle$ is the direct sum of the differentials on each term of the sum.

Lemma 3.3. *The residue* $\mathrm{Res}_l^{\cdot}$ *is a morphism of complexes*

$$\mathrm{Res}_l^{\cdot}\colon W_l\Lambda_X^{\cdot}\langle\log Q\rangle \to \mathrm{Res}_l^{\cdot}\Lambda_X\langle\log Q\rangle \tag{3.28}$$

Proof. By the definition of the filtration W (formula (3.10)) we conclude that $\mathrm{Res}_l^{\cdot}$ sends $W_l\Lambda_X^{\cdot}\langle\log Q\rangle$ to $\mathrm{Res}_l^{\cdot}\Lambda_X\langle\log Q\rangle$. Let us prove that it commutes with d. Using formulas (3.2) and (3.24) we compute

$$\mathrm{Res}_l^k(d(\omega,\sigma,\theta)) = (\mathrm{Res}_l^k(d\omega), \mathrm{Res}_l^k(d\sigma), \mathrm{Res}_{l+1}^{k-1}(d\theta + (-1)^k(i^*\omega - q^*\sigma)))$$

Now $\omega \in W_l\mathcal{E}_{\tilde{X}}^k\langle\log\tilde{Q}\rangle$, so that by chapter 1 $i^*\omega \in W_l\mathcal{E}_{\tilde{E}}^k\langle\log\tilde{M}\rangle$; by lemma 3.2 it follows $\mathrm{Res}_{l+1}^{k-1}(i^*\omega) = 0$. In the same way we find $\mathrm{Res}_{l+1}^{k-1}(q^*\sigma) = 0$. Hence

$$\mathrm{Res}_l^k(d(\omega,\sigma,\theta)) = (d\mathrm{Res}_l^k(\omega), d\mathrm{Res}_l^k(d\sigma), d\mathrm{Res}_{l+1}^{k-1}(\theta))$$

because we know that in the smooth case the residue commutes with d (lemma 1.1 of chapter 1). This proves the lemma. □

By lemma 3.2, and the definition (3.27) we obtain

Proposition 3.1. $\mathrm{Res}_l^{\cdot}$ *induces a morphism:*

$$\mathrm{Res}_l^{\cdot}\colon \frac{W_l\Lambda_X^{\cdot}\langle\log Q\rangle}{W_{l-1}\Lambda_X^{\cdot}\langle\log Q\rangle} \to \mathrm{Res}_l^{\cdot}\Lambda_X\langle\log Q\rangle \tag{3.29}$$

or

$$\mathrm{Res}_l^{\cdot}\colon \frac{W_l\mathcal{E}_X^{\cdot}\langle\log Q\rangle}{W_{l-1}\mathcal{E}_X^{\cdot}\langle\log Q\rangle} \to \mathcal{E}_{\tilde{Q}^{[l]}}^{\cdot}(-l) \oplus \mathcal{E}_{M^{[l]}}^{\cdot}(-l) \oplus \mathcal{E}_{\tilde{M}^{[l+1]}}^{\cdot}(-(l+2)) \tag{3.30}$$

(we keep the same notation for the quotient mapping).

Hence we obtain a morphism of cohomology sheaves. Then

Lemma 3.4. Res_l^k *induces an isomorphism of the cohomology sheaves*

$$\underline{\mathrm{Res}_l^k}\colon \mathcal{H}^k\left(\frac{W_l\Lambda_X^{\cdot}\langle\log Q\rangle}{W_{l-1}\Lambda_X^{\cdot}\langle\log Q\rangle}\right) \to \mathcal{H}^k\left(\mathrm{Res}_l^{\cdot}\Lambda_X\langle\log Q\rangle\right) \tag{3.31}$$

Proof. We have

$$\frac{W_l\Lambda^k_X\langle\log Q\rangle}{W_{l-1}\Lambda^k_X\langle\log Q\rangle} = \tag{3.32}$$
$$= \frac{W_l\mathcal{E}^k_{\tilde{X}}\langle\log\tilde{Q}\rangle}{W_{l-1}\mathcal{E}^k_{\tilde{X}}\langle\log\tilde{Q}\rangle} \oplus \frac{W_l\mathcal{E}^k_{E}\langle\log M\rangle}{W_{l-1}\mathcal{E}^k_{E}\langle\log M\rangle} \oplus \frac{W_{l+1}\mathcal{E}^{k-1}_{\tilde{E}}\langle\log\tilde{M}\rangle}{W_{l}\mathcal{E}^{k-1}_{\tilde{E}}\langle\log\tilde{M}\rangle}$$

and from the proof of lemma 3.3 the differential of an element $([\omega],[\sigma],[\theta])$ of $\frac{W_l\Lambda^{\cdot}_X\langle\log Q\rangle}{W_{l-1}\Lambda^{\cdot}_X\langle\log Q\rangle}$ is the term-by-term differential $(d[\omega],d[\sigma],d[\theta])$. On the other hand by (3.23), (3.24) and (3.26) the differential of the residue complex in (3.31) is also a direct sum of differentials. The lemma follows from the corresponding result in the smooth case (theorem 1.7 of chapter 1). □

As an immediate consequence of the above lemma we obtain

Theorem 3.2. *For fixed l, the morphism $\underline{\mathrm{Res}}^k_l$ induces natural isomorphisms*

$$\underline{\mathrm{Res}}^k_l : H^k\left(X,\frac{W_l\Lambda^{\cdot}_X\langle\log Q\rangle}{W_{l-1}\Lambda^{\cdot}_X\langle\log Q\rangle}\right) \to H^k(X,\mathrm{Res}^{\cdot}_l\Lambda_X\langle\log Q\rangle) = \tag{3.33}$$
$$= H^{k-l}(\tilde{Q}^{[l]},\mathcal{E}^{\cdot}_{\tilde{Q}^{[l]}}) \oplus H^{k-l}(M^{[l]},\mathcal{E}^{\cdot}_{M^{[l]}}) \oplus H^{k-1-(l+1)}(\tilde{M},\mathcal{E}^{\cdot}_{\tilde{M}^{[l+1]}})$$

Proof. Taking the cohomology of the sheaves in both members of (3.31) we obtain isomorphisms

$$H^q(X,\mathcal{H}^p\left(\frac{W_l\Lambda^{\cdot}_X\langle\log Q\rangle}{W_{l-1}\Lambda^{\cdot}_X\langle\log Q\rangle}\right)) \to H^q(X,\mathcal{H}^p\left(\mathrm{Res}^{\cdot}_l\Lambda_X\langle\log Q\rangle\right)) \tag{3.34}$$

By part I, chapter 3 the first member (resp. the second member) of (3.34) is the term E_2 of a spectral sequence converging to the first member (resp. the second member) of (3.33), and the isomorphisms in (3.34) are compatible with the maps in the spectral sequences. Then in the limit (3.33) we get an isomorphism. □

3.5 Residues and mixed Hodge structures (quasi-smooth case)

3.5.1 Hodge filtrations and residues

Let D be one of the smooth manifolds $\tilde{Q}$, M or $\tilde{M}$. We recall the meaning of $D^{[l]}$, as in (3.21).

We define on the De Rham complexes of the manifolds $D^{[l]}$, the usual Hodge filtration

$$F^p\mathcal{E}^{\cdot}_{D^{[l]}}$$

which induces a shifted filtration on the shifted complex:

$$^{(-l)}F^p\mathcal{E}^k_{D^{[l]}}(-l) = F^{p-l}\mathcal{E}^{k-l}_{D^{[l]}} \tag{3.35}$$

and we define the conjugate filtration

$$^{(-l)}\bar{F}^q\mathcal{E}^k_{D^{[l]}}(-l) = \bar{F}^{q-l}\mathcal{E}^{k-l}_{D^{[l]}} \tag{3.36}$$

If $\tilde{X}$, E, $\tilde{E}$ are compact Kähler manifolds, each $D^{[l]}$ is also compact Kähler and the Hodge filtration of the De Rham complex induces a pure Hodge structure on the cohomology of $D^{[l]}$. Indeed, one has

$$H^k(D^{[l]},\mathbb{C}) = \bigoplus_{a+b=k} F^a\bar{F}^bH^k(D^{[l]},\mathbb{C}) \tag{3.37}$$

This means that the shifted filtrations of (3.35), (3.36) induce also a pure Hodge structure, namely (3.37) can be rewritten as

$$H^k(D^{[l]},\mathbb{C}) = \bigoplus_{a+b=k} {}^{(-l)}F^{a+l}\ {}^{(-l)}\bar{F}^{b+l}H^k(D^{[l]},\mathbb{C}) \tag{3.38}$$

Renaming the indices, we obtain

$$H^{k-l}(D^{[l]},\mathbb{C}) = \bigoplus_{a+b=k+l} {}^{(-l)}F^a\ {}^{(-l)}\bar{F}^bH^{k-l}(D^{[l]},\mathbb{C}) \tag{3.39}$$

where the direct sum of (3.39) is taken over pairs (a,b) with $a+b=k+l$ (instead of $k-l$ as usual). That is, $H^{k-l}(D^{[l]},\mathbb{C})$ equipped with the $(-l)$-shifted filtrations, acquires a pure Hodge structure of weight $k+l$.

We come now to the definition of Hodge filtrations F^p and $\bar{F}^q$ on the logarithmic complex $\Lambda^{\cdot}_X\langle\log Q\rangle$ and on the residue complex $\mathrm{Res}^{\cdot}_l\Lambda_X\langle\log Q\rangle$ defined in (3.26). We define

$$\begin{aligned} {}^{(-l)}F^a\Lambda^k_X\langle\log Q\rangle &= F^a\mathcal{E}^k_{\tilde{X}}\langle\log\tilde{Q}\rangle\oplus \\ &\oplus F^a\mathcal{E}^k_E\langle\log M\rangle\oplus F^a\mathcal{E}^{k-1}_{\tilde{E}}\langle\log\tilde{M}\rangle \end{aligned} \tag{3.40}$$

$$\begin{aligned} {}^{(-l)}\bar{F}^b\Lambda^k_X\langle\log Q\rangle &= \bar{F}^{b-l}\mathcal{E}^k_{\tilde{X}}\langle\log\tilde{Q}\rangle\oplus \\ &\oplus\bar{F}^{b-l}\mathcal{E}^k_E\langle\log M\rangle\oplus\bar{F}^{b-l-1}\mathcal{E}^{k-1}_{\tilde{E}}\langle\log\tilde{M}\rangle \end{aligned} \tag{3.41}$$

In the second members of (3.40), (3.41) $F^a,\bar{F}^b$ denote the Hodge filtrations defined in chapter 1. On the other hand taking into account (3.27) we put

$$\begin{aligned} {}^{(-l)}F^a\mathrm{Res}^p_l\Lambda_X\langle\log Q\rangle &= F^{a-l}\mathcal{E}^{p-l}_{\tilde{Q}^{[l]}}\oplus \\ &\oplus F^{a-l}\mathcal{E}^{p-l}_{M^{[l]}}\oplus F^{a-l-1}\mathcal{E}^{p-1-(l+1)}_{\tilde{M}^{[l+1]}} \end{aligned} \tag{3.42}$$

$$\begin{aligned} {}^{(-l)}\bar{F}^b\mathrm{Res}^p_l\Lambda_X\langle\log Q\rangle &= \bar{F}^{b-l}\mathcal{E}^{p-l}_{\tilde{Q}^{[l]}}\oplus \\ &\oplus\bar{F}^{b-l}\mathcal{E}^{p-l}_{M^{[l]}}\oplus\bar{F}^{b-l-1}\mathcal{E}^{p-1-(l+1)}_{\tilde{M}^{[l+1]}} \end{aligned} \tag{3.43}$$

Remark. The filtrations defined in (3.40), (3.41) are not conjugate.

With the above notations, we deduce immediately

Lemma 3.5. *For any l, the residues induce morphisms of filtered spaces for the shifted filtrations*

$$\mathrm{Res}_l^{\cdot}: {}^{(-l)}F^p W_l \Lambda_X^{\cdot}\langle\log Q\rangle \to {}^{(-l)}F^p \mathrm{Res}_l^{\cdot} \Lambda_X\langle\log Q\rangle$$

$$\mathrm{Res}_l^{\cdot}: {}^{(-l)}F^p \left(\frac{W_l \Lambda_X^{\cdot}\langle\log Q\rangle}{W_{l-1}\Lambda_X^{\cdot}\langle\log Q\rangle}\right) \to {}^{(-l)}F^p \mathrm{Res}_l^{\cdot} \Lambda_X\langle\log Q\rangle$$

and also for the $\bar{F}$ filtrations

$$\mathrm{Res}_l^{\cdot}: {}^{(-l)}\bar{F}^q W_l \Lambda_X^{\cdot}\langle\log Q\rangle \to {}^{(-l)}\bar{F}^q \mathrm{Res}_l^{\cdot} \Lambda_X\langle\log Q\rangle$$

$$\mathrm{Res}_l^{\cdot}: {}^{(-l)}\bar{F}^q \left(\frac{W_l \Lambda_X^{\cdot}\langle\log Q\rangle}{W_{l-1}\Lambda_X^{\cdot}\langle\log Q\rangle}\right) \to {}^{(-l)}\bar{F}^q \mathrm{Res}_l^{\cdot} \Lambda_X\langle\log Q\rangle$$

Lemma 3.6. *1) The filtrations defined in* (3.40), (3.41), (3.42), (3.43), *are preserved by the corresponding differentials.*

2) Let us suppose that the manifolds $\tilde{X}$, E and $\tilde{E}$ in the diagram (3.1) *are compact Kähler manifolds. Then the differential d_0 is a strict morphism for the filtrations F and $\bar{F}$.*

Proof. The property 1) is clear.
By (3.19)

$$E_0^{l,k}(X) = E_0^{l,k}(\tilde{X}) \oplus E_0^{l,k}(E) \oplus E_0^{l+1,k-1}(\tilde{E}) \tag{3.44}$$

where for example $E_0^{l,k}(\tilde{X})$ is the spectral sequence on $\tilde{X}$ corresponding to the complex $\mathcal{E}_{\tilde{X}}^{\cdot}\langle\log M\rangle$, equipped with its filtration W, as in chapter 1. The differential d_0 is the direct sum of differentials on each term, which are strict by chapter 2, theorem 2.8; property 2) follows. □

3.5.2 Pure Hodge structure on $E_1^{l,k}(X)$

In the sequel we suppose that the manifolds $\tilde{X}$, E and $\tilde{E}$ in the diagram (3.1) *are compact Kähler manifolds.*

Let us consider the spectral sequence $E_r^{l,k}(X)$ with first term

$$E_1^{l,k} = H^k\left(X, \frac{W_l \Lambda_X^{\cdot}\langle\log Q\rangle}{W_{l-1}\Lambda_X^{\cdot}\langle\log Q\rangle}\right)$$

(see (3.16)).

Proposition 3.2. *The first term $E_1^{l,k}(X)$ of the spectral sequence has a pure Hodge structure of weight $k+l$ induced by the shifted filtrations ${}^{(-l)}F$, ${}^{(-l)}\bar{F}$*

$$E_1^{l,k}(X) = \bigoplus_{p+q=k+l} {}^{(-l)}F^p \; {}^{(-l)}\bar{F}^q E_1^{l,k}(X) \tag{3.45}$$

Moreover the residue

$$\underline{\mathrm{Res}}_l^k : E_1^{l,k}(X) \to H^k(X, \mathrm{Res}_l^{\cdot}\Lambda_X\langle \log Q\rangle) \tag{3.46}$$

induces an isomorphism of pure Hodge structures on $E_1^{l,k}(X)$ *and on* $H^k(X, \mathrm{Res}_l^{\cdot}\Lambda_X\langle \log Q\rangle)$ *for the shifted Hodge filtrations* ${}^{(-l)}F$, ${}^{(-l)}\bar{F}$, *where*

$$\begin{gathered} H^k(X, \mathrm{Res}_l^{\cdot}\Lambda_X\langle \log Q\rangle) = \\ H^{k-l}(\tilde{Q}^{[l]}, \mathbb{C}) \oplus H^{k-l}(M^{[l]}, \mathbb{C}) \oplus H^{k-1-(l+1)}(\tilde{M}^{[l+1]}, \mathbb{C}) \end{gathered} \tag{3.47}$$

Proof. By (3.32) and (3.16) we have

$$E_1^{l,k}(X) = E_1^{l,k}(\tilde{X}) \oplus E_1^{l,k}(E) \oplus E_1^{l+1,k-1}(\tilde{E}) \tag{3.48}$$

The filtrations ${}^{(-l)}F^p$ and ${}^{(-l)}\bar{F}^q$ on $E_1^{l,k}(X)$ induce in each summand of (3.48) their own shifted filtrations ${}^{(-l)}F^p$ and ${}^{(-l)}\bar{F}^q$, so that by proposition 1.1 of chapter 1 $E_1^{l,k}(\tilde{X})$, $E_1^{l,k}(E)$ and $E_1^{l+1,k-1}(\tilde{E})$ carry pure Hodge structures of the same weight $k+l = k-1+l+1$. Then (3.48) implies that $E_1^{l,k}(X)$ carries a pure Hodge structure of weight $k+l$. The equality (3.47) follows from the formula (3.26). Since by (3.24) the residue on X is the direct sum of residues on $\tilde{X}$, E and $\tilde{E}$, the proposition follows from proposition 1.1 of chapter 1. □

3.5.3 The differential d_1

The relevant differentials

$$d_1 : E_1^{l,k}(X) \to E_1^{l-1,k+1}(X) \tag{3.49}$$

occur for $l = -1, 0, 1$. Let $[[\pi]]_1$ be an element of $E_1^{l,k}(X)$, which means that one has a triple

$$\pi = (\omega, \sigma, \theta) \in W_l\mathcal{E}^k_{\tilde{X}}\langle \log \tilde{Q}\rangle \oplus W_l\mathcal{E}^k_E\langle \log M\rangle \oplus W_{l+1}\mathcal{E}^{k-1}_{\tilde{E}}\langle \log \tilde{M}\rangle$$

such that

$$d(\omega, \sigma, \theta) \in W_{l-1}\mathcal{E}^{k+1}_{\tilde{X}}\langle \log \tilde{Q}\rangle \oplus W_{l-1}\mathcal{E}^{k+1}_E\langle \log M\rangle \oplus W_l\mathcal{E}^k_{\tilde{E}}\langle \log \tilde{M}\rangle$$

Then:

$$d_1([[\pi]]_1) = [[d(\omega, \sigma, \theta)]]_1 \in E_1^{l-1,k+1}(X)$$

Lemma 3.7.

1) *The differential* d_1 *is a morphism of pure Hodge structures defined by* (3.45).

2) We define $\hat{d}_1$ as the differential d_1 at the level of residues, namely, such that the following diagram is commutative

$$\begin{array}{ccc} E_1^{l,k}(X) & \xrightarrow{\quad d_1 \quad} & E_1^{l-1,k+1}(X) \\ \Big\downarrow {\scriptstyle \underline{\mathrm{Res}}_l^k} & & \Big\downarrow {\scriptstyle \underline{\mathrm{Res}}_{l-1}^{k+1}} \\ H^k\left(X, \mathrm{Res}_l^{\cdot}\Lambda_X\langle\log Q\rangle\right) & \xrightarrow{\hat{d}_1} & H^{k+1}\left(X, \mathrm{Res}_{l-1}^{\cdot}\Lambda_X\langle\log Q\rangle\right) \end{array}$$

Then, $\hat{d}_1$ is a morphism of pure Hodge structures for the shifted filtrations ${}^{(-l)}F$, ${}^{(-l)}\bar{F}$ (3.42), (3.43) on $H^k\left(X, \mathrm{Res}_l^{\cdot}\Lambda_X\langle\log Q\rangle\right)$ and ${}^{(-l+1)}F$, ${}^{(-l+1)}\bar{F}$ on $H^{k+1}\left(X, \mathrm{Res}_{l-1}^{\cdot}\Lambda_X\langle\log Q\rangle\right)$ and is given by the formula (3.57) below.

Proof. Proof of 2). We consider the only relevant $l = -1, 0, 1$. Let $[\alpha] = ([\alpha_1], [\alpha_2], [\alpha_3]) \in H^k\left(X, \mathrm{Res}_l^{\cdot}\Lambda_X\langle\log Q\rangle\right)$, where by (3.47) α_1, α_2, α_3 are d-closed forms on $\tilde{Q}^{[l]}$, $M^{[l]}$, $\tilde{M}^{[l+1]}$, of degrees $k-l$, $k-l$ and $k-l-2$ respectively. We apply lemma 1.5 of chapter 1 separately to α_1, α_2, α_3: there exists $(\pi_1, \pi_2, \pi_3) \in W_l\mathcal{E}_{\tilde{X}}^{\cdot}\langle\log \tilde{Q}\rangle \oplus W_l\mathcal{E}_E^{\cdot}\langle\log M\rangle \oplus W_{l+1}\mathcal{E}_{\tilde{E}}^{\cdot}\langle\log \tilde{M}\rangle$ such that the residue of π_i is α_i.

Hence, π_1, π_2, π_3 induce elements

$$[[\pi_1]]_1 \in E_1^{l,k}(\tilde{X}),\ [[\pi_2]]_1 \in E_1^{l,k}(E),\ [[\pi_3]]_1 \in E_1^{l+1,k-1}(\tilde{E}) \tag{3.50}$$

and thus by (3.48) an element

$$[[\pi]]_1 \in E_1^{l,k}(X) = E_1^{l,k}(\tilde{X}) \oplus E_1^{l,k}(E) \oplus E_1^{l+1,k-1}(\tilde{E}) \tag{3.51}$$

Note in particular that

$$d\pi_1 \in W_{l-1}\mathcal{E}_{\tilde{X}}^{\cdot}\langle\log \tilde{Q}\rangle,\ d\pi_2 \in W_{l-1}\mathcal{E}_E^{\cdot}\langle\log M\rangle,\ d\pi_3 \in W_l\mathcal{E}_{\tilde{E}}^{\cdot}\langle\log \tilde{M}\rangle \tag{3.52}$$

Then, by definition,

$$\hat{d}_1[\alpha] = \underline{\mathrm{Res}}_{l-1}^{k+1} d_1[[\pi]]_1 \tag{3.53}$$

Now

$$d_1[[\pi]]_1 = [[d\pi]]_1$$

and from (3.2)

$$d(\pi_1, \pi_2, \pi_3) = (d\pi_1, d\pi_2, d\pi_3 + (-1)^k(i^*\pi_1 - q^*\pi_2) \tag{3.54}$$

From chapter 1, $i^*\pi_1 - q^*\pi_2 \in W_l\mathcal{E}_{\tilde{E}}^k\langle\log \tilde{M}\rangle$, hence by (3.52) the second member of (3.54) is in the space $W_{l-1}\Lambda_X^{\cdot}\langle\log Q\rangle$. Moreover, $[[d\pi_1]]_1$, $[[d\pi_2]]_1$, $[[d\pi_3]]_1$, define elements in $E_1^{l-1,k+1}(\tilde{X})$, $E_1^{l-1,k+1}(E)$ and $E_1^{l,k}(\tilde{E})$ respectively, and

$$d_1[[\pi_j]]_1 = [[d\pi_j]]_1,\ j = 1, 2, 3$$

So using the results of proposition 1.1 of chapter 1 we have:

$$\underline{\mathrm{Res}}_{l-1}^{k+1} d_1[[\pi_j]]_1 = \hat{d}_1[\alpha_j] = \delta_l^1 \gamma_j[\alpha_j], \quad j = 1, 2 \tag{3.55}$$

where we have denoted by γ_1 (resp. γ_2) the Gysin map for the pair $(\tilde{X}, \tilde{Q})$ (resp.(E, M)), δ_l^1 being the Kronecker symbol, and

$$\underline{\mathrm{Res}}_l^k d_1[[\pi_3]]_1 = \hat{d}_1[\alpha_3] = \delta_l^0 \gamma_3[\alpha_3] \tag{3.56}$$

γ_3 being the Gysin map for the pair $(\tilde{E}, \tilde{M})$. In (3.53), the residue morphism is the direct sum of the residues in each component of $d_1[[\pi]]_1$, so that finally using formula (3.24) for Res_{l-1}^{k+1} we obtain

$$\hat{d}_1[\alpha] = \left(\delta_l^1 \gamma_1[\alpha_1], \delta_l^1 \gamma_2[\alpha_2], \delta_l^0 \gamma_3[\alpha_3] + \left[\underline{\mathrm{Res}}_l^k (i^*\pi_1 - q^*\pi_2)\right]\right) \tag{3.57}$$

It is clear from (3.57) that $\hat{d}_1$ is a morphism

$$H^k\left(X, \mathrm{Res}_l^{\cdot} \Lambda_X \langle \log Q \rangle\right) \to H^{k+1}\left(X, \mathrm{Res}_{l-1}^{\cdot} \Lambda_X \langle \log Q \rangle\right) \tag{3.58}$$

which is a morphism of filtered spaces for the filtrations ${}^{(-l)}F$, ${}^{(-l)}\bar{F}$, and ${}^{(-l+1)}F$, ${}^{(-l+1)}\bar{F}$, respectively. This is a consequence of (3.42), (3.43) defining these filtrations, from the fact that γ_j, $j = 1, 2, 3$ is a morphism of filtered spaces, and the fact that if the α_j have given types, one can choose the π_j with a well defined type (as in lemma 1.5 of chapter 1), that the pull-back i^* and q^* respect types, and the residue also.
Proof of 1). $\hat{d}_1$ is a morphism of filtered spaces for the Hodge filtration and the residues are isomorphisms of filtered spaces. So also d_1 is a morphism of filtered spaces and thus it is a morphism of pure Hodge structures. Taking into account that $\underline{\mathrm{Res}}^k(i^*\pi_1) = i^* \underline{\mathrm{Res}}^k \pi_1 = i^*\alpha_1$ and the analogous formula for $\underline{\mathrm{Res}}^k(q^*\pi_2)$, the formula (3.57) becomes more explicitly

$$\hat{d}_1[\alpha] = \begin{cases} (0, 0, \gamma_3[\alpha_3] + [(i^*\pi_1 - q^*\pi_2)]) & \text{for } l = 0 \\ (\gamma_1[\alpha_1], \gamma_2[\alpha_2], [i^*\alpha_1 - q^*\alpha_2)]) & \text{for } l = 1 \\ 0 \quad \text{for } l \neq 0, 1 \end{cases} \tag{3.59}$$

where

$$\begin{cases} E_1^{-1,k} = H^k(\tilde{E}, \mathbb{C}) \\ E_1^{0,k} = H^k(\tilde{X}, \mathbb{C}) \oplus H^k(E, \mathbb{C}) \oplus H^{k-2}(\tilde{M}, \mathbb{C}) \\ E_1^{1,k} = H^{k-1}(\tilde{Q}, \mathbb{C}) \oplus H^{k-1}(M, \mathbb{C}) \end{cases} \tag{3.60}$$

□

3.5.4 The conjugate complex

The complex $\Lambda_X^{\cdot}\langle \log Q \rangle$ is not closed under conjugation. We can define *the conjugate complex* $\Lambda_X^{\cdot}\langle \overline{\log Q} \rangle$ by

$$\Lambda_X^{\cdot}\langle \overline{\log Q} \rangle = \mathcal{E}_{\tilde{X}}^{\cdot}\langle \overline{\log \tilde{Q}} \rangle \oplus \mathcal{E}_E^{\cdot}\langle \overline{\log M} \rangle \oplus \mathcal{E}_{\tilde{E}}^{\cdot}\langle \overline{\log \tilde{M}} \rangle(-1)$$

where $\mathcal{E}^{\cdot}_{\tilde{X}}\langle\overline{\log \tilde{Q}}\rangle$, $\mathcal{E}^{\cdot}_{E}\langle\overline{\log M}\rangle$ and $\mathcal{E}^{\cdot}_{\tilde{E}}\langle\overline{\log \tilde{M}}\rangle$ have been defined in chapter 1.

There is a conjugation for logarithmic forms, such that $\omega \in \Lambda^k_X\langle\log Q\rangle$ if and only if $\overline{\omega} \in \Lambda^k_X\langle\overline{\log Q}\rangle$.
Moreover

$$\overline{d\omega} = d\overline{\omega} \tag{3.61}$$

On the conjugate complex we can also define:

- The conjugate weight filtration $\overline{W}_l$:

$$\omega \in \overline{W}_l\Lambda^k_X\langle\overline{\log Q}\rangle \iff \overline{\omega} \in W_l\Lambda^k_X\langle\log Q\rangle$$

- The conjugate residues $\overline{\mathrm{Res}^k_l}$:

$$\overline{\mathrm{Res}^k_l} : \overline{W}_l\Lambda^k_X\langle\overline{\log Q}\rangle \to \mathrm{Res}^k_l\Lambda_X\langle\log Q\rangle \tag{3.62}$$

  i.e. the conjugate residues have the same target $\mathrm{Res}^k_l\Lambda_X\langle\log Q\rangle$ as the residues Res^k_l.

Using (3.61), (3.62) it is easy to express the results in lemma 3.3, proposition 3.1, theorem 3.2, lemma 3.6 replacing $W_l\Lambda^k_X\langle\log Q\rangle$ with $\overline{W}_l\Lambda^k_X\langle\overline{\log Q}\rangle$. In particular let us note the following immediate consequence.

Proposition 3.3. *The spectral sequence with respect to the filtration $\overline{W}_l$ of the complex $\Gamma(X, \Lambda^k_X\langle\overline{\log Q}\rangle)$ coincides, for $r \geq 1$, with the spectral sequence $E^{l,k}_r(X)$ with respect to the filtration W_l of the complex $\Gamma(X, \Lambda^k_X\langle\log Q\rangle)$. The conjugation $\omega \to \overline{\omega}$ from $\Gamma(X, \Lambda^k_X\langle\log Q\rangle)$ to $\Gamma(X, \Lambda^k_X\langle\overline{\log Q}\rangle)$ induces a conjugation on each term $E^{l,k}_r(X)$, $r \geq 1$; moreover the differentials d_r commute to conjugations.*

The Hodge filtrations on $\Lambda^k_X\langle\overline{\log Q}\rangle$ are defined, following (3.40), (3.41):

$$\begin{aligned} {}^{(-l)}F^b\Lambda^k_X\langle\overline{\log Q}\rangle &= F^{b-l}\Lambda^k_X\langle\overline{\log Q}\rangle \\ {}^{(-l)}\bar{F}^a\Lambda^k_X\langle\overline{\log Q}\rangle &= \bar{F}^a\Lambda^k_X\langle\overline{\log Q}\rangle \end{aligned} \tag{3.63}$$

In the second members of (3.63) $F^b, \bar{F}^a$ denote the Hodge filtrations. Then it is clear that

Proposition 3.4. *The conjugation $\omega \to \overline{\omega}$ transforms ${}^{(-l)}F^a\Lambda^{\cdot}_X\langle\log Q\rangle$ to ${}^{(-l)}\bar{F}^a\Lambda^{\cdot}_X\langle\overline{\log Q}\rangle$ and ${}^{(-l)}\bar{F}^b\Lambda^{\cdot}_X\langle\log Q\rangle$ to ${}^{(-l)}F^b\Lambda^{\cdot}_X\langle\overline{\log Q}\rangle$*

3.5.5 Degeneration of the spectral sequence

In this section, *we suppose that $\tilde{X}$, E and $\tilde{E}$ are compact Kähler manifolds.* We prove that the spectral sequence $E^{l,k}_r(X)$ degenerates at the term E_2. We study carefully the term $E^{l,k}_2(X)$. In fact by (3.60) the only nonzero terms occur for $l = -1, 0, 1$, and the only nontrivial d_2 occurs for $l = 1$.

On each term $E^{l,k}_2(X)$ of the spectral sequence of the filtration W_l there are two kinds of filtrations:

- The *direct filtrations* ${}^{(-l)}F_1$, ${}^{(-l)}\bar{F}_1$, induced by the filtrations ${}^{(-l)}F$ and ${}^{(-l)}\bar{F}$ of the complex $\Lambda^{\cdot}_X\langle\log Q\rangle$ (see (3.40) and (3.41)); here $E_2^{l,k}(X)$ must be considered as a quotient of the subspace $Z_2^{l,k} \subset W_l\Lambda^k_X\langle\log Q\rangle$; precisely $\alpha \in {}^{(-l)}F_1^a E_2^{l,k}(X)$ means: there exists $x \in {}^{(-l)}F_1^a W_l\Lambda^k_X\langle\log Q\rangle$, with $dx \in W_{l-2}\Lambda^{k+1}_X\langle\log Q\rangle$, so that $x \in Z_2^{l,k}$, and the class of x in $E_2^{l,k}(X)$ is α (and analogously for ${}^{(-l)}\bar{F}_1$.)

- The *recursive filtrations* ${}^{(-l)}F_2$, ${}^{(-l)}\bar{F}_2$ induced on $E_2^{l,k}(X)$ considered as the cohomology of the complex $(E_1^{l,k}(X), d_1)$; precisely $\alpha \in {}^{(-l)}F_2^a E_2^{l,k}(X)$ means: there exists $y \in {}^{(-l)}F^a E_1^{l,k}(X)$, with $d_1(y) = 0$ and the cohomology class of y in $E_2^{l,k}(X)$ (as the d_1-cohomology of $E_1^{l,k}(X)$) is α (and the same for ${}^{(-l)}\bar{F}_2$.)

Moreover it is easy to prove that $F_1 \subset F_2$, $\bar{F}_1 \subset \bar{F}_2$: if $x \in {}^{(-l)}F_1^a Z_2^{l,k}$ as above, then $x \in {}^{(-l)}F^a Z_1^{l,k}$ and $d_1[x]_1 = 0$; hence $y = [x]_1 \in {}^{(-l)}F^a E_1^{l,k}(X) \cap \ker d_1$ and its class in $E_2^{l,k}(X)$ is α.

Lemma 3.8. *The filtrations* ${}^{(-l)}F_1$ *and* ${}^{(-l)}\bar{F}_1$*, as well as* ${}^{(-l)}F_2$ *and* ${}^{(-l)}\bar{F}_2$*, are conjugate.*

The statement for ${}^{(-l)}F_1$, ${}^{(-l)}\bar{F}_1$ follows from the propositions 3.3 and 3.4. For ${}^{(-l)}F_2$ and ${}^{(-l)}\bar{F}_2$ it is enough to prove it in the case of $E_1^{l,k}(X)$, where it is obvious because $F_1 = F_2$, $\bar{F}_1 = \bar{F}_2$ in $E_1^{l,k}(X)$.

Theorem 3.3.

1) The differentials d_r *of the spectral sequence are* 0 *for* $r \geq 2$.

2) d_2 *is a morphism of the filtrations* ${}^{(-l)}F_1$ *and* ${}^{(-l)}\bar{F}_1$*:*

$$d_2\colon {}^{(-l)}F_1^a E_r^{l,k}(X) \to {}^{(-l+r)}F_1^a E_2^{l-2,k+1}(X) \tag{3.64}$$
$$d_2\colon {}^{(-l)}\bar{F}_1^b E_2^{l,k}(X) \to {}^{(-l+2)}\bar{F}_1^b E_2^{l-2,k+1}(X) \tag{3.65}$$

3) The filtrations ${}^{(-l)}F_2$, ${}^{(-l)}\bar{F}_2$ *induce pure Hodge structures on* $E_2^{l,k}(X)$

$$E_2^{l,k}(X) = \bigoplus_{a+b=k+l} {}^{(-l)}F_2^a \, {}^{(-l)}\bar{F}_2^b E_r^{l,k}(X) \tag{3.66}$$

4) On $E_2^{l,k}(X)$ *the filtration* ${}^{(-l)}F_1$ *(resp.* ${}^{(-l)}\bar{F}_1$*) is identical to the filtration* ${}^{(-l)}F_2$ *(resp.* ${}^{(-l)}\bar{F}_2$*). We shall denote simply by* ${}^{(-l)}F$, ${}^{(-l)}\bar{F}$*, these filtrations.*

Proof. Property 2) is clear, since the differential d preserves the filtrations ${}^{(-l)}F$ and ${}^{(-l)}\bar{F}$ of the complex $\Lambda^{\cdot}_X\langle\log Q\rangle$ (lemma 3.6, 1).
Property 3): d_1 preserves ${}^{(-l)}F$ and ${}^{(-l)}\bar{F}$ on $E_1^{l,k}(X)$ so that its cohomology

$E_2^{l,k}(X)$ carries a pure Hodge structure as in (3.66).
Property 1) for $r \geq 3$ is obvious for degree reasons. Then we note that 2), 3), 4) imply 1) for $r = 2$. In fact if we know that the direct and the recursive filtrations on $E_2^{l,k}(X)$ coincide, by (3.64) and (3.65) $d_2\colon E_2^{l,k}(X) \to E_r^{l-2,k+1}(X)$ becomes a morphism of pure Hodge structures of respective weights $k+l$, $k+l-1$, which are different, so that $d_2 = 0$ by proposition 1.4 of part I, chapter 1.
Finally, let us show 4). By lemma 3.8 it is sufficient to treat the case of ${}^{(-l)}F_1$ and ${}^{(-l)}F_2$.

We already know ${}^{(-l)}F_1 \subset {}^{(-l)}F_2$, so we need to show ${}^{(-l)}F_2 \subset {}^{(-l)}F_1$. For simplicity we write $L^k = \Lambda_X^k\langle\log Q\rangle$.

Let $\alpha \in F_2^a E_2^{l,k}$; there is a representative in $F^a E_1^{l,k} \cap Ker d_1$, i.e. an element $x \in F^a Z_1^{l,k} \subset F^a W_l L^k$ with $d_1[x]_1 = 0$; we have $dx \in W_{l-1}L^{k+1} \cap F^a W_l L^{k+1}$ so that

$$dx \in F^a W_{l-1} L^{k+1} \tag{3.67}$$

$d_1[x]_1 = 0$ gives

$$dx = dy + z, \quad y \in Z_0^{l-1,k} = W_{l-1}L^k, \quad z \in Z_0^{l-2,k+1} = W_{l-2}L^{k+1} \tag{3.68}$$

We obtain an element $[y]_0 \in E_0^{l-1,k}$, and from (3.68) we get $d_0[y]_0 \in F^a E_0^{l-1,k}$. Since d_0 is strict by lemma 3.6, we find $y' \in F^a W_{l-1} L^k$ with $d_0[y]_0 = d_0[y']_0$ or

$$dy = dy' + z', \quad z' \in W_{l-2}L^{k+1} \tag{3.69}$$

From (3.67), (3.68), (3.69) we obtain

$$x - y' \in F^a W_l L^k$$

and

$$d(x - y') \in W_{l-2} L^{k+1}$$

so that $x - y' \in F^a Z_2^{l,k}$ and its class in $E_2^{l,k}$ is α. This proves 4). □

Therefore we have given a proof of the following result:

Theorem 3.4. *Let us suppose that $\tilde{X}$, E, $\tilde{E}$ are compact Kähler manifolds. Then the cohomology $H^k(X \setminus Q, \mathbb{C})$, provided with the weight filtration W shifted by $-k$, carries a mixed Hodge structure. More precisely, the graded spaces $\frac{W_l H^k(X \setminus Q,\mathbb{C})}{W_{l-1}H^k(X\setminus Q,\mathbb{C})}$ are isomorphic to $E_2^{l,k}(X)$ and thus have a pure Hodge structure as in* (3.66).

(The shift of W by $-k$ is needed to normalize the weight in (3.66); in the shifted filtration $W_l' = W_{l-k}$ the quotient $\frac{W_l' H^k(X\setminus Q,\mathbb{C})}{W_{l-1}' H^k(X\setminus Q,\mathbb{C})}$ has weight l, as expected).

Remark. We point out that the statement of theorem 3.4 is not yet complete. More precisely, the graded spaces $\frac{W_l H^k(X\setminus Q,\mathbb{C})}{W_{l-1}H^k(X\setminus Q,\mathbb{C})}$ are isomorphic to $E_2^{l,k}(X)$, which carry a pure Hodge structure of weight $k+l$ for the filtration F induced by $E_1^{l,k}(X)$ ($E_2^{l,k}(X)$ is a quotient of a subspace of $E_1^{l,k}(X)$). On the other hand, the filtration F on $H^k(X \setminus Q, \mathbb{C})$ induces a filtration on the quotient $\frac{W_l H^k(X\setminus Q,\mathbb{C})}{W_{l-1}H^k(X\setminus Q,\mathbb{C})}$. We have not shown that the two filtrations, under the isomorphism $E_2^{l,k}(X) \simeq \frac{W_l H^k(X\setminus Q,\mathbb{C})}{W_{l-1}H^k(X\setminus Q,\mathbb{C})}$ coincide (and the same for $\bar{F}$). This will be made clear later in the general case (chapter 4).

Chapter 4

Mixed Hodge structures on noncompact singular spaces

4.1 Introduction

In this chapter, we introduce logarithmic differential forms in the most general case of a singular open analytic space $X \setminus Q$ which is the complement of a subspace Q of a compact analytic space X.

We take a desingularization of X:

$$\begin{array}{ccc} \tilde{E} & \xrightarrow{i} & \tilde{X} \\ {\scriptstyle q}\downarrow & & \downarrow{\scriptstyle \pi} \\ E & \xrightarrow{j} & X \end{array}$$

where $E \subset X$ is a nowhere dense closed subspace with $\operatorname{sing}(X) \subset E$, and $\tilde{X}$ is a smooth manifold.

Let

$$\tilde{Q} = \pi^{-1}(Q), \; M = E \cap Q, \; \tilde{M} = \tilde{E} \cap \tilde{Q}.$$

Then we define the relevant objects on X (logarithmic differential forms, weight and Hodge filtrations, residues...) as triples of objects on $\tilde{X}$, E, $\tilde{E}$. The results about $\tilde{X}$ are known by the theory of the manifold case (chapter 2); the definitions and results about E and $\tilde{E}$ are known by induction on the dimension ($\dim E, \dim \tilde{E} < \dim X$). The weight filtration on the complex of logarithmic forms on X with poles along Q is also defined combining the order of poles together with the rank of the space in the hypercovering of the complex. Though one should not expect that the theory on X is a "direct sum" of the theories on $\tilde{X}$, E, $\tilde{E}$, it turns out (as in the compact case of Part II chapter 3) that this is not far from being true. In fact let us consider the spectral sequence $E_r^{m,k}$ attached to the weight filtration (the main ingredient for the mixed Hodge theory). Then (under the right assumption on $X \setminus Q$, in particular if $X \setminus Q$ is quasi-projective):

1) $E_0^{m,k}$, $d_0 \colon E_0^{m,k} \to E_0^{m,k}$, $E_1^{m,k}$ are direct sums sums of the corresponding objects on $\tilde{X}$, E and $\tilde{E}$;

2) the spectral sequence degenerates at the level 2: $d_r = 0$, hence $E_r^{m,k} = E_2^{m,k}$, for $r \geq 2$.

3) the $E_2^{m,k}$ carry a pure Hodge structure, and they are isomorphic to the graded quotients $\frac{W_m H^k(X\setminus Q,\mathbb{C})}{W_{m-1}H^k(X\setminus Q,\mathbb{C})}$ of the cohomology $H^k(X \setminus Q, \mathbb{C})$ with respect to the weight filtration.

So we must add, to the separate knowledge of the properties of $\tilde{X}$, E and $\tilde{E}$, only the computation of $d_1 \colon E_1^{m,k} \to E_1^{m-1,k}$. Moreover, one can find an explicit formula for d_1 (formula (4.57).)

We can also define the notion of residue in the above general context. As in the smooth case, the residues live on certain smooth compact manifolds, so that they allow us to transport classical Hodge theory on compact manifolds and induce mixed Hodge theory on open singular spaces. The first term $E_1^{m,k}$ of the spectral sequence appears to be the cohomology of another complex, the residue complex.

4.2 Logarithmic complexes on singular spaces

In this section we show how to define complexes of forms with logarithmic poles at infinity on any complex space.

4.2.1 Logarithmic forms

By a *pair* (of complex spaces) (X, Q) we mean the data of a complex space X and of a closed, nowhere dense complex subspace Q. Let $g \colon Q \to X$ and $\rho \colon X \setminus Q \to X$ be the natural embeddings. A *morphism of pairs* $f \colon (X, Q) \to (Y, R)$ is a morphism $f \colon X \to Y$ such that $f(Q) \subset R$ and $f(X\setminus Q) \subset Y\setminus Q$. For a pair (X, Q) we define a family of complexes of fine sheaves $\mathcal{R}(X\langle\log Q\rangle) = \{\Lambda^{\cdot}_X\langle\log Q\rangle\}$ and for every morphism $f \colon (X, Q) \to (Y, R)$ a family $\mathcal{R}(f)$ of morphisms of complexes between the $\Lambda^{\cdot}_Y\langle\log R\rangle \in \mathcal{R}(Y\langle\log R\rangle)$ and some of the $\Lambda^{\cdot}_X\langle\log Q\rangle \in \mathcal{R}(X\langle\log Q\rangle)$, more precisely morphisms $\Lambda^{\cdot}_Y\langle\log R\rangle \to f_*\Lambda^{\cdot}_X\langle\log Q\rangle$ which we simply denote $\Lambda^{\cdot}_Y\langle\log R\rangle \to \Lambda^{\cdot}_X\langle\log Q\rangle$ and call *(admissible) pullback* with the following properties.

(I) The restriction $\Lambda^{\cdot}_X\langle\log Q\rangle|_{X\setminus Q} = \Lambda^{\cdot}_{X\setminus Q}$ of $\Lambda^{\cdot}_X\langle\log Q\rangle$ to $X \setminus Q$ belongs to $\mathcal{R}(X\setminus Q)$ (as defined in Part II, chapter 2), and the natural morphism of complexes on X

$$\Lambda^{\cdot}_X\langle\log Q\rangle \to \rho_*\Lambda^{\cdot}_{X\setminus Q}$$

induces isomorphisms in cohomology:

$$\mathcal{H}^k(\Lambda^{\cdot}_X\langle\log Q\rangle) \to \mathcal{H}^k\rho_*\Lambda^{\cdot}_{X\setminus Q} \tag{4.1}$$

in particular for every open set $W \subset X$ one has

$$H^k(W, \Lambda^{\cdot}_X\langle \log Q\rangle) = H^k(W, \rho_* \Lambda^{\cdot}_{X\setminus Q}) = H^k(W \setminus Q, \mathbb{C})$$

(II) For $k > 2dimX$, $\Lambda^k_X\langle \log Q\rangle = 0$.

(III) If X is smooth and Q is a divisor with normal crossing, the logarithmic De Rham complex $\mathcal{E}^{\cdot}_X\langle \log Q\rangle$ belongs to $\mathcal{R}(X\langle \log Q\rangle)$; for every morphism $f\colon (X,Q) \to (Y,R)$ with Y smooth and R a divisor with normal crossing the ordinary De Rham pullback $f^*\colon \mathcal{E}^{\cdot}_Y\langle \log R\rangle \to f_*\mathcal{E}^{\cdot}_X\langle \log Q\rangle$ is an admissible pullback.

The family of pullback will satisfy the following properties.

(C) (Composition). The composition of two pullback is a pullback.

(EP) (Existence of pullback). Let $f\colon (X,Q) \to (Y,R)$ be a morphism of pairs, and let us fix $\Lambda^{\cdot}_Y\langle \log R\rangle \in \mathcal{R}(Y\langle \log R\rangle)$; then there exists a $\Lambda^{\cdot}_X\langle \log Q\rangle \in \mathcal{R}(X\langle \log Q\rangle)$ and a pullback $\Lambda^{\cdot}_Y\langle \log R\rangle \to \Lambda^{\cdot}_X\langle \log Q\rangle$.

(U) (Uniqueness of pullback). Let $f\colon (X,Q) \to (Y,R)$ be a morphism of pairs, and $\alpha\colon \Lambda^{\cdot}_Y\langle \log R\rangle \to \Lambda^{\cdot}_X\langle \log Q\rangle$, $\beta\colon \Lambda^{\cdot}_Y\langle \log R\rangle \to \Lambda^{\cdot}_X\langle \log Q\rangle$ two pullback corresponding to f; then $\alpha = \beta$.

(F) (Filtering). If $\Lambda^{\cdot,1}_X\langle \log Q\rangle$, $\Lambda^{\cdot,2}_X\langle \log Q\rangle \in \mathcal{R}(X\langle \log Q\rangle)$, there exist two pullback $\Lambda^{\cdot,1}_X\langle \log Q\rangle \to \Lambda^{\cdot}_X\langle \log Q\rangle$, $\Lambda^{\cdot,2}_X\langle \log Q\rangle \to \Lambda^{\cdot}_X\langle \log Q\rangle$ corresponding to the identity, with the same target $\Lambda^{\cdot}_X\langle \log Q\rangle \in \mathcal{R}(X\langle \log Q\rangle)$.

It is useful to define $\Lambda^p_X\langle \log Q\rangle = 0$ at points $x \in X$ where $Q = X$ (in a neighborhood of x).

We sketch now the construction of the family $\mathcal{R}(X\langle \log Q\rangle)$ for any pair (X,Q). We omit the proofs, because they follow almost word by word (*mutatis mutandis*) the proofs in part II, chapter 2.

Step 1). Let X be smooth, and Q any subspace. Let

$$\begin{array}{ccc} \tilde{Q} & \longrightarrow & \tilde{X} \\ \downarrow & & \downarrow{\scriptstyle \pi} \\ Q & \xrightarrow{\;g\;} & X \end{array} \tag{4.2}$$

be a proper modification,where $\tilde{X}$ is smooth, $\tilde{Q}$ is a divisor with normal crossing and π is an isomorphism outside $\tilde{Q}$; we define

$$\mathcal{E}^{\cdot}_X\langle \log Q\rangle = \pi_*(\mathcal{E}^{\cdot}_{\tilde{X}}\langle \log \tilde{Q}\rangle)$$

The above complex, which depends on the choice of the diagram (4.2),will be called a *logarithmic De Rham complex of the pair* (X,Q) *associated to the diagram* (4.2); the reader should keep in mind that this gives new complexes even when Q is a divisor with normal crossing.

Step 2). Let (X, Q) be any pair, $E \subset X$ a nowhere dense closed subspace, with $\operatorname{sing}(X) \subset E$, and consider a diagram

$$\begin{array}{ccc} \tilde{E} & \xrightarrow{\ i\ } & \tilde{X} \\ {\scriptstyle q}\downarrow & & \downarrow{\scriptstyle \pi} \\ E & \xrightarrow{\ j\ } & X \end{array} \tag{4.3}$$

Let

$$\tilde{Q} = \pi^{-1}(Q), \ M = E \cap Q, \ \tilde{M} = \tilde{E} \cap \tilde{Q}.$$

Let $\Lambda^{\cdot}_E\langle \log M\rangle \in \mathcal{R}(E\langle \log M\rangle)$ (which exists by induction on $dim(X)$). Then we can find $\Lambda^{\cdot}_{\tilde{E}}\langle \log \tilde{M}\rangle \in \mathcal{R}(\tilde{E}\langle \log \tilde{M}\rangle)$, a logarithmic De Rham complex $\mathcal{E}^{\cdot}_{\tilde{X}}\langle \log \tilde{Q}\rangle$ as in step 1), a pullback $\phi \in \mathcal{R}(q)$

$$\phi \colon \Lambda^{\cdot}_E\langle \log M\rangle \to \Lambda^{\cdot}_{\tilde{E}}\langle \log \tilde{M}\rangle \tag{4.4}$$

a pullback $\psi \in \mathcal{R}(i)$

$$\psi \colon \mathcal{E}^{\cdot}_{\tilde{X}}\langle \log \tilde{Q}\rangle \to \Lambda^{\cdot}_{\tilde{E}}\langle \log \tilde{M}\rangle$$

so that we define the complex

$$\Lambda^k_X\langle \log Q\rangle = \pi_* \mathcal{E}^k_{\tilde{X}}\langle \log \tilde{Q}\rangle \oplus j_* \Lambda^k_E\langle \log M\rangle \oplus (j \circ q)_* \Lambda^{k-1}_{\tilde{E}}\langle \log \tilde{M}\rangle \tag{4.5}$$

whose differential is by definition

$$d(\omega, \sigma, \theta) = (d\omega, d\sigma, d\theta + (-1)^k(\psi(\omega) - \phi(\sigma)). \tag{4.6}$$

The above complex induces the isomorphism of cohomology sheaves (4.1).

Proof of the surjectivity in (4.1). Let $x \in Q$ and $\rho = (\omega, \sigma, \theta) \in \Lambda^k_X(U \setminus Q)$ be a d-closed section of $\rho_* \Lambda^k_{X \setminus Q}$ on an open neighborhood U of x. Since $d\sigma = 0$, by induction on dimensions there are $\sigma' \in \Lambda^k_E\langle \log M\rangle(U \cap E)$ and $\sigma'' \in \Lambda^{k-1}_E((U \cap E) \setminus M)$ with $\sigma - d\sigma'' = \sigma'$ (after possibly shrinking U); replacing ρ by $\rho - d(0, \sigma'', 0)$ we can suppose that $\sigma \in \Lambda^k_E\langle \log M\rangle(U \cap E)$. A similar argument (using theorem 2.4 of chapter 2) shows that we can also suppose $\omega \in \mathcal{E}^k_{\tilde{X}}\langle \log \tilde{Q}\rangle(\tilde{U})$. Since $(-1)^k(\psi(\omega) - \phi(\sigma)) = -d\theta$, by induction (we use the injectivity on $\tilde{E}$) there is $\theta' \in \Lambda^{k-1}_{\tilde{E}}\langle \log \tilde{M}\rangle(q^{-1}(U \cap E))$ such that $(-1)^k(\psi(\omega) - \phi(\sigma)) = -d\theta'$. Again by induction , $d(\theta - \theta') = 0$ implies that we can write $\theta - \theta' = d\tilde{\theta} + \theta''$, where $\theta'' \in \Lambda^{k-1}_{\tilde{E}}\langle \log \tilde{M}\rangle(q^{-1}(U \cap E))$ and $\tilde{\theta} \in \Lambda^{k-2}_{\tilde{E}}(q^{-1}(U \cap E) \setminus \tilde{M}))$. Finally

$$(\omega, \sigma, \theta) - d(0, 0, \tilde{\theta}) = (\omega, \sigma, \theta' + \theta'')$$

which ends the proof of the surjectivity in (4.1).

Proof of the injectivity in (4.1). Let U be an open neighborhood of $x \in Q$ and $\mu = (\omega, \sigma, \theta) \in \Lambda^k_X\langle\log Q\rangle(U)$ such that there exists $(\omega', \sigma', \theta') \in \Lambda^{k-1}_X(U \setminus Q)$ with $(\omega, \sigma, \theta) = d(\omega', \sigma', \theta')$. Then in particular $\sigma = d\sigma'$ so that (by the injectivity on E) there is a $\sigma'' \in \Lambda^{k-1}_E\langle\log M\rangle(U \cap E)$ with $\sigma = d\sigma''$ (after possibly shrinking U); we can replace μ with $\mu - d(0, \sigma'', 0)$, so that we can suppose that $\sigma = 0$. A similar argument on $\tilde{X}$ (starting with $d\omega = 0$) leads us to suppose also $\omega = 0$. Then we have $\theta = d\theta'$ and the surjectivity on $\tilde{E}$ gives a $\theta'' \in \Lambda^{k-2}_{\tilde{E}}\langle\log \tilde{M}\rangle(q^{-1}(U \cap E))$ with $\theta = d\theta''$. The conclusion follows. □

Note that $\Lambda^k_X\langle\log Q\rangle$ is a fine sheaf defined on all of X. A section of $\Lambda^k_X\langle\log Q\rangle$ on an open set $U \subset X$ will be called a *logarithmic differential form* of degree k on U.

Notation. From now on we will write for simplicity

$$\Lambda^{\cdot}_X\langle\log Q\rangle = \mathcal{E}^{\cdot}_{\tilde{X}}\langle\log \tilde{Q}\rangle \oplus \Lambda^{\cdot}_E\langle\log M\rangle \oplus \Lambda^{\cdot}_{\tilde{E}}\langle\log \tilde{M}\rangle(-1) \tag{4.7}$$

instead of

$$\Lambda^{\cdot}_X\langle\log Q\rangle = \pi_*\mathcal{E}^{\cdot}_{\tilde{X}}\langle\log \tilde{Q}\rangle \oplus j_*\Lambda^{\cdot}_E\langle\log M\rangle \oplus (j \circ q)_*\Lambda^{\cdot}_{\tilde{E}}\langle\log \tilde{M}\rangle(-1)$$

Theorem 4.1. *One has a natural isomorphism induced by the morphism of inclusion*

$$H^p(W, \Lambda^{\cdot}_X\langle\log Q\rangle) = H^p(W, \rho_*\Lambda^{\cdot}_{X\setminus Q}) = H^p(W \setminus Q, \mathbb{C}) \tag{4.8}$$

or in other words the cohomology of $W \setminus D$ can be calculated as the cohomology of the complex of sections $(\Gamma(W, \Lambda^{\cdot}_X\langle\log Q\rangle), d)$ of $\Lambda^{\cdot}_X\langle\log Q\rangle$.

Proof. First we note that the second isomorphism in the above formula (4.8), is a consequence of theorem 2.1 of part II, chapter 2 for the complex $\Lambda^{\cdot}_{X\setminus Q}$. Each of the complexes $\Lambda^{\cdot}_X\langle\log Q\rangle$ and $\rho_*\Lambda^{\cdot}_{X\setminus Q}$ induces a spectral sequence of hypercohomology on W:

$$H^p(W, \mathcal{H}^q\Lambda^{\cdot}_X\langle\log Q\rangle) \Longrightarrow H^{p+q}(W, \Lambda^{\cdot}_X\langle\log Q\rangle)$$
$$H^p(W, \mathcal{H}^q\rho_*\Lambda^{\cdot}_{X\setminus Q}) \Longrightarrow H^{p+q}(W, \rho_*\Lambda^{\cdot}_{X\setminus Q})$$

By the isomorphism (4.1) the cohomology sheaves $\mathcal{H}^q\Lambda^{\cdot}_X\langle\log Q\rangle$ and $\mathcal{H}^q\rho_*\Lambda^{\cdot}_{X\setminus Q}$ are isomorphic. So in the limit we find that the cohomologies $H^p(W, \Lambda^{\cdot}_X\langle\log Q\rangle)$ and $H^p(W, \rho_*\Lambda^{\cdot}_{X\setminus Q})$ are isomorphic. □

From the construction of $\Lambda^{\cdot}_X\langle\log Q\rangle$ it follows that there is a uniquely determined family $((X_a, Q_a), h_a)_{a\in A}$ of pairs (X_a, Q_a) (X_a smooth and Q_a a divisor with normal crossing in X_a) and proper maps of pairs $h_a\colon (X_a, Q_a) \to (X, Q)$ such that

$$\Lambda^k_X\langle\log Q\rangle = \bigoplus_{a\in A} (h_a)_*\mathcal{E}^{k-q(a)}_{X_a}\langle\log Q_a\rangle \tag{4.9}$$

where $q(a) = q_X(a)$ is a nonnegative integer, which depends only on $a \in A$ and not on k; moreover, there exist mappings $h_{ab}\colon (X_a, Q_a) \to (X_b, Q_b)$, commuting with h_a and h_b, such that the differential $\Lambda^k_X\langle\log Q\rangle \to \Lambda^{k+1}_X\langle\log Q\rangle$ is given by .

$$d(\omega_a) = (d\omega_a + \sum_b \epsilon^{(k)}_{ab} h^*_{ab}\omega_b) \tag{4.10}$$

where $\epsilon^{(k)}_{ab}$ can take the values $0, \pm 1$.

The family $((X_a, Q_a), h_a)_{a\in A}$ will be called the *hypercovering* of (X, Q) associated to $\Lambda^{\cdot}_X\langle\log Q\rangle$, and $q_X(a)$ will be the **rank** of (X_a, Q_a).

Clearly if $h_{ab}\colon (X_a, Q_a) \to (X_b, Q_b)$ exists, one has

$$q_X(a) = q_X(b) + 1 \tag{4.11}$$

Remark.

1. In the situation of the diagram (4.3) and of the complex (4.5), we notice that $(\tilde{X}, \tilde{Q})$ is not in general a pair (X_a, Q_a) of the hypercovering. In fact the logarithmic complex $\mathcal{E}^{\cdot}_{\tilde{X}}\langle\log\tilde{Q}\rangle$ is given by a $\phi_*\mathcal{E}^{\cdot}_{\tilde{X}'}\langle\log\tilde{Q}'\rangle$ where $\tilde{X}'$ is a manifold, $\tilde{Q}'$ a divisor with normal crossing, and $\phi\colon (\tilde{X}', \tilde{Q}') \to (\tilde{X}, \tilde{Q})$ is a modification. In this case, $(\tilde{X}', \tilde{Q}')$ is a pair of the hypercovering, and the corresponding morphism of pairs is $\pi \circ \phi\colon (\tilde{X}', \tilde{Q}') \to (X, Q)$.

2. Notice also, for future use, that when one restricts $\Lambda^{\cdot}_X\langle\log Q\rangle$ to $X \setminus Q$, one obtains a complex $\Lambda^{\cdot}_{X\setminus Q}$ whose associated hypercovering is $(X_a \setminus Q_a)$ (with the same rank $q_X(a)$).

So, among the pairs (X_a, Q_a) of the hypercovering of $\Lambda^{\cdot}_X\langle\log Q\rangle$, one can distinguish three types

i) the pair $(\tilde{X}', \tilde{Q}')$ which is the modification of $(\tilde{X}, \tilde{Q})$, whose rank is 0;

ii) pairs (X_b, Q_b) corresponding to the hypercovering of $\Lambda^{\cdot}_E\langle\log M\rangle$ whose rank is
$$q_X(b) = q_E(b)$$

iii) pairs (X_c, Q_c) corresponding to the hypercovering of $\Lambda^{\cdot}_{\tilde{E}}\langle\log\tilde{M}\rangle$ whose rank is
$$q_X(c) = q_{\tilde{E}}(c) + 1$$

At the level of hypercoverings, a pullback

$$\phi\colon \Lambda^{\cdot}_Y\langle\log R\rangle \to \Lambda^{\cdot}_X\langle\log Q\rangle$$

is also given by

$$\phi\colon \bigoplus_{b\in B} \mathcal{E}^{k-q_Y(b)}_{Y_b}\langle\log R_b\rangle \to \bigoplus_{a\in A} \mathcal{E}^{k-q_X(a)}_{X_a}\langle\log Q_a\rangle \tag{4.12}$$

and can be realized as follows

(i) for any $a \in A$, there exists at most a unique $b = b(a) \in B$ with $q_Y(b) = q_X(a)$ and a morphism of pairs

$$f_{ab}\colon (X_a, Q_a) \to (Y_b, R_b)$$

inducing a standard De Rham pullback

$$f_{ab}^*\colon \mathcal{E}^{\cdot}_{Y_b}\langle \log R_b \rangle \to \mathcal{E}^{\cdot}_{X_a}\langle \log Q_a \rangle$$

We put $f_{ab}^* = 0$ if f_{ab} is not defined as a morphism of pairs.

(ii) Then ϕ of (4.12) is given by

$$\phi\left(\oplus_{b\in B}\omega_b\right) = \oplus_{a\in A}(f_{ab}^*\omega_b) \tag{4.13}$$

(this is well defined because for any $a \in A$, f_{ab}^* is unique, and maybe 0).

Definition 4.1. The hypercovering (X_a, Q_a) associated to $\Lambda^{\cdot}_X\langle \log Q \rangle$ is called a Kähler hypercovering if each X_a is a Kähler manifold.

Arguing as in theorem 2.10 of part II, chapter 2, we obtain

Theorem 4.2.

(i) Let X be a compact (B)-Kähler space, $Q \subset X$ a closed subspace. Then there exists a $\Lambda^{\cdot}_X\langle \log Q \rangle \in \mathcal{R}(X\langle \log Q \rangle)$ with a Kähler hypercovering.

(ii) Let $f\colon (X, Q) \to (Y, R)$ be a morphism of pairs, where X and Y are compact (B)-Kähler spaces; let $\Lambda^{\cdot}_Y\langle \log R \rangle \in \mathcal{R}(Y\langle \log R \rangle)$ with a Kähler hypercovering. Then there exists a $\Lambda^{\cdot}_X\langle \log Q \rangle \in \mathcal{R}(X\langle \log Q \rangle)$ with a Kähler hypercovering and a pullback $\Lambda^{\cdot}_Y\langle \log R \rangle \to \Lambda^{\cdot}_X\langle \log Q \rangle$.

(iii) If X is a compact (B)-Kähler space and $\Lambda^{\cdot,1}_X\langle \log Q \rangle$, $\Lambda^{\cdot,2}_X\langle \log Q \rangle$ admit a Kähler hypercovering, there exist two pullback $\Lambda^{\cdot,1}_X\langle \log Q \rangle \to \Lambda^{\cdot}_X\langle \log Q \rangle$, $\Lambda^{\cdot,2}_X\langle \log Q \rangle \to \Lambda^{\cdot}_X\langle \log Q \rangle$ corresponding to the identity, where the common target $\Lambda^{\cdot}_X\langle \log Q \rangle$ has a Kähler hypercovering.

4.3 The weight filtration W

4.3.1 On a logarithmic De Rham complex when X is smooth

Let (X, Q) be a pair with X a manifold, Q a (closed nowhere dense) subspace and $\pi\colon (\tilde{X}, \tilde{Q}) \to (X, Q)$ a modification with $\tilde{X}$ a manifold and $\tilde{Q}$ a divisor with normal crossing. We have a logarithmic De Rham complex

$$\mathcal{E}^{\cdot}_X\langle \log Q \rangle = \pi_*\mathcal{E}^{\cdot}_{\tilde{X}}\langle \log \tilde{Q} \rangle$$

where $\mathcal{E}^{\cdot}_{\tilde{X}}\langle\log\tilde{Q}\rangle$ is the standard logarithmic De Rham complex defined in chapter 2. We define an increasing filtration

$$W_m\mathcal{E}^{\cdot}_X\langle\log Q\rangle = \pi_* W_m\mathcal{E}^{\cdot}_{\tilde{X}}\langle\log\tilde{Q}\rangle \tag{4.14}$$

where $W_m\mathcal{E}^{\cdot}_{\tilde{X}}\langle\log\tilde{Q}\rangle$ is the filtration by the order of poles as defined in chapter 2.

4.3.2 On a general logarithmic complex

If $\Lambda^{\cdot}_X\langle\log Q\rangle$ is a logarithmic complex as in (4.7) (where we have skipped the symbols of direct images of sheaves), we define the filtration

$$\begin{gathered} W_m\Lambda^{\cdot}_X\langle\log Q\rangle = \\ = W_m\mathcal{E}^{\cdot}_{\tilde{X}}\langle\log\tilde{Q}\rangle \oplus W_m\Lambda^{\cdot}_E\langle\log M\rangle \oplus^{(1)} W_m\Lambda^{\cdot}_{\tilde{E}}\langle\log\tilde{M}\rangle(-1) \end{gathered} \tag{4.15}$$

where ${}^{(1)}W_m$ is the filtration shifted by + 1, namely

$${}^{(1)}W_m\Lambda^k_{\tilde{E}}\langle\log\tilde{M}\rangle(-1) = W_{m+1}\Lambda^{k-1}_{\tilde{E}}\langle\log\tilde{M}\rangle \tag{4.16}$$

In (4.15) $W_m\Lambda^{\cdot}_E\langle\log M\rangle$ and $W_m\Lambda^{\cdot}_{\tilde{E}}\langle\log\tilde{M}\rangle$ are defined by recursion on the dimension and $W_m\mathcal{E}^{\cdot}_{\tilde{X}}\langle\log\tilde{Q}\rangle$ is defined as above by (4.14).

Lemma 4.1.

1) $\Lambda^{\cdot}_X\langle\log Q\rangle, d)$ *is a filtered complex for* W_m*:*

$$d(\ W_m\Lambda^k_X\langle\log Q\rangle)\ \subset W_m\Lambda^{k+1}_X\langle\log Q\rangle \tag{4.17}$$

and is given, at the level of hypercovering by

$$W_m\Lambda^{\cdot}_X\langle\log Q\rangle = \bigoplus_{a\in A} {}^{(q_X(a))}W_m\mathcal{E}^{\cdot}_{X_a}\langle\log Q_a\rangle(-q_X(a)) \tag{4.18}$$

2) W_m *is preserved by pullback.*

Proof. We prove first (4.18) by recursion on the dimension. We start from (4.15). One of the pairs (X_a, Q_a) is $(\tilde{X}', \tilde{Q}')$ which is a modification of $(\tilde{X}, \tilde{Q})$, whose rank is 0. For this pair, $W_m\mathcal{E}^{\cdot}_{\tilde{X}}\langle\log\tilde{Q}\rangle$ coincides with $W_m\mathcal{E}^{\cdot}_{\tilde{X}'}\langle\log\tilde{Q}'\rangle$ (forgetting the direct image symbols).

For the pairs (X_b, Q_b) of the hypercovering of $\Lambda^{\cdot}_E\langle\log M\rangle$ one has

$$W_m\Lambda^{\cdot}_E\langle\log M\rangle = \bigoplus_b {}^{(q_E(b))}W_m\mathcal{E}^{\cdot}_{X_b}\langle\log Q_b\rangle(-q_E(b))$$

where we know that $q_E(b) = q_X(b)$. For the pairs (X_c, Q_c) belonging the hypercovering of $\Lambda^{\cdot}_{\tilde{E}}\langle\log\tilde{M}\rangle$ one has

$$W_m\Lambda^{\cdot}_{\tilde{E}}\langle\log\tilde{M}\rangle = \bigoplus_c {}^{(q_{\tilde{E}}(c))}W_m\mathcal{E}^{\cdot}_{X_c}\langle\log Q_c\rangle(-q_{\tilde{E}}(c))$$

But $q_{\tilde{E}}(c) = q_X(c) - 1$ so that

$$^{(1)}W_m\Lambda^{\cdot}_{\tilde{E}}\langle\log \tilde{M}\rangle(-1) = \bigoplus_c {}^{(q_X(c))}W_m\mathcal{E}^{\cdot}_{X_c}\langle\log Q_c\rangle(-q_X(c))$$

These formulas, together with the definition (4.7), prove (4.18).

Now, we consider a pullback as in (4.12), (4.13). We know that the nonzero f^*_{ab} are standard De Rham pullback between pairs of same rank

$$f^*_{ab} \colon \mathcal{E}^{\cdot}_{Y_b}\langle\log R_b\rangle \to \mathcal{E}^{\cdot}_{X_a}\langle\log Q_a\rangle, \quad q_X(a) = q_Y(b)$$

On the standard logarithmic De Rham complexes, W_m is the filtration by the order of poles and so it is preserved by standard De Rham pullback. So using (4.13) and (4.18) we see that a pullback preserves the filtration W_m.

Finally, we prove that d preserves the filtration W_m. We use the equations (4.18) and the definition (4.10) of d. Assuming that for each a

$$\omega_a \in {}^{(q_X(a))}\, W_m\mathcal{E}^{k-q_X(a)}_{X_a}\langle\log Q_a\rangle$$

we see that

$$d\omega_a \in {}^{(q_X(a))}\, W_m\mathcal{E}^{k+1-q_X(a)}_{X_a}\langle\log Q_a\rangle \tag{4.19}$$

and

$$h^*_{ab}\omega_b \in {}^{(q_X(b))}\, W_m\mathcal{E}^{k-q_X(b)}_{X_a}\langle\log Q_a\rangle$$

But $q_X(b) = q_X(a) - 1$, so that

$$\sum_b \epsilon^{(k)}_{ab}\, h^*_{ab}\omega_b \in {}^{(q_X(a))}\, W_{m-1}\mathcal{E}^{k+1-q_X(a)}_{X_a}\langle\log Q_a\rangle \tag{4.20}$$

The equation (4.20), together with (4.19), prove that d preserves the filtration W. □

Moreover from (4.20) and the definition of d in (4.10), we obtain

Lemma 4.2. *If* $\oplus_{a\in A}\omega_a \in W_m\Lambda^{\cdot}_X\langle\log Q\rangle$, *then*

$$d(\oplus_{a\in A}\omega_a) = \oplus_{a\in A}(d\omega_a) \qquad \mathrm{mod}\ W_{m-1}\Lambda^{\cdot}_X\langle\log Q\rangle$$

4.3.3 Filtration of the cohomology and spectral sequence

Let $V \subset X$ be an open subset. The filtration W_m induced on the complex $\Gamma(V, \Lambda^{\cdot}_X\langle\log Q\rangle)$ gives rise to a spectral sequence whose first term is

$$E_1^{m,k}(V) = H^k\left(V, \frac{W_m\Lambda^{\cdot}_X\langle\log Q\rangle}{W_{m-1}\Lambda^{\cdot}_X\langle\log Q\rangle}\right) = H^k\left(V, \frac{W_m\Lambda^{\cdot}_X\langle\log Q\rangle}{W_{m-1}\Lambda^{\cdot}_X\langle\log Q\rangle}\right)$$

The second equality in the above formula holds because the $\frac{W_m\Lambda^k_X\langle\log Q\rangle}{W_{m-1}\Lambda^k_X\langle\log Q\rangle}$ are fine sheaves, so that we have isomorphisms

$$\frac{\Gamma(V, W_m\Lambda^k_X\langle\log Q\rangle)}{\Gamma(V, W_{m-1}\Lambda^k_X\langle\log Q\rangle)} = \Gamma\left(V, \frac{W_m\Lambda^k_X\langle\log Q\rangle}{W_{m-1}\Lambda^k_X\langle\log Q\rangle}\right)$$

By (4.18) we have

$$\frac{W_m\Lambda^{\cdot}_X\langle\log Q\rangle}{W_{m-1}\Lambda^{\cdot}_X\langle\log Q\rangle} = \bigoplus_{a\in A}(h_a)_*\left(\frac{{}^{(q(a))}W_m\mathcal{E}^{\cdot}_{X_a}\langle\log Q_a\rangle(-q(a))}{{}^{(q(a))}W_{m-1}\mathcal{E}^{\cdot}_{X_a}\langle\log Q_a\rangle(-q(a))}\right)$$

so that

$$\Gamma\left(V, \frac{W_m\Lambda^{\cdot}_X\langle\log Q\rangle}{W_{m-1}\Lambda^{\cdot}_X\langle\log Q\rangle}\right) = = \bigoplus_{a\in A}\Gamma\left(V, (h_a)_*\frac{W_{m+q(a)}\mathcal{E}^{\cdot}_{X_a}\langle\log Q_a\rangle(-q(a))}{W_{m+q(a)-1}\mathcal{E}^{\cdot}_{X_a}\langle\log Q_a\rangle(-q(a))}\right) \tag{4.21}$$

Lemma 4.3.

1) *The differential d_0 of the graded complex $\Gamma\left(V, \frac{W_m\Lambda^{\cdot}_X\langle\log Q\rangle}{W_{m-1}\Lambda^{\cdot}_X\langle\log Q\rangle}\right)$, under the identification (4.21), is given by*

$$d_0\left([\oplus_{a\in A}\omega_a]\right) = \left(\oplus_{a\in A}d_0[\omega_a]\right) \tag{4.22}$$

where [ ] in the left hand side denotes the equivalence class modulo $W_{m-1}\Lambda^{\cdot}_X\langle\log Q\rangle$, and [ ] in the right hand side the equivalence class modulo ${}^{(q(a))}W_{m-1}\mathcal{E}^{\cdot}_{X_a}\langle\log Q_a\rangle(-q(a))$.

2) *The first term of the spectral sequence is*

$$E_1^{m,k}(V) = \bigoplus_{a\in A}H^k\left(V, (h_a)_*\left(\frac{{}^{(q(a))}W_m\mathcal{E}^{\cdot}_{X_a}\langle\log Q_a\rangle(-q(a))}{{}^{(q(a))}W_{m-1}\mathcal{E}^{\cdot}_{X_a}\langle\log Q_a\rangle(-q(a))}\right)\right) \tag{4.23}$$

Proof.

1) is immediate from lemma 4.2

2) $E_1^{m,k}(V)$ is the cohomology of the complex on the left hand side of (4.21) for the d_0 of (4.22). Hence the equality (4.21) and the formula (4.22) prove (4.23). □

Notice that by chapter 2 and (4.23)

$$E_1^{m,k}(V) = \bigoplus_{a\in A}E_1^{m+q(a),k-q(a)}(h_a^{-1}(V)) \tag{4.24}$$

(where the terms E_1 in the second members are the E_1 of the spectral sequence of chapter 2 for each (X_a, Q_a)).

If $V = X$ the above formula (4.24) can also be written as

$$E_1^{m,k}(X) = E_1^{m,k}(\tilde{X}) \oplus E_1^{m,k}(E) \oplus E_1^{m+1,k-1}(\tilde{E}) \tag{4.25}$$

where $E_r^{m,k}(\tilde{X})$, $E_r^{m,k}(E)$, $E_r^{m,k}(\tilde{E})$ are the spectral sequences arising from the weight filtrations on the complexes $\Gamma(\tilde{X}, \mathcal{E}^{\cdot}_{\tilde{X}}\langle\log \tilde{Q}\rangle)$, $\Gamma(\tilde{E}, \Lambda^{\cdot}_{E}\langle\log M\rangle)$ and $\Gamma(\tilde{E}, \Lambda^{\cdot}_{\tilde{E}}\langle\log \tilde{M}\rangle)$ respectively.

4.4 Residues

4.4.1 Residues on the graded complexes

For each pair (X_a, Q_a) of the hypercovering, we can define residues as in chapter 2 for $\mathcal{E}^{\cdot}_{X_a}\langle\log Q_a\rangle$ and we deduce mappings

$$\mathrm{Res}^k_{a,m}\colon (h_a)_* \left({}^{(q(a))}W_m\mathcal{E}^k_{X_a}\langle\log Q_a\rangle(-q(a))\right) \to (h_a)_*\mathcal{E}^{k-m-2q(a)}_{Q_a{}^{[m+q(a)]}} \tag{4.26}$$

$$\mathrm{Res}^k_{a,m}\colon (h_a)_* \left(\frac{{}^{(q(a))}W_m\mathcal{E}^{\cdot}_{X_a}\langle\log Q_a\rangle(-q(a))}{{}^{(q(a))}W_{m-1}\mathcal{E}^{\cdot}_{X_a}\langle\log Q_a\rangle(-q(a))}\right) \to (h_a)_*\mathcal{E}^{k-m-2q(a)}_{Q_a{}^{[m+q(a)]}} \tag{4.27}$$

with the conventions:
$Q_a{}^{[0]} = X_a$ and then $\mathrm{Res}^k_{a,m} =$ identity;
$Q_a{}^{[r]} = \emptyset$ if $r < 0$ and the corresponding $\mathrm{Res}^k_{a,m} = 0$ (indeed $W_{m+q(a)} = 0$ if $m + q(a) < 0$).

We can define the residue as a mapping which is the direct sum of $\mathrm{Res}^k_{a,m}$:

$$\mathrm{Res}^k_m\colon W_m\Lambda^k_X\langle\log Q\rangle \to \bigoplus_{a\in A}(h_a)_*\mathcal{E}^{k-m-2q(a)}_{Q_a{}^{[m+q(a)]}} \tag{4.28}$$

$$\mathrm{Res}^k_m\colon \frac{W_m\Lambda^k_X\langle\log Q\rangle}{W_{m-1}\Lambda^k_X\langle\log Q\rangle} \to \bigoplus_{a\in A}(h_a)_*\mathcal{E}^{k-m-2q(a)}_{Q_a{}^{[m+q(a)]}} \tag{4.29}$$

These residues induce residues in cohomology. Indeed by lemma 2.2 of chapter 2 applied to the residue mapping $\mathrm{Res}^k_{a,m}$ on open sets $h_a^{-1}(V)$, we deduce an isomorphism of local cohomologies:

Lemma 4.4. *One has an isomorphism of local cohomologies*

$$\underline{\mathrm{Res}}^p_{a,m}\colon \mathcal{H}^p\left((h_a)_*\left(\frac{{}^{(q(a))}W_m\mathcal{E}^{\cdot}_{X_a}\langle\log Q_a\rangle(-q(a))}{{}^{(q(a))}W_{m-1}\mathcal{E}^{\cdot}_{X_a}\langle\log Q_a\rangle(-q(a))}\right)\right) \to$$
$$\to \mathcal{H}^p\left((h_a)_*\mathcal{E}^{\cdot}_{Q_a{}^{[m+q(a)]}}(-m-2q(a))\right)$$

Then we deduce, using the spectral sequence of local cohomologies:

Theorem 4.3. *For any open set $V \subset X$, one has an isomorphism of cohomologies:*

$$\underline{\mathrm{Res}}^k_m : H^k\left(V, \frac{W_m\Lambda^{\cdot}_X\langle\log Q\rangle}{W_{m-1}\Lambda^{\cdot}_X\langle\log Q\rangle}\right) \rightarrow \bigoplus_{a\in A} H^{k-m-2q(a)}\left(h_a^{-1}(V)\cap {Q_a}^{[m+q(a)]}, \mathbb{C}\right) \tag{4.30}$$

We define the **residue complex**

$$\mathrm{Res}^{\cdot}_m\Lambda_X\langle\log Q\rangle = \bigoplus_{a\in A}(h_a)_*\mathcal{E}^{\cdot}_{{Q_a}^{[m+q(a)]}}(-m-2q(a)) \tag{4.31}$$

whose differential d is the direct sum of the differentials on each term of the sum. Then the cohomology of this complex is exactly the second member of (4.30). Thus:

Corollary 4.1. $\underline{\mathrm{Res}}^k_m$ *induces an isomorphism*

$$\underline{\mathrm{Res}}^k_m : H^k\left(V, \frac{W_m\Lambda^{\cdot}_X\langle\log Q\rangle}{W_{m-1}\Lambda^{\cdot}_X\langle\log Q\rangle}\right) \rightarrow H^k(V, \mathrm{Res}^{\cdot}_m\Lambda_X\langle\log Q\rangle)$$

Proposition 4.1. *The residue mapping*

$$\mathrm{Res}^k_m : W_m\Lambda^k_X\langle\log Q\rangle \rightarrow \mathrm{Res}^k_m\Lambda_X\langle\log Q\rangle$$

commutes with the respective differentials.

Proof. We consider a form

$$\oplus_a\omega_a \in \bigoplus_a (h_a)_*\left({}^{(q(a))}W_m\mathcal{E}^k_{X_a}\langle\log Q_a\rangle(-q(a))\right)$$

So we have:

$$d\mathrm{Res}^k_m(\oplus_a\omega_a) = \oplus_a\mathrm{Res}^{k+1}_m d\omega_a \in \bigoplus_a (h_a)_*\mathcal{E}^{k-m-2q(a)+1}_{{Q_a}^{[m+q(a)+1]}}$$

because the residue commutes with the differential on any $\mathcal{E}^k_{X_a}\langle\log Q_a\rangle$. On the other hand

$$d(\oplus_a\omega_a) = \oplus_a(d\omega_a + \sum_b \epsilon^{(k)}_{ab}h^*_{ab}\omega_b) \tag{4.32}$$

where $q(b) = q(a) - 1$

$$\mathrm{Res}^{k+1}_m(d(\oplus_a\omega_a)) = \oplus_a\mathrm{Res}^{k+1}_m\left(d\omega_a + \sum_b \epsilon^{(k)}_{ab}h^*_{ab}\omega_b\right) \tag{4.33}$$

We now distinguish three types of indices $a \in A$

(i) If a is such that $m+q(a)>0$, or $m-1+q(a)\geq 0$, by (4.20) the form

$$\sum_b \epsilon_{ab}^{(k)} h_{ab}^* \omega_b \in^{(q(a))} W_{m-1}\mathcal{E}_{X_a}^{k+1-q(a)}\langle \log Q_a\rangle$$

is smooth on X_a and its residue Res_m^{k+1} is 0 on ${Q_a}^{[m+q(a)]}$, so that for this type of a

$$\mathrm{Res}_m^{k+1}(d\omega_a + \sum_b \epsilon_{ab}^{(k)} h_{ab}^* \omega_b) = \mathrm{Res}_m^{k+1} d\omega_a \tag{4.34}$$

(ii) If a is such that $m+q(a)=0$, then the residue

$$\mathrm{Res}_m^{k+1}(d\omega_a + \sum_b \epsilon_{ab}^{(k)} h_{ab}^* \omega_b) = d\omega_a + \sum_b \epsilon_{ab}^{(k)} h_{ab}^* \omega_b$$

because ${Q_a}^{[0]}=X_a$ and the corresponding residue is the identity. But

$$\omega_b \in {}^{(q(b))}W_m\mathcal{E}_{X_b}^k\langle \log Q_b\rangle(-q(b))$$

Since $m+q(b)=-1$ and $W_m\mathcal{E}_{X_b}^k\langle \log Q_b\rangle = 0$ for $m=-1$, $\omega_b=0$ and

$$\mathrm{Res}_m^{k+1}(d\omega_a + \sum_b \epsilon_{ab}^{(k)} h_{ab}^* \omega_b) = d\omega_a = \mathrm{Res}_m^{k+1} d\omega_a \tag{4.35}$$

(iii) Finally if a is such that $m+q(a)<0$, the corresponding residues are identically 0:

$$\mathrm{Res}_m^{k+1}(d\omega_a + \sum_b \epsilon_{ab}^{(k)} h_{ab}^* \omega_b) = 0 = \mathrm{Res}_m^{k+1} d\omega_a \tag{4.36}$$

From (4.32), ..., (4.36), we deduce

$$\mathrm{Res}_m^{k+1}(d(\oplus_a\omega_a)) = \oplus_a \mathrm{Res}_m^{k+1} d\omega_a = d(\mathrm{Res}_m^k(\oplus_a\omega_a)$$

which is exactly the statement of proposition 4.1. □

4.4.2 Global residues on the filtered complexes

We have a residue morphism on global sections, induced by (4.28)

$$\mathrm{Res}_m^k : \Gamma(V, W_m\Lambda_X^k\langle \log Q\rangle) \to \bigoplus_{a\in A}\Gamma\left(h_a^{-1}(V)\cap {Q_a}^{[m+q(a)]}, \mathcal{E}_{{Q_a}^{[m+q(a)]}}^{k-m-2q(a)}\right)$$

From proposition 4.1 we deduce a mapping

$$\begin{aligned}&\mathrm{RES}_m^k : W_mH^k(V, \Lambda_X^{\cdot}\langle \log Q\rangle) \to \\ &\bigoplus_{a\in A} H^{k-m-2q(a)}\left(h_a^{-1}(V)\cap {Q_a}^{[m+q(a)]}, \mathbb{C}\right)\end{aligned} \tag{4.37}$$

Using theorem 4.3, we prove exactly as in lemma 2.4 and theorem 2.3 of chapter 2

Lemma 4.5. *The kernel of* RES^k_m *is* $W_{m-1}H^k(V, \Lambda^{\cdot}_X\langle\log Q\rangle)$.

Theorem 4.4. *Let* $[\omega]$ *be a class in* $W_m H^k(V, \Lambda^{\cdot}_X\langle\log Q\rangle)$, *and* $l < m$. *Then* $[\omega] \in W_l H^k(V, \Lambda^{\cdot}_X\langle\log Q\rangle)$ *if and only if*

$$\mathrm{RES}^k_m[\omega] = \mathrm{RES}^k_{m-1}[\omega] = \cdots = \mathrm{RES}^k_{l+1}[\omega] = 0$$

4.5 Residues and mixed Hodge structure (singular case)

From now on we assume that all the manifolds X_a *of the hypercovering of* X *are compact Kähler manifolds.*

4.5.1 Shifted Hodge filtrations and residues

We come back to (4.26) which we rewrite as:

$$\begin{aligned}\mathrm{Res}^k_{a,m} \colon (h_a)_* \left({}^{(q(a))}W_m \mathcal{E}^k_{X_a}\langle\log Q_a\rangle(-q(a))\right) \to \\ \to (h_a)_* \mathcal{E}^k_{{Q_a}^{[m+q(a)]}}(-m - 2q(a))\end{aligned} \tag{4.38}$$

Notice that in the right hand side the shift by $-m - 2q(a)$ corresponds to a shift $-q(a)$ due to the rank of (X_a, Q_a) plus a shift $-m-q(a)$ due to the order of poles. So we define shifted Hodge filtrations for the ${Q_a}^{[r]}$ in the following way

$${}^{(-r)}F^p\mathcal{E}^k_{{Q_a}^{[r]}}(-r - q(a)) = F^{p-r}\mathcal{E}^{k-r-q(a)}_{{Q_a}^{[r]}} \tag{4.39}$$

$${}^{(-r)}\bar{F}^q\mathcal{E}^k_{{Q_a}^{[r]}}(-r - q(a)) = {}^{(-r)}\bar{F}^{q-r}\mathcal{E}^{k-r-q(a)}_{{Q_a}^{[r]}} \tag{4.40}$$

Then, we have a shifted Hodge decomposition for the cohomology of ${Q_a}^{[m+q(a)]}$

$$\begin{aligned} &H^{k-m-2q(a)}({Q_a}^{[m+q(a)]}, \mathbb{C}) = \\ &= \bigoplus_{p+q=k+m} {}^{(-m-q(a))}F^p\, {}^{(-m-q(a))}\bar{F}^q H^{k-m-2q(a)}({Q_a}^{[m+q(a)]}, \mathbb{C})\end{aligned} \tag{4.41}$$

Finally, one can define shifted Hodge filtrations ${}^{(-m)}F^p$ and ${}^{(-m)}\bar{F}^q$ on the residue complex (4.31) by

$$\begin{aligned}&{}^{(-m)}F^p\mathrm{Res}^k_m\Lambda^{\cdot}_X\langle\log Q\rangle = \\ &= \bigoplus_{a\in A}(h_a)_*^{(-m-q(a))}F^p\mathcal{E}^k_{{Q_a}^{[m+q(a)]}}(-m - 2q(a))\end{aligned} \tag{4.42}$$

$$\begin{aligned}&{}^{(-m)}\bar{F}^q\mathrm{Res}^k_m\Lambda^{\cdot}_X\langle\log Q\rangle = \\ &= \bigoplus_{a\in A}(h_a)_*^{(-m-q(a))}\bar{F}^q\mathcal{E}^k_{{Q_a}^{[m+q(a)]}}(-m - 2q(a))\end{aligned} \tag{4.43}$$

Because of corollary 4.1 the cohomology of the complex of global sections of (4.31) is the direct sum of the $H^{k-m-2q(a)}({Q_a}^{[m+q(a)]}, \mathbb{C})$ and thus, using (4.41) we see that it carries a pure Hodge structure of weight $k+m$:

$$H^k(X, \mathrm{Res}^{\cdot}_m \Lambda_X\langle \log Q\rangle) = \bigoplus_a H^{k-m-2q(a)}({Q_a}^{[m+q(a)]}, \mathbb{C})$$

$$= \bigoplus_{p+q=k+m} {}^{(-m)}F^p \; {}^{(-m)}\bar{F}^q H^k(X, \mathrm{Res}^{\cdot}_m \Lambda_X\langle \log Q\rangle) \qquad (4.44)$$

where we have defined

$${}^{(-m)}F^p \; {}^{(-m)}\bar{F}^q H^k(X, \mathrm{Res}^{\cdot}_m \Lambda_X\langle \log Q\rangle) =$$
$$= \bigoplus_{a\in A} {}^{(-m-q(a))}F^p \; {}^{(-m-q(a))}\bar{F}^q H^{k-m-2q(a)}({Q_a}^{[m+q(a)]}, \mathbb{C}) \qquad (4.45)$$

We also define Hodge filtrations on complexes of forms with poles, namely:

$${}^{(-m-q(a))}F^p \; {}^{(q(a))}\mathcal{E}^k_{X_a}\langle \log Q_a\rangle(-q(a)) =$$
$$= F^p \; {}^{(q(a))}\mathcal{E}^k_{X_a}\langle \log Q_a\rangle(-q(a)) \qquad (4.46)$$

$${}^{(-m-q(a))}\bar{F}^q \; {}^{(q(a))}\mathcal{E}^k_{X_a}\langle \log Q_a\rangle(-q(a)) =$$
$$= \bar{F}^{q-m-q(a)} \; {}^{(q(a))}\mathcal{E}^k_{X_a}\langle \log Q_a\rangle(-q(a)) \qquad (4.47)$$

and then, we define Hodge filtrations on the logarithmic complexes:

$${}^{(-m)}F^p\Lambda^k_X\langle \log Q\rangle = \bigoplus_{a\in A}(h_a)_* F^p \; {}^{(q(a))}\mathcal{E}^k_{X_a}\langle \log Q_a\rangle(-q(a)) \qquad (4.48)$$

$${}^{(-m)}\bar{F}^q\Lambda^k_X\langle \log Q\rangle = \bigoplus_{a\in A}(h_a)_* \bar{F}^{q-m-q(a)} \, {}^{(q(a))}\mathcal{E}^k_{X_a}\langle \log Q_a\rangle(-q(a)) \qquad (4.49)$$

Remark. The filtrations defined in (4.48), (4.49) are not conjugate. Moreover the role of ${}^{(-m)}F^a$, ${}^{(-m)}\bar{F}^b$ is not symmetric; in particular ${}^{(-m)}F^a$ does not depend on m, so it is a true filtration on the complex $\Lambda^{\cdot}_X\langle \log Q\rangle$, while ${}^{(-m)}\bar{F}^b$ depends on m and is adapted to the subcomplex $W_m\Lambda^{\cdot}_X\langle \log Q\rangle$ and to the quotient complex $\frac{W_m\Lambda^{\cdot}_X\langle \log Q\rangle}{W_{m-1}\Lambda^{\cdot}_X\langle \log Q\rangle}$.

Lemma 4.6.

1) *The filtrations defined in* (4.42), (4.43), (4.46), (4.47), (4.48), (4.49) *are preserved by the corresponding differentials. Hence they induce filtrations on the cohomology* $H^k(X \setminus Q, \mathbb{C})$ *under the isomorphism* (4.8) *of theorem 4.1.*

2) *Let us suppose that every* X_a *in the hypercovering of* X *is a compact Kähler manifold. Then the differential* d_0 *of the graded complex:*

$$\frac{\Gamma(X, W_m\Lambda^{\cdot}_X\langle \log Q\rangle)}{\Gamma(X, W_{m-1}\Lambda^{\cdot}_X\langle \log Q\rangle)} = \Gamma\left(X, \frac{W_m\Lambda^{\cdot}_X\langle \log Q\rangle}{W_{m-1}\Lambda^{\cdot}_X\langle \log Q\rangle}\right)$$

is a strict morphism for the filtration F.

Proof. Property 1) is a consequence of the formula (4.10).
By (4.21) we have

$$\Gamma\left(X, \frac{W_m\Lambda^{\cdot}_X\langle\log Q\rangle}{W_{m-1}\Lambda^{\cdot}_X\langle\log Q\rangle}\right) = \bigoplus_{a\in A}\Gamma\left(X_a, \frac{W_{m+q(a)}\mathcal{E}^{\cdot}_{X_a}\langle\log Q_a\rangle(-q(a))}{W_{m+q(a)-1}\mathcal{E}^{\cdot}_{X_a}\langle\log Q_a\rangle(-q(a))}\right)$$

hence by lemma 4.3, formula (4.22) the differential d_0 is a direct sum of differentials d_0 for the pairs (X_a, Q_a), where X_a is smooth and Q_a is a divisor with normal crossings, which are strict by theorem 2.8 in chapter 2; property 2) follows. □

Moreover

Lemma 4.7. *For any m, the residues induce morphisms of filtered spaces for the shifted Hodge filtrations*

$$\mathrm{Res}^{\cdot}_m\colon\ {}^{(-m)}F^pW_m\Lambda^{\cdot}_X\langle\log Q\rangle \to {}^{(-m)}F^p\mathrm{Res}^{\cdot}_m\Lambda_X\langle\log Q\rangle$$

$$\mathrm{Res}^{\cdot}_m\colon\ {}^{(-m)}\bar{F}^qW_m\Lambda^{\cdot}_X\langle\log Q\rangle \to {}^{(-m)}\bar{F}^q\mathrm{Res}^{\cdot}_m\Lambda_X\langle\log Q\rangle$$

as well as on the graded complexes for W_m

$$\mathrm{Res}^{\cdot}_m\colon\ {}^{(-m)}F^p\left(\frac{W_m\Lambda^{\cdot}_X\langle\log Q\rangle}{W_{m-1}\Lambda^{\cdot}_X\langle\log Q\rangle}\right) \to {}^{(-m)}F^p\mathrm{Res}^{\cdot}_m\Lambda_X\langle\log Q\rangle \quad (4.50)$$

$$\mathrm{Res}^{\cdot}_m\colon\ {}^{(-m)}\bar{F}^q\left(\frac{W_m\Lambda^{\cdot}_X\langle\log Q\rangle}{W_{m-1}\Lambda^{\cdot}_X\langle\log Q\rangle}\right) \to {}^{(-m)}\bar{F}^q\mathrm{Res}^{\cdot}_m\Lambda_X\langle\log Q\rangle \quad (4.51)$$

Proof. The residues are defined as direct sums of the residues $\mathrm{Res}^k_{a,m}$. It is clear that for each a:

$$\mathrm{Res}^k_{a,m}\colon (h_a)_*F^p\,{}^{(q(a))}W_m\mathcal{E}^k_{X_a}\langle\log Q_a\rangle(-q(a)) \to (h_a)_*F^{p-m-q(a)}\mathcal{E}^{k-m-2q(a)}_{Q_a{}^{[m+q(a)]}}$$

$$\mathrm{Res}^k_{a,m}\colon (h_a)_*\bar{F}^{q-m-q(a)}\,{}^{(q(a))}W_m\mathcal{E}^k_{X_a}\langle\log Q_a\rangle(-q(a)) \to (h_a)_*\bar{F}^{q-m-q(a)}\mathcal{E}^{k-m-2q(a)}_{Q_a{}^{[m+q(a)]}}$$

Indeed, if $m+q(a)>0$, $\mathrm{Res}^k_{a,m}$ is an actual residue, and the result follows from chapter 2. If $m+q(a)=0$, $\mathrm{Res}^k_{a,m}$ is the identity, so the result is trivial, and if $m+q(a)<0$, $\mathrm{Res}^k_{a,m}=0$.

Using the definitions (4.48) and (4.49) of the filtrations on $\Lambda^{\cdot}_X\langle\log Q\rangle$, and the definition (4.44) and (4.45) of the filtrations on the residue complex $\mathrm{Res}^{\cdot}_m\Lambda_X\langle\log Q\rangle$ we deduce easily the mappings (4.50), (4.51). □

4.5.2 Pure Hodge structure on $E_1^{m,k}$

Lemma 4.8.

1) The first term $E_1^{m,k}(X)$ of the spectral sequence has a pure Hodge structure of weight $k+m$ induced by the shifted filtrations ${}^{(-m)}F$, ${}^{(-m)}\bar{F}$

$$E_1^{m,k}(X) = \bigoplus_{p+q=k+m} {}^{(-m)}F^p\, {}^{(-m)}\bar{F}^q E_1^{m,k}(X) \qquad (4.52)$$

2) The residue defined as in (4.30)

$$\begin{aligned}\underline{\mathrm{Res}}_m^k : E_1^{m,k}(X) = H^k\left(X, \frac{W_m\Lambda_X^{\cdot}\langle\log Q\rangle}{W_{m-1}\Lambda_X^{\cdot}\langle\log Q\rangle}\right) \to \\ H^k\left(X, \mathrm{Res}_m^{\cdot}\Lambda_X\langle\log Q\rangle\right) = \bigoplus_a H^{k-m-2q(a)}({Q_a}^{[m+q(a)]}, \mathbb{C})\end{aligned} \qquad (4.53)$$

is an isomorphism of pure Hodge structures induced by the shifted Hodge filtrations ${}^{(-m)}F$, ${}^{(-m)}\bar{F}$.

Proof. First by lemma 4.7, we see that Res_m^k induces the isomorphism $\underline{\mathrm{Res}}_m^k$ defined by (4.53) at the level of the cohomologies, which is a morphism of filtered spaces for the filtrations ${}^{(-m)}F$, ${}^{(-m)}\bar{F}$. So we have morphisms

$$\begin{aligned}\underline{\mathrm{Res}}_m^k : {}^{(-m)}F^p\, {}^{(-m)}\bar{F}^q E_1^{m,k}(X) \to \\ {}^{(-m)}F^p\, {}^{(-m)}\bar{F}^q H^k\left(X, \mathrm{Res}_m^{\cdot}\Lambda_X\langle\log Q\rangle\right)\end{aligned} \qquad (4.54)$$

which are injective (because by corollary 4.1 $\underline{\mathrm{Res}}_m^k$ is an isomorphism of vector spaces between $E_1^{m,k}(X)$ and $H^k\left(X, \mathrm{Res}_m^{\cdot}\Lambda_X\langle\log Q\rangle\right)$. But Res_m^k is the direct sum of $\mathrm{Res}_{a,m}^k$ for all a and by formula (4.24)

$$E_1^{m,k}(X) = \bigoplus_a E_1^{m+q(a),k-q(a)}(X_a)$$

Using definition (4.48) and (4.49) for the shifted Hodge filtrations, we have

$$\begin{aligned}&{}^{(-m)}F^p\, {}^{(-m)}\bar{F}^q E_1^{m,k}(X) = \\ &= \bigoplus_a {}^{(-m-q(a))}F^p\, {}^{(-m-q(a))}\bar{F}^q E_1^{m+q(a),k-q(a)}(X_a)\end{aligned} \qquad (4.55)$$

and we know that

$$\begin{aligned}&\underline{\mathrm{Res}}_{a,m}^k : {}^{(-m-q(a))}F^p\, {}^{(-m-q(a))}\bar{F}^q E_1^{m+q(a),k-q(a)}(X_a) \to \\ &\to {}^{(-m-q(a))}F^p\, {}^{(-m-q(a))}\bar{F}^q H^{k-m-2q(a)}({Q_a}^{[m+q(a)]}, \mathbb{C})\end{aligned}$$

is an isomorphism because of the results of chapter 2 (proposition 2.3) applied to the manifold X_a and the divisor with normal crossing Q_a. But using (4.45) and (4.55), we see that $\underline{\mathrm{Res}}_m^k$ in (4.54) is an isomorphism of each graded space. So the filtrations ${}^{(-m)}F$, ${}^{(-m)}\bar{F}$, induce on $E_1^{m,k}(X)$ a pure Hodge structure isomorphic to the pure Hodge structure defined by (4.44) on $H^k(X, \mathrm{Res}_m^{\cdot}\Lambda_X\langle\log Q\rangle)$, the isomorphism being $\underline{\mathrm{Res}}_m^k$. □

4.5.3 The differential d_1

As usual, one defines

$$d_1 \colon E_1^{m,k}(X) \to E_1^{m-1,k+1}(X)$$

in the following manner. If $[[\pi]]_1$ is an element of $E_1^{m,k}(X)$, one has a collection $\pi = \oplus_{a \in A} \pi_a$ with

$$\oplus_a \pi_a \in {}^{(q(a))}W_m \mathcal{E}^{k}_{X_a} \langle \log Q_{\dot{a}} \rangle (-q(a))$$

and

$$d(\oplus_a \pi_a) \in {}^{(q(a))}W_{m-1} \mathcal{E}^k_{X_a} \langle \log Q_a \rangle (-q(a))$$

Then:

$$d_1([[\pi]]_1) = [[d(\oplus_a \pi_a)]]_1 \in E_1^{m-1,k+1}(X)$$

Lemma 4.9.

1) The differential d_1 is a morphism of pure Hodge structures defined by (4.52).

2) We define $\hat{d}_1$ as the differential d_1 at the level of residues, namely, such that the following diagram is commutative

$$\begin{array}{ccc} E_1^{m,k}(X) & \xrightarrow{\quad d_1 \quad} & E_1^{m-1,k+1}(X) \\ \Big\downarrow {\scriptstyle \underline{\mathrm{Res}}_m^k} & & \Big\downarrow {\scriptstyle \underline{\mathrm{Res}}_{m-1}^{k+1}} \\ H^k\left(X, \mathrm{Res}_m^{\cdot} \Lambda_X \langle \log Q \rangle\right) & \xrightarrow{\hat{d}_1} & H^{k+1}\left(X, \mathrm{Res}_{m-1}^{\cdot} \Lambda_X \langle \log Q \rangle\right) \end{array}$$

Then, $\hat{d}_1$ is a morphism of pure Hodge structures for the shifted filtrations ${}^{(-m)}F$, ${}^{(-m)}\bar{F}$ ((4.42), (4.44)) on $H^k\left(X, \mathrm{Res}_m^{\cdot} \Lambda_X \langle \log Q \rangle\right)$ and ${}^{(-m+1)}F$, ${}^{(-m+1)}\bar{F}$, on $H^{k+1}\left(X, \mathrm{Res}_{m-1}^{\cdot} \Lambda_X \langle \log Q \rangle\right)$ and is given by the formula

$$(\hat{d}_1[\alpha])_a = \gamma_a[\alpha_a] + \left[\underline{\mathrm{Res}}_{m+q(a)-1}^{k-q(a)+1} \sum_b \epsilon_{ab}^{(k)} h_{ab}^{*} \pi_b \right] \tag{4.56}$$

where we have denoted by γ_a the generalized Gysin morphism constructed in chapter 2 (proposition 2.3) for the pair (X_a, Q_a).

Proof.
Proof of 2). We consider an element $\alpha \in H^k\left(X, \mathrm{Res}_m^{\cdot} \Lambda_X \langle \log Q \rangle\right)$, $[\alpha] = \oplus_a [\alpha_a] \in \bigoplus_a H^{k-m-2q(a)}(Q_a{}^{[m+q(a)]}, \mathbb{C})$. Each α_a is itself a collection of d-closed forms

$$\alpha_{a,I} \in H^{k-m-2q(a)}(Q_{a,I}, \mathbb{C}),\ |I| = m + q(a)$$

For each $a \in A$, and thus for each pair (X_a, Q_a) of the hypercovering, we follow the construction of proposition 2.3 and lemma 2.7 of chapter 2. Namely, we construct

$$\pi_a \in W_{m+q(a)}\mathcal{E}^{k-q(a)}_{X_a}\langle \log Q_a\rangle$$

whose corresponding residue is α_a. Thus

$$\pi_a \in^{(q(a))} W_m\mathcal{E}^k_{X_a}\langle \log Q_a\rangle(-q(a))$$
$$\mathrm{Res}^k_{a,m}\pi_a = \alpha_a$$

Thus, each π_a induces an element

$$[[\pi_a]]_1 \in E_1^{m+q(a),k-q(a)}(X)$$

and thus the collection $\oplus_a \pi_a$ induces an element

$$[[\pi]]_1 = \oplus_a[[\pi_a]]_1 \in E_1^{m,k}(X) = \bigoplus_a E_1^{m+q(a),k-q(a)}(X)$$

(see (4.24)). Then, we define

$$\hat{d}_1[\alpha] = \underline{\mathrm{Res}}^{k+1}_{m-1} d_1[[\pi]]_1 \tag{4.57}$$

Now

$$d_1[[\pi]]_1 = [[d\pi]]_1$$

and from (4.10)

$$(d\pi)_a = d\pi_a + \sum_b \epsilon^{(k)}_{ab} h^*_{ab}\pi_b$$

where the sum is on indices $b \in A$ with

$$q(a) = q(b) + 1$$

From lemma 4.2, we know that this sum is in the space $W_{m-1}\Lambda^{\cdot}_X\langle \log Q\rangle$. Moreover, π_a defines an element in $E_1^{m+q(a),k-q(a)}(X)$ and

$$d_1[[\pi_a]]_1 = [[d\pi_a]]_1$$

So using the results of proposition 2.3 of chapter 2, in the case of the pair (X_a, Q_a), we have:

$$\underline{\mathrm{Res}}^{k-q(a)+1}_{m+q(a)-1} d_1[[\pi_a]]_1 = \hat{d}_1[\alpha_a] = \gamma_a[\alpha_a]$$

where we have denoted by γ_a is the generalized Gysin morphism for the pair (X_a, Q_a). In (4.57), the residue morphism is the direct sum of the residues in each component of $d_1[[\pi]]_1$, so that finally

$$(\hat{d}_1[\alpha])_a = \gamma_a[\alpha_a] + \left[\underline{\mathrm{Res}}^{k-q(a)+1}_{m+q(a)-1} \sum_b \epsilon^{(k)}_{ab} h^*_{ab}\pi_b\right] \tag{4.58}$$

It is clear from (4.58) that $\hat{d}_1$ is a morphism

$$H^k\left(X, \mathrm{Res}^{\cdot}_m \Lambda_X\langle \log Q\rangle\right) \to H^{k+1}\left(X, \mathrm{Res}^{\cdot}_{m-1}\Lambda_X\langle \log Q\rangle\right)$$

which is a morphism of filtered spaces for the filtrations ${}^{(-m)}F$, ${}^{(-m)}\bar{F}$, and ${}^{(-m+1)}F$, ${}^{(-m+1)}\bar{F}$, respectively. This is a consequence of (4.45), defining these filtrations, as shifted standard Hodge filtration on the cohomologies of the ${Q_a}^{[l]}$, from the fact that γ_a is a morphism of filtered spaces, and the fact that if $[\alpha]_a$ have given types, one can choose the π_a with a well defined type (as in lemma 1.5 of chapter 2), that the pullback respects types, and the residue also.

Proof of 1). $\hat{d}_1$ is a morphism of filtered spaces for the Hodge filtration and the residues are isomorphisms of filtered spaces. So d_1 is a morphism of filtered spaces also and thus it is a morphism of pure Hodge structures. □

4.5.4 The conjugate complex

The complex $\Lambda^{\cdot}_X\langle \log Q\rangle$ is not closed under conjugation. We can define *the conjugate complex* $\Lambda^{\cdot}_X\langle\overline{\log Q}\rangle$ by

$$\Lambda^{\cdot}_X\langle\overline{\log Q}\rangle = \mathcal{E}^{\cdot}_{\tilde{X}}\langle\overline{\log \tilde{Q}}\rangle \oplus \Lambda^{\cdot}_E\langle\overline{\log M}\rangle \oplus \Lambda^{\cdot}_{\tilde{E}}\langle\overline{\log \tilde{M}}\rangle(-1)$$

where $\mathcal{E}^{\cdot}_{\tilde{X}}\langle\overline{\log \tilde{Q}}\rangle$ is the conjugate of $\mathcal{E}^{\cdot}_{\tilde{X}}\langle\log \tilde{Q}\rangle$, and $\Lambda^{\cdot}_E\langle\overline{\log M}\rangle$, $\Lambda^{\cdot}_{\tilde{E}}\langle\overline{\log \tilde{M}}\rangle$ can be defined by induction on the dimension of X. Also, by (4.9)

$$\Lambda^k_X\langle\overline{\log Q}\rangle = \bigoplus_{a\in A}(h_a)_*\mathcal{E}^{k-q(a)}_{X_a}\langle\overline{\log Q_a}\rangle \tag{4.59}$$

There is a conjugation for logarithmic forms, such that $\omega \in \Lambda^k_X\langle\log Q\rangle$ if and only if $\overline{\omega} \in \Lambda^k_X\langle\overline{\log Q}\rangle$.

Moreover

$$\overline{d\omega} = d\,\overline{\omega} \tag{4.60}$$

as it follows from the formula (4.10) for d and the property that the pullback h^*_{ab} commute to conjugation.

On the conjugate complex we can also define:

- The conjugate weight filtration $\overline{W}$:

$$\omega \in \overline{W}_m\Lambda^k_X\langle\overline{\log Q}\rangle \iff \overline{\omega} \in W_m\Lambda^k_X\langle\log Q\rangle$$

- The conjugate residues $\overline{\mathrm{Res}^k_m}$:

$$\overline{\mathrm{Res}^k_m} : \overline{W}_m\Lambda^k_X\langle\overline{\log Q}\rangle \to \bigoplus_{a\in A}(h_a)_*\mathcal{E}^{k-m-2q(a)}_{{Q_a}^{[m+q(a)]}} = \mathrm{Res}^k_m\Lambda_X\langle\log Q\rangle \tag{4.61}$$

i.e. the conjugate residues have the same target $\mathrm{Res}^k_m\Lambda_X\langle\log Q\rangle$ as the residues Res^k_m.

Using (4.59), (4.60), (4.61), it is easy to express the results in lemma 4.4, theorem 4.3, corollary 4.1 and proposition 4.1, replacing $W_m\Lambda^k_X\langle\log Q\rangle$ with $\overline{W}_m\overline{\Lambda}^k_X\langle\overline{\log Q}\rangle$. In particular let us note the following immediate consequence.

Proposition 4.2. *The spectral sequence with respect to the filtration* $\overline{W}_m$ *of the conjugate complex* $\Gamma(X, \Lambda^k_X\langle\overline{\log Q}\rangle)$ *coincides, for* $r \geq 1$, *with the spectral sequence* $E_r^{m,k}(X)$ *with respect to the filtration* W_m *of the complex* $\Gamma(X, \Lambda^k_X\langle\log Q\rangle)$. *The map* $\omega \to \overline{\omega}$ *from* $\Gamma(X, \Lambda^k_X\langle\log Q\rangle)$ *to* $\Gamma(X, \Lambda^k_X\langle\overline{\log Q}\rangle)$ *induces a conjugation on each term* $E_r^{m,k}(X)$, $r \geq 1$; *moreover the differentials* d_r *commute to conjugations.*

The shifted Hodge filtrations on $\Lambda^k_X\langle\overline{\log Q}\rangle$ are defined, following (4.39), (4.40):

$$\begin{aligned} {}^{(-r)}F^b\Lambda^k_X\langle\overline{\log Q}\rangle &= F^{b-r}\Lambda^k_X\langle\overline{\log Q}\rangle \\ {}^{(-r)}\bar{F}^a\Lambda^k_X\langle\overline{\log Q}\rangle &= \bar{F}^a\Lambda^k_X\langle\overline{\log Q}\rangle \end{aligned} \tag{4.62}$$

Then it is clear that

Proposition 4.3. *The conjugation* $\omega \to \overline{\omega}$ *transforms* ${}^{(-r)}F^a\Lambda^{\cdot}_X\langle\log Q\rangle$ *to* ${}^{(-r)}\bar{F}^a\Lambda^{\cdot}_X\langle\overline{\log Q}\rangle$ *and* ${}^{(-r)}\bar{F}^b\Lambda^{\cdot}_X\langle\log Q\rangle$ *to* ${}^{(-r)}F^b\Lambda^{\cdot}_X\langle\overline{\log Q}\rangle$.

4.6 Degeneration of the spectral sequence

We recall that we are supposing that every X_a *is a compact Kähler manifold.* On each term $E_r^{m,k}(X)$ of the spectral sequence of the filtration W_m there are two kinds of filtrations:

- The *direct filtrations* ${}^{(-m)}F_1$, ${}^{(-m)}\bar{F}_1$, induced by the filtrations ${}^{(-m)}F$ and ${}^{(-m)}\bar{F}$ of the complex $\Gamma(X, \Lambda^{\cdot}_X\langle\log Q\rangle)$ (see (4.48) and (4.49)) precisely here $E_r^{m,k}(X)$ must be considered as a quotient of the subspace $Z_r^{m,k} \subset \Gamma(X, W_m\Lambda^k_X\langle\log Q\rangle)$.

- The *recursive filtrations* ${}^{(-m)}F_2$, ${}^{(-m)}\bar{F}_2$ induced recursively on $E_r^{m,k}(X)$ considered as the cohomology of the complex $(E_{r-1}^{m,k}(X), d_{r-1})$.

It is clear that on $E_1^{m,k}(X)$ the filtration ${}^{(-m)}F_1$ (resp. ${}^{(-m)}\bar{F}_1$) is identical to the filtration ${}^{(-m)}F_2$ (resp. ${}^{(-m)}\bar{F}_2$). Moreover it is easy to prove that $F_1 \subset F_2$, $\bar{F}_1 \subset \bar{F}_2$.

Lemma 4.10. *The filtrations* ${}^{(-m)}F_1$ *and* ${}^{(-m)}\bar{F}_1$, *as well as* ${}^{(-m)}F_2$ *and* ${}^{(-m)}\bar{F}_2$, *are conjugate.*

The statement for ${}^{(-m)}F_1$, ${}^{(-m)}\bar{F}_1$ follows from the propositions 4.2 and 4.3. For ${}^{(-m)}F_2$ and ${}^{(-m)}\bar{F}_2$ it is enough to prove it in the case of $E_1^{m,k}(X)$, where it is obvious because $F_1 = F_2$, $\bar{F}_1 = \bar{F}_2$ in $E_1^{m,k}(X)$.

Theorem 4.5. *We suppose that all the manifolds X_a of the hypercovering of X are compact Kähler manifolds.*

1) The differentials d_r of the spectral sequence are 0 for $r \geq 2$.

2) d_r is a morphism of the filtrations ${}^{(-m)}F_1$ and ${}^{(-m)}\bar{F}_1$:

$$d_r\colon {}^{(-m)}F_1^a E_r^{m,k}(X) \to {}^{(-m+r)} F_1^a E_r^{m-r,k+1}(X) \tag{4.63}$$
$$d_r\colon {}^{(-m)}\bar{F}_1^b E_r^{m,k}(X) \to {}^{(-m+r)} \bar{F}_1^b E_2^{m-r,k+1}(X) \tag{4.64}$$

3) The filtrations ${}^{(-m)}F_2$, ${}^{(-m)}\bar{F}_2$ induce pure Hodge structures of weight $k+m$ on $E_r^{m,k}(X)$

$$E_r^{m,k}(X) = \bigoplus_{a+b=k+m} {}^{(-m)}F_2^a\ {}^{(-m)}\bar{F}_2^b E_r^{m,k}(X) \tag{4.65}$$

4) On $E_r^{m,k}(X)$ the filtration ${}^{(-m)}F_1$ (resp. ${}^{(-m)}\bar{F}_1$) is identical to the filtration ${}^{(-m)}F_2$ (resp. ${}^{(-m)}\bar{F}_2$).

Proof. The property 2) is clear, since the differential d preserves the filtrations ${}^{(-m)}F$ and ${}^{(-m)}\bar{F}$ of the complex $\Lambda_X^{\cdot}\langle \log Q\rangle$ (lemma 4.6, 1). We prove 1), 3), 4) by induction on r; for $r = 1$ the property 3) is included in lemma 4.8, and the properties 1) and 4) are trivial. The property 3) for r is a consequence of 1), 2), 3) and 4) for $r-1$: d_{r-1} preserves ${}^{(-m)}F_2 = {}^{(-m)}F_1$ and ${}^{(-m)}\bar{F}_2 = {}^{(-m)}\bar{F}_1$ on $E_{r-1}^{m,k}(X)$ so that its cohomology $E_r^{m,k}(X)$ carries a pure Hodge structure of weight $k+m$ as in (4.65).

Then we note that 2), 3), 4) imply 1). In fact if we know that the direct and the recursive filtrations coincide, by (4.63) and (4.64) $d_r\colon E_r^{m,k}(X) \to E_r^{m-r,k+1}(X)$ become a morphism of pure Hodge structures of respective weights $k+m$, $k+m-r+1$, which are different for $r \geq 2$, so that $d_r = 0$ in that case (proposition 1.4 of part I, chapter 1).

Finally, let us suppose that 1), 3), 4) are true for $s \leq r$ and let us prove 4) for $r+1$ (≥ 2).

By lemma 4.10 it is sufficient to treat the case of ${}^{(-m)}F_1$ and ${}^{(-m)}F_2$. We already know $F_1 \subset F_2$, so we need to show $F_2 \subset F_1$. We shall denote by ${}^{(-m)}F$, the filtrations ${}^{(-m)}F_1$, ${}^{(-m)}F_2$ when they coincide. For simplicity we write $L^k = \Gamma(X, \Lambda_X^k\langle \log Q\rangle)$.

Let $\alpha \in F_2^a E_{r+1}^{m,k}$; there is a representative in $F^a E_r^{m,k} \cap Ker d_r$, i.e. an element $x \in F^a Z_r^{m,k} \subset F^a W_m L^k$ with $d_r[x]_r = 0$; we have $dx \in W_{m-r}L^{k+1} \cap F^a W_m L^{k+1}$ so that

$$dx \in F^a W_{m-r} L^{k+1} \tag{4.66}$$

$d_r[x]_r = 0$ gives

$$dx = dx_1 + z_1, \quad x_1 \in Z_{r-1}^{m-1,k}, z_1 \in Z_{r-1}^{m-r-1,k+1} \tag{4.67}$$

Then successively we get by 1) $d_{r-1}[x_1]_{r-1} = 0$ so that

$$dx_1 = dx_2 + z_2, \quad x_2 \in Z_{r-2}^{m-2,k}, z_2 \in Z_{r-2}^{m-r-1,k+1} \tag{4.68}$$

$$\dots$$

$$dx_{r-2} = dx_{r-1} + z_{r-1}, \quad x_{r-1} \in Z_1^{m-r+1,k}, z_{r-1} \in Z_1^{m-r-1,k+1} \tag{4.69}$$

From the above equalities we find that

$$dx_{r-1} = dx \qquad \mathrm{mod}\ Z_0^{m-r-1,k+1} = W_{m-r-1}L^{k+1} \tag{4.70}$$

Since $x_{r-1} \in Z_1^{m-r+1,k}$, we can compute $d_1[x_{r-1}]_1$. It follows from (4.70) and (4.66) that $d_1[x_{r-1}]_1 \in F^a E_1^{m-r,k}$. But d_1 is a morphism of pure Hodge structures (lemma 4.9), hence it is strict for F. Therefore there exists $x'_{r-1} \in F^a Z_1^{m-r+1,k}$ with $d_1[x'_{r-1}]_1 = d_1[x_{r-1}]_1$, that is

$$dx_{r-1} = dx'_{r-1} + dy + z', \quad y \in Z_0^{m-r,k}, z' \in Z_0^{m-r-1,k+1} \tag{4.71}$$

Note that

$$dx'_{r-1} \in F^a W_{m-r+1} L^{k+1} \cap W_{m-r} L^{k+1} \subset F^a W_{m-r} L^{k+1} \tag{4.72}$$

We obtain an element $[y]_0 \in E_0^{m-r,k}$, and from (4.70), (4.71), (4.72) we get $d_0[y]_0 \in F^a E_0^{m-r,k}$. Since d_0 is strict by lemma 4.6, we find $y' \in F^a W_{m-r} L^k$ with $d_0[y]_0 = d_0[y']_0$ or

$$dy = dy' + z'', \quad z'' \in W_{m-r-1} L^{k+1} \tag{4.73}$$

From (4.66)...(4.73) we obtain

$$x - x'_{r-1} - y' \in F^a W_m L^k$$

and

$$d(x - x'_{r-1} - y') \in W_{m-r-1} L^{k+1}$$

so that $x - x'_{r-1} - y' \in F^a Z_{r+1}^{m,k}$ and its class in $E_{r+1}^{m,k}$ is α. This proves 4) for $r+1$. □

It follows that there are natural isomorphisms

$$E_2^{m,k} \simeq \frac{W_m H^k(X \setminus Q, \mathbb{C})}{W_{m-1} H^k(X \setminus Q, \mathbb{C})} \tag{4.74}$$

The cohomology space $H^k(X \setminus Q, \mathbb{C})$ carries a Hodge filtration F_c given by (4.48) and lemma 4.6,1). More precisely

$$F_c^p H^k(X \setminus Q, \mathbb{C}) = F^p H^k(X, \Lambda_X^{\cdot}\langle \log Q \rangle) \tag{4.75}$$

Theorem 4.6. *Under the same assumptions as in theorem 4.5, let F_d be the direct (or the recursive) filtration on $E_2^{m,k}$, and F_c be the filtration induced on $E_2^{m,k}$, under the isomorphism (4.74), by the filtration F_c given by (4.75). Then $F_d = F_c$.*

Proof. As in the proof of the previous theorem, let $L^k = \Gamma(X, \Lambda_X^k\langle \log Q\rangle)$. The isomorphism (4.74) means the following. Let $x \in Z_2^{m,k}$; then there exists $x' \in W_m L^k$ with $dx' = 0$ and

$$x' = x + d\tilde{x} + z, \quad \tilde{x} \in Z_1^{m-1,k-1}, \quad z \in Z_1^{m-1,k}$$

or

$$x' \equiv x + z$$

where the symbol $\equiv$ means cohomologous.

Let $\alpha \in F_c^p E_2^{m,k}$; there exists $x \in F^p W_m L^k$ with $dx = 0$, inducing α in $E_2^{m,k} \simeq \frac{W_m H^k(X\setminus Q, \mathbb{C})}{W_{m-1} H^k(X\setminus Q, \mathbb{C})}$. It is clear that $x \in Z_2^{m,k}$, so that $x \in F^p W_m L^k \cap Z_2^{m,k} = F^p Z_2^{m,k}$. Hence $\alpha \in F_d^p E_2^{m,k}$. This proves $F_c \subset F_d$.

Conversely, if $\alpha \in F_d^p E_2^{m,k}$ there exists $x \in F^p Z_2^{m,k}$ inducing α. By the isomorphism (4.74) there exists $x' \in W_m L^k$ with $dx' = 0$ and

$$x' \equiv x + z, \quad z \in Z_1^{m-1,k} \tag{4.76}$$

let $[z]_1 \in E_1^{m-1,k}$ be the class of z. By (4.76) we have

$$dz = -dx \in F^p W_m L^{k+1}$$

and since $dz \in W_{m-2} L^{k+1}$ we obtain

$$dz \in W_{m-2} L^{k+1} \cap F^p W_m L^{k+1} \subset F^p W_{m-2} L^{k+1}$$

Thus $d_1[z]_1 \in F^p E_1^{m-2,k+1}$. Since d_1 is a strict morphism for F, we find $x_1 \in F^p Z_1^{m-1,k}$ with $d_1[z]_1 = d_1[x]_1$, or $d_1[z - x_1]_1 = 0$, that is, $z - x_1$ gives a class in $E_2^{m-1,k}$. By the isomorphism (4.74) for $m-1$, there is $x'' \in W_{m-1} L^k$, $dx'' = 0$, with $x'' \equiv z - x_1 - z_1$, $\quad z_1 \in Z_1^{m-2,k}$, or

$$z \equiv x'' + x_1 + z_1$$

and by (4.76)

$$x' \equiv x + x'' + x_1 + z_1$$

The cohomology class of x'' belongs to $W_{m-1} H^k$ (where $H^k = H^k(X \setminus Q, \mathbb{C})$), so that it vanishes in the quotient (4.74); thus we write

$$x' \equiv (x + x_1) + z_1 \qquad \text{mod } W_{m-1} H^k$$

We note that $x + x_1 \in F^p W_m L^k$. We proceed as above, finding that $d_1[z_1]_1 \in F^p E_1^{m-3,k+1}$ and we obtain

$$x' \equiv (x + x_1 + x_2) + z_2 \qquad \text{mod } W_{m-1} H^k, \quad x_2 \in F^p Z_1^{m-2,k}, \; z_2 \in Z_1^{m-3,k}$$

Going on, because $W_s L^{\cdot} = 0$ for $s << 0$ we finally write

$$x' \equiv x + x_1 + x_2 + \cdots + x_s \qquad \mathrm{mod}\ W_{m-1}H^k$$

with $x_j \in F^p Z_1^{m-j,k}$ so that $x' \in F^p W_m L^k$ and x', through $W_m H^k$, induces α. This proves $F_d \subset F_c$. □

Therefore we have given a proof of the Deligne result:

Theorem 4.7. *Let us suppose that all the manifolds X_a of the hypercovering of X are compact Kähler manifolds. We provide the cohomology $H^k(X \setminus Q, \mathbb{C})$ with the weight filtration W shifted by $-k$, the Hodge filtration F induced by the complex of global sections of $\Lambda^{\cdot}_X\langle \log Q\rangle$, and the filtration $\bar{F}$ conjugate to F. Then $H^k(X \setminus Q, \mathbb{C})$ carries a mixed Hodge structure. More precisely, the quotients $\frac{W_m H^k(X \setminus Q, \mathbb{C})}{W_{m-1} H^k(X \setminus Q, \mathbb{C})}$ are isomorphic to $E_2^{m,k}(X)$ and thus have a pure Hodge structure of weight $k+m$. The filtration induced by F on $E_2^{m,k}(X)$ coincides with the direct and the recursive filtration.*

Remark. The shift of W by $-k$ is needed to normalize the weights in the quotients. In the shifted filtration $W'_m = W_{m-k}$ we obtain

$$\frac{W'_m H^k(X \setminus Q, \mathbb{C})}{W'_{m-1} H^k(X \setminus Q, \mathbb{C})} \simeq E_2^{m-k,k}(X)$$

that is, the quotient $\frac{W'_m H^k(X, \mathbb{C})}{W'_{m-1} H^k(X, \mathbb{C})}$ has weight m, as expected.

Theorem 4.8 (Uniqueness of the mixed Hodge structures). *Let X be a compact (B)-Kähler space, $Q \subset X$ a subspace, and $\Lambda^{\cdot,1}_X\langle \log Q\rangle$, $\Lambda^{\cdot,2}_X\langle \log Q\rangle$ $\in \mathcal{R}(X\langle \log Q\rangle$, such that the associated hypercoverings are Kähler. Then $\Lambda^{\cdot,1}_X\langle \log Q\rangle$ and $\Lambda^{\cdot,2}_X\langle \log Q\rangle$ induce identical mixed Hodge structures on the cohomology of $X \setminus Q$.*

Proof. By the property of filtering for Kähler hypercoverings (theorem 4.2 (iii)) there exists a third $\Lambda^{\cdot}_X\langle \log Q\rangle$ whose associated hypercovering is Kähler, and two pullback $\Lambda^{\cdot,1}_X\langle \log Q\rangle \to \Lambda^{\cdot}_X\langle \log Q\rangle$, $\Lambda^{\cdot,2}_X\langle \log Q\rangle \to \Lambda^{\cdot}_X\langle \log Q\rangle$ corresponding to the identity. Both pullback induce the identity in the cohomology $H^k(X \setminus Q, \mathbb{C})$. Since a morphism of mixed Hodge structures which is an isomorphism of vector spaces (in our case: the identity) is an isomorphism of mixed Hodge structures (proposition 1.5 of part I, chapter 1), we conclude that the mixed Hodge structures induced in cohomology by $\Lambda^{\cdot,1}_X\langle \log Q\rangle$ and $\Lambda^{\cdot,2}_X\langle \log Q\rangle$ coincide with the mixed Hodge structure induced by $\Lambda^{\cdot}_X\langle \log Q\rangle$, hence they are the same. □

Theorem 4.9 (Functoriality of the mixed Hodge structures). *Let $f\colon (X, Q) \to (Y, R)$ be a morphism of pairs, where X and Y are compact (B)-Kähler spaces, and $\Phi\colon \Lambda^{\cdot}_Y\langle \log R\rangle \to \Lambda^{\cdot}_X\langle \log Q\rangle$ a pullback. We suppose that the hypercoverings associated to $\Lambda^{\cdot}_Y\langle \log R\rangle$ and $\Lambda^{\cdot}_X\langle \log Q\rangle$ are Kähler. Then Φ induces on* *cohomology the natural pullback* $f^*\colon H^k(Y\setminus R,\mathbb{C})\to H^k(X\setminus Q,\mathbb{C})$*, and* f^* *is a morphism of mixed Hodge structures.*

Proof. We know that Φ induces on cohomology the pullback f^* and respects both the filtrations W_m and F^p. Hence also f^* respects the filtrations W_m and F^p. □

4.7 Strictness of d

The goal of the present section is to prove the

Theorem 4.10. *Let us suppose that all the manifolds* X_a *of the hypercovering associated to* $\Lambda^{\cdot}_X\langle\log Q\rangle$ *are compact Kähler manifolds. The differential*

$$d\colon \Gamma(X,\Lambda^k_X\langle\log Q\rangle)\to\Gamma(X,\Lambda^{k+1}_X\langle\log Q\rangle)$$

is strict for the Hodge filtration F *defined in* (4.48).

By theorem 1.1 of part I, chapter 1, the above theorem is equivalent to

Theorem 4.11. *Let* $L^{\cdot}=\Gamma(X,\Lambda^{\cdot}_X\langle\log Q\rangle)$*. The spectral sequence* $E_r^{m,k}(L^{\cdot},F)$ *degenerates at* E_1.

In order to prove the theorem 4.11 we need some preliminary results on spectral sequences.

Lemma 4.11. *Let* $L^{\cdot}$ *be a complex of* $\mathbb{C}$*-vector spaces provided with a filtration* W*. Let* $E_r^{m,k}$ *be the corresponding spectral sequence. We suppose that the terms* $E_1^{m,k}$ *are finite-dimensional. Then the terms* $E_r^{m,k}$ *are finite-dimensional for* $r\geq 1$ *and*

$$\sum_m \dim E_r^{m,k}\geq \dim H^k(L^{\cdot}) \tag{4.77}$$

Moreover the equalities in (4.77) *hold if and only if* $E_r^{m,k}=E_\infty^{m,k}$ *for all* (m,k).

Proof. Let r_0 be an integer (which always exists) such that $E_{r_0}^{m,k}=E_\infty^{m,k}$ for all (m,k); this is equivalent to

$$d_s\equiv 0\quad\text{for } s\geq r_0-1 \tag{4.78}$$

Then $E_{r_0}^{m,k}=\frac{W_mH^k(L^{\cdot})}{W_{m-1}H^k(L^{\cdot})}$, so that

$$\sum_m \dim E_{r_0}^{m,k}=\sum_m \dim\frac{W_mH^k(L^{\cdot})}{W_{m-1}H^k(L^{\cdot})}=\dim H^k(L^{\cdot})$$

If $r \geq r_0$, $E_r^{m,k} = E_{r_0}^{m,k}$ so that equality in (4.77) holds for the $E_r^{m,k}$.
Let $1 \leq r \leq r_0$. Since $E_{r+1}^{m,k}$ is a quotient of a subspace of $E_r^{m,k}$, it follows that $\dim E_r^{m,k} \geq \dim E_{r+1}^{m,k}$, and $\dim E_r^{m,k} = \dim E_{r+1}^{m,k}$ for all (m,k) if and only if $d_r \colon E_r^{m,k} \to E_r^{m-r,k+1}$ is 0 for all (m,k). From the above statements we find that

$$\sum_m \dim E_r^{m,k} \geq \sum_m \dim E_{r_0}^{m,k} = \dim H^k(L^\cdot)$$

and in the left the equality holds if and only if $d_r \equiv \cdots \equiv d_{r_0-1} \equiv 0$, i.e., after (4.78),

$$d_s \equiv 0 \quad \text{for } s \geq r$$

which means $E_r^{m,k} = E_\infty^{m,k}$ for all (m,k). □

Lemma 4.12. *Let $L^\cdot$ be a complex of $\mathbb{C}$-vector spaces provided with a filtration F, such that the differential d is strict for F. The natural morphism*

$$\mathrm{Gr}_{-p}^F(H^k(L^\cdot)) = \frac{F^pH^k(L^\cdot)}{F^{p+1}H^k(L^\cdot)} \to H^k(\mathrm{Gr}_{-p}^F L^\cdot) = H^k\left(\frac{F^pL^\cdot}{F^{p+1}L^\cdot}\right) \tag{4.79}$$

is an isomorphism.

Proof. It is easy to see that the morphism (4.79) is injective (with no strictness assumption).
A class α in $H^k\left(\frac{F^pL^\cdot}{F^{p+1}L^\cdot}\right)$ is represented by $y \in F^pL^k$, with $dy \in F^{p+1}L^{k+1}$. Since d is strict, there exists $z \in F^{p+1}L^k$ such that $dz = dy$. We obtain $x = y - z$ with $dx = 0$ which determines a class in $\frac{F^pH^k(L^\cdot)}{F^{p+1}H^k(L^\cdot)}$ whose image in $H^k\left(\frac{F^pL^\cdot}{F^{p+1}L^\cdot}\right)$ is α. □

Proposition 4.4. *Let $L^\cdot$ be a complex of $\mathbb{C}$-vector spaces provided with two filtrations F (decreasing) and W (increasing). Let $E_r^{m,k}(L^\cdot, W)$ be the spectral sequence of $L^\cdot$ with respect to W, and $E_r^{m,k}(\mathrm{Gr}_{-p}^F L^\cdot, W)$ the similar spectral sequence of the graded complex $\mathrm{Gr}_{-p}^F L^\cdot$. If $d_j \colon E_r^{m,k}(\mathrm{Gr}_{-p}^F L^\cdot, W) \to E_r^{m-j,k+1}(\mathrm{Gr}_{-p}^F L^\cdot, W)$ are strict for the recursive filtration induced by F, then there are natural isomorphisms of complexes*

$$\mathrm{Gr}_{-p}^F E_r^{m,\cdot}(L^\cdot, W) \to E_r^{m,\cdot}(\mathrm{Gr}_{-p}^F L^\cdot, W) \tag{4.80}$$

The proof is by induction on r. For $r = 0$ the two sides in (4.80) are respectively $\mathrm{Gr}_{-p}^F \mathrm{Gr}_m^W L^k$ and $\mathrm{Gr}_m^W \mathrm{Gr}_{-p}^F L^k$, which are isomorphic by Zassenhaus lemma (proposition 2.6 of chapter 2.)

Let us suppose that (4.80) holds for $r-1$; then we have the following identities, commuting with differentials:

$$\begin{gathered}\mathrm{Gr}_{-p}^F E_r^{m,k}(L^\cdot, W) = \mathrm{Gr}_{-p}^F H^k(E_{r-1}^{m,\cdot}(L^\cdot, W)) = (\text{by lemma 4.12}) \\ = H^k(\mathrm{Gr}_{-p}^F E_{r-1}^{m,\cdot}(L^\cdot, W)) = (\text{by induction}) = H^k(E_{r-1}^{m,\cdot}(\mathrm{Gr}_{-p}^F L^\cdot, W)) = \\ = E_r^{m,k}(\mathrm{Gr}_{-p}^F L^\cdot, W).\end{gathered}$$

Proof of theorem 4.11. Let W be the weight filtration on $L^k = \Gamma(X, \Lambda_X^k\langle\log Q\rangle)$ By lemma 4.11 we must show that $\dim E_1^{p,k}(L^\cdot, F)$ is finite and the equality

$$\sum_p \dim E_1^{p,k}(L^\cdot, F) = \dim H^k(L^\cdot) \tag{4.81}$$

holds.

Let us recall that

$$E_1^{p,k}(L^\cdot, F) = \sum_p \dim H^k(\mathrm{Gr}_{-p}^F L^\cdot) \tag{4.82}$$

We apply the inequality (4.77) to the spectral sequence $E_r^{m,k}(\mathrm{Gr}_{-p}^F L^\cdot, W)$ and $r = 2$.

By theorem 4.5 the spectral sequence $E_r^{m,k}(L^\cdot, W)$ degenerates at E_2. Since the spaces $E_1^{m,k}(L^\cdot, W)$ are finite dimensional, we can apply lemma 4.11 for $r = 2$ and get the equality

$$\sum_p \dim E_2^{m,k}(L^\cdot, W) = \dim H^k(L^\cdot) \tag{4.83}$$

By lemma 4.6, the differential $d_0\colon E_0^{m,k}(L^\cdot, W) \to E_0^{m,k+1}(L^\cdot, W)$ is strict; and $d_1\colon E_1^{m,k}(L^\cdot, W) \to E_1^{m,k+1}(L^\cdot, W)$, as a morphism of pure Hodge structures (lemma 4.9) is strict too. Thus we can apply the proposition 4.4 to $E_2^{m,k}(L^\cdot, W)$, obtaining an isomorphism

$$\mathrm{Gr}_{-p}^F E_r^{m,k}(L^\cdot, W) \simeq E_r^{m,k}(\mathrm{Gr}_{-p}^F L^\cdot, W) \ (r = 1, 2) \tag{4.84}$$

For $r = 1$ the formula (4.84) insures that $E_1^{m,k}(\mathrm{Gr}_{-p}^F L^\cdot, W)$ is finite dimensional, so that the lemma 4.11, formula (4.77), for the spectral sequence $E_r^{m,k}(\mathrm{Gr}_{-p}^F L^\cdot, W)$ and $r = 2$ gives the inequality

$$\sum_m \dim E_2^{m,k}(\mathrm{Gr}_{-p}^F L^\cdot, W) \geq \dim H^k(\mathrm{Gr}_{-p}^F L^\cdot) \tag{4.85}$$

Finally we use step by step (4.82), (4.85), (4.84) (for $r = 2$), (4.83) obtaining

$$\sum_p \dim E_1^{p,k}(L^\cdot, F) = \sum_p \dim H^k(\mathrm{Gr}_{-p}^F L^\cdot) \leq \sum_{p,m} \dim E_2^{m,k}(\mathrm{Gr}_{-p}^F L^\cdot, W) =$$
$$= \sum_{p,m} \dim \mathrm{Gr}_{-p}^F E_2^{m,k}(L^\cdot, W) = \sum_m E_2^{m,k}(L^\cdot, W) = \dim H^k(L^\cdot)$$

This, together with lemma 4.11, gives the expected equality (4.81). □

Under the assumptions of theorem 4.10 the differential d is strict also with respect to the filtration W_m, up to a shift by $+1$:

Theorem 4.12. *Let us suppose that all the manifolds X_a of the hypercovering associated to $\Lambda^{\cdot}_X\langle\log Q\rangle$ are compact Kähler manifolds. Let $\omega \in \Gamma(X, \Lambda^k_X\langle\log Q\rangle)$ such that $d\omega \in W_m\Gamma(X, \Lambda^{k+1}_X\langle\log Q\rangle)$; there exists $\theta \in W_{m+1}\Gamma(X, \Lambda^k_X\langle\log Q\rangle)$ with $d\theta = d\omega$.*

Proof. Let $L^k = \Gamma(X, \Lambda^k_X\langle\log Q\rangle)$, and

$$E_r^{m,k} = \frac{Z_r^{m,k}}{dZ_{r-1}^{m+r-1,k-1} + Z_{r-1}^{m-1,k}}$$

(with differentials d_r) be the spectral sequence associated to the filtration W_m on $L^{\cdot}$, which degenerates at E_2 by theorem 4.5. We define another increasing filtration W' by

$$W'_m L^k = Z_1^{m-k,k}$$

Since $dZ_1^{m-k,k} \subset Z_1^{m-k-1,k+1}$, the filtration W' is preserved by d.

Let $(E'^{m,k}_r, d'_r)$ be the spectral sequence associated to the filtration W' on $L^{\cdot}$. An easy computation shows that $E'^{m,k}_r = E^{m,k}_{r+1}$ and $d'_r = d_{r+1}$. Hence the spectral sequence $E'^{m,k}_r$ degenerates at E_1, or equivalently the differential

$$d\colon \Gamma(X, \Lambda^k_X\langle\log Q\rangle) \to \Gamma(X, \Lambda^{k+1}_X\langle\log Q\rangle)$$

is strict for the filtration W'.

Let $\omega \in L^k$ with $d\omega \in W_m L^{k+1}$; since $d(d\omega) = 0$, $d\omega \in Z_1^{m,k+1} = W'_{m+k+1}L^{k+1}$, so that there exists $\theta \in W'_{m+k+1}L^k$ with $d\theta = d\omega$. But $W'_{m+k+1}L^k = Z_1^{m+1,k} \subset W_{m+1}L^k$, hence $\theta \in W_{m+1}L^k$. □

References

[A] D. Arapura, Building mixed Hodge structures, The arithmetic and geometry of algebraic cycles, *CRM Proc. Lecture Notes* 24, Amer. Math. Soc., Providence, RI, 2000.

[A1] D. Arapura, Hodge theory with local coefficients on compact varieties, *MSRI publications* (1989).

[AG0] V. Ancona, B. Gaveau, Families of differential forms on complex spaces, *Ann. Scuola Norm. Sup. Pisa Cl. Sci.* (5) 1, 119–150 (2002).

[AG1] V. Ancona, B. Gaveau, Le complexe de De Rham d'un espace analytique normal, *C. R. Académie des Sciences de Paris* 315, 577–581 (1992).

[AG2] V. Ancona, B. Gaveau, Intégrales de formes différentielles et théorèmes de De Rham sur un espace analytique, *C. R. Académie des Sciences de Paris* 317, 65–70 (1993).

[AG3] V. Ancona, B. Gaveau, Théorèmes de De Rham sur un espace analytique, *Revue Roumaine de Math. pures et appliqués* tome XXXVIII n. 7–8, 579–594 (1993).

[AG4] V. Ancona, B. Gaveau, The De Rham complex of a reduced complex space, in: *Contribution to complex analysis and analytic geometry* (H. Skoda and J.M. Trépreau eds.) Vieweg 1994.

[AG5] V. Ancona, B. Gaveau, Théorie de Hodge et complexe de De Rham holomorphe pour certains espaces analytiques, *Bull. des Sciences Mathématiques* 122, 163–201 (1998).

[AG6] V. Ancona, B. Gaveau, Suites spectrales, formes holomorphes et théorie de Hodge de certains espaces analytiques, *Proceedings of Coimbra conference on Partial differential equations and Mathematical Physics* (J. Vaillant and J. Carvalho e Silva editors) Textos Mat. Ser. B, 24, 1–14 (2000).

[AG7] V. Ancona, B. Gaveau, Residues on complex spaces (to appear)

[B] J. I. Burgos, a C^∞ log complex. *Compositio Math.* 92, 61–86 (1994).

[BL] M. Berger, A. Lascoux, variétés kählériennes compactes, *Lecture Notes in Math.* 154, Berlin, Heidelberg, New York, Springer 1970.

[BH] T. Bloom, M. Herrera, De Rham cohomology of an analytic space, *Inventiones math.* 7, 275–296 (1969).

[BHae] A. Borel, A. Haefliger, La classe d'homologie fondamentale d'un espace analytique, *Bull. Soc. Math. France* 89, 461–513 (1961).

[BM] E. Bierstone, P.D. Milman, Canonical desingularization in characteristic zero by blowing up the maximum strata of a local invariant, *Inventiones math.* 128, 207-302, (1997).

[BMO] A. Borel, J.C. Moore, Homology theory for locally compact spaces, *Mich. math. J.* 7, 137–159 (1960).

[Br] G. Bredon, Sheaf theory, New York, Mac Graw-Hill 1967.

[C] J. Carlson, Polyhedral resolutions of algebraic varieties, *Trans. Am. Math. Soc.* 292, 595–612 (1985).

[D1] P. Deligne Théorème de Lefschetz et critères de dégénérescence des suites spectrales. *Publ. Math. IHES* 35, 259–278 (1968).

[D2] P. Deligne, Théorie de Hodge I, *Actes du Congrès International des Mathématiciens* (Nice, 1970), Tome 1, 425–430, Gauthier-Villars, Paris 1971.

[D] P. Deligne, Théorie de Hodge II et III. *Publ. Math. IHES* 40, 5–58 (1971) et 44, 5–77 (1974).

[DP] P. Dolbeault, J.B. Poly, Differential forms with subanalytic singularities, integral cohomology, residues, *Proceedings of Symposia in Pure Mathematics* 30, 255–261 (1977).

[DR] G. De Rham, Variétés différentiables. Formes, courants, formes harmoniques. *Publications de l'Institut de Mathématique de l'Université de Nancago* III. Paris, Hermann 1973.

[E] F. Elzein, Mixed Hodge structures, *Trans. Am. Math. Soc.* 275, 71–106 (1983).

[F] A Fujiki,Closedness of the Douady space of compact Kähler spaces, *Publ. RIMS Kyoto* 14, 1–52 (1978).

[GF] H. Grauert, K. Fritzsche, Several complex variables, *Graduate texts in mathematics* 38, Berlin, Heidelberg, New York, Springer 1976.

[GK] H. Grauert, H. Kerner, Deformationen von Singularitäten komplexer Räume, *Math. Ann.* 153, 236–260 (1964).

[G] R. Godement, Topologie algébrique et théorie des faisceaux, Paris, Hermann 1958.

[GI] B. Giesecke, Simpliziale zerlegungen abzälbare analytischer Räume, *Math. Zeitschrift* 83, 177–213 (1964).

[GNPP1] F. Guillèn, V. Navarro Aznar, P. Pascual Gainza, P. Puertas, Théorie de Hodge via schémas cubiques, Notes polycopiées, Univ. Politecnica de Catalunya (1982).

[GNPP2] F. Guillèn, V. Navarro Aznar, P. Pascual Gainza, P. Puertas, Hyperrésolutions cubiques et descente cohomologique, *Lecture Notes in Math.* 1335, Berlin, Heidelberg, New York, Springer 1988.

[GS] P. Griffiths, W. Schmid, Recent developments in Hodge theory: a discussion of techniques and results, *Proceedings of the International Colloquium on Discrete subgroups of Lie groups* (Bombay 1973), Oxford university Press 1975.

[GN] F. Guillèn, V. Navarro Aznar, Un critère d' extension d'un foncteur défini sur les schémas lisses. Publ. Math. IHES, 95, 1–95 (2002).

[GR] H. Grauert, R. Remmert, Coherent analytic sheaves, *Grundlehren der math. Wissenshaften* 265, Berlin-Heidelberg-New York, Springer 1984.

[GP] P. A. Griffiths, On the periods of certain rational integrals. I, II, *Ann. of Math.* (2) 90, 460–495 (1969); ibid. (2) 90, 496–541, (1969)

[H1] H. Hironaka, resolution of singularities of an algebraic variety over a field of characteristic zero, *Ann. of Math.* 79, 109-326 (1964).

[H2] H. Hironaka, Flattening theorem in complex-analytic geometry, *Amer. J. of Math.* 97, 503–547 (1975).

[H3] H. Hironaka, Introduction to real-analytic sets and real-analytic maps, *Quaderni dei gruppi di ricerca matematica del CNR* Pisa 1973.

[He] M. Herrera, Integration on a semianalytic set, *Bull. Soc.Math. France* 94, 81–180 (1959).

[L] J. Leray, Le calcul différentiel et intégral sur une variété analytique complexe (Problème de Cauchy III), *Bull. Soc. Math. France* 87, 141–180, (1967).

[Le] P. Lelong, Intégration sur un ensemble analytique complexe, *Bull. Soc.Math. France* 87, 81–180 (1959).

[Ło] S. Łojasievicz, Triangulation of semianalytic sets, *Ann. Scuola Norm. Sup. Pisa Cl. Sci.* (3) 18, 449–474 (1964).

[M] B. G. Moishezon, On n-dimensional compact varieties with n algebraically independent meromorphic functions, I,II,III, *AMS Translations* 63, 51–177 (1967).

[Ma] B. Malgrange, Ideals of differentiable functions, *Studies in Mathematics* No. 3 Tata Institute of Fundamental Research, Bombay; Oxford University Press, London 1967.

[MK] J. Morrow, K. Kodaira, Complex manifolds, Holt, Rinehart and Winston, Inc., New York-Montreal, Que.-London, 1971.

[N] V. Navarro Aznar, Sur la théorie de Hodge-Deligne. *Inventiones math.* 90, 11–76 (1987).

[R] H. J. Reiffen, Das Lemma von Poincaré für holomorphe Differentialformen auf komplexen Räumen, *Math. Zeitschrift* 101, 269–284 (1964).

[S] D. Sullivan, Infinitesimal computations in topology, *Publ. Math. IHES* 47, 269–331 (1977).

[SD] B.Saint-Donat, Techniques de descente cohomologique (d'après des notes de P. Deligne) S.G.A. 4 V bis, *Lecture Notes in Math.* 270, 83–162 , Berlin-Heidelberg-New York, Springer 1972.

[V] J. Varouchas, Sur l'image d'une varieté kälerienne compacte, in Fonctions de plusieurs variables complexes V, 245–259, *Lecture Notes in Math.* 1188, Berlin-Heidelberg-New York, Springer 1986.

[Ve] A. Verona, Triangulation of stratified fibre bundles. *Manuscripta Math.* 30 425–445 (1979/80)

[W] A. Weil, Introduction à l'étude des variétés kählériennes, *Publications de l'Institut de Mathematique de l'Université de Nancago,* VI. Paris Hermann 1958.

Index